33e Symposium International FESF Strasbourg

AF864672

Social Strategies

Monographien zur Soziologie und Gesellschaftspolitik
Monographs on Sociology and Social Policy
Vol. 40

Editor: Paul Trappe

Peter Lang
Bern · Berlin · Bruxelles · Frankfurt am Main · New York · Oxford · Wien

Edmundas Kazimieras Zavadskas (ed.)

33e Symposium International FESF Strasbourg

Des développements récents dans la protection de l'environnement

Recent Developments in Environmental Protection

Neuere Entwicklungen im Umweltschutz

Peter Lang
Bern · Berlin · Bruxelles · Frankfurt am Main · New York · Oxford · Wien

Bibliographic information published by Die Deutsche Bibliothek
Die Deutsche Bibliothek lists this publication in the Deutsche Nationalbibliografie; detailed bibliographic data is available on the Internet at ‹http://dnb.ddb.de›.

British Library and Library of Congress Cataloguing-in-Publication Data:
A catalogue record for this book is available from *The British Library,* Great Britain, and from *The Library of Congress,* USA

ISSN 1424-0467
ISBN 3-03910-612-0
US-ISBN 0-8204-7161-5

© Peter Lang AG, European Academic Publishers, Bern 2005
Hochfeldstrasse 32, Postfach 746, CH-3000 Bern 9, Switzerland
info@peterlang.com, www.peterlang.com, www.peterlang.net

All rights reserved.
All parts of this publication are protected by copyright.
Any utilisation outside the strict limits of the copyright law, without the permission of the publisher, is forbidden and liable to prosecution.
This applies in particular to reproductions, translations, microfilming, and storage and processing in electronic retrieval systems.

Printed in Germany

Inhaltsverzeichnis

Preface

An interesting phenomenon has been existing in Europe for the last 25 years – Faculté Européenne des Sciences du Foncier (FESF), the European Faculty of Land Use and Development, Europäische Fakultät für Bodenordnung, an organization that unites scientists from different fields of science.

Dr. Paul Trappe, President of the Faculty and Professor of Basle University, besides of his function as the leader of this international organization, for the last several decades he also has been working as an editor of a series of monographs entitled Social Strategies (SoStra – Monographien zur Soziologie und Gesellschaftspolitik, Monographs on Sociology and Social Policy).

Regrettably, it was not until 2001 that my colleague Dr. Vida Maliené finally took part in a symposium organized by this faculty. It was the 31st symposium and it took place in Strasbourg.

For the past several years we have been working successfully with Professor Erich Weiss of University of Bonn, who is also an Honorary Doctor of Vilnius Gediminas Technical University. I am very grateful to him for suggesting Vilnius Gediminas Technical University as the venue for the 33rd Symposium.

"Recent Developments in Environmental Protection" was the main subject discussed at the symposium. This issue is the question of the day in Lithuania as well.

We were glad that representatives from twelve countries took part in the symposium: six were from EU member countries (Austria, France, Finland, Sweden, Germany, and Greece), two were from non-EU countries (Switzerland and Russia) and four – from those who that are about to join EU this year (Latvia, Lithuania, Poland, and Slovenia).

In total, 19 participants of the symposium were from other countries and they read 16 reports (Germany – 4, France – 2, Poland – 2, Switzerland – 2, and Austria, Latvia, Finland, Slovenia, Sweden, Russia – each one lecture).

Lithuanian specialists did not miss such a great opportunity and got to know many prominent foreign guests. They also presented their reports on various topical issues both to Lithuania and other countries of the world. Lithuania has been represented by 19 participants.

The symposium was unique in that the reports were read in three languages – 13 reports were delivered in English, 9 in German, and 4 in French.

Sociologists, economists, managers, lawyers, geology and ecology professionals, land use managers, surveyors and construction engineers took part in the work of this symposium.

It is interesting that despite the fact that the participants were from most diverse scientific fields, they fully understood each other maybe because they were all interested in the important issues discussed at the symposium, i.e. the recent developments in environmental protection.

The opening speeches were given by Professor Dr. Raimundas Kirvaitis, Vice-Rector of Vilnius Gediminas Technical University, Professor Dr. Albertas Vasiliauskas, Vice-Minister of Environment and Urban Development and a forestry scientist. The greeting speech in French, German and English was given by FESF President – Professor Dr. Paul Trappe.

Four sessions were held under the chairmanship of these professors: Prof. Dr. A. Kaklauskas, Prof. Dr. H. Mattson, Assoc. Prof. Dr. M. Subic Kovac, Assoc. Prof. Dr. C. Aksamitauskas.

Sessions were held in the University Senate Hall. The participants stayed in Sauls Žiedas Hotel situated nearby the University building. The hotel offered good accommodation for our guests and its vicinity to our university made it possible to save time. Participants actively participated in various activities. After reading the reports speakers took part in lively discussions with the audience. Unfortunately, the symposium lasted for only a few days. Nevertheless, the symposium organizers, the scientists of Vilnius Gediminas Technical University, did their best to make it possible for the participants to learn about cultural life of Vilnius. The participants visited the Old Town that has been enlisted on UNESCO's list of cultural heritage, the ancient Lithuanian capital of Trakai with its unique castle of grand dukes built on an island of a lake. On the last evening the participants attended a concert of the State Symphony Orchestra.

During the symposium the sincere and constructive relationship prevailed, therefore we hope that in future the participants will maintain contacts and will willingly take part in the forthcoming FESF undertakings. Despite the fact that only 12 countries were represented in the symposium, geographically speaking the participants came from most various parts of the world: from Scandinavia and Greece, from France and Tomsk, a city in Siberia.

Eight participants from Vilnius Gediminas Technical University took part in the conference, also 11 colleagues from Lithuanian University of Agriculture, various ministries, departments, and research institutions. Colleagues from Vilnius Gediminas Technical University took part by preparing of 7 presentations.

Six major presentations were especially interesting. Ten presentations on recent achievements in 8 European states were of great informative value.

Lithuanian representatives provided presentations on 10 topics which covered a wide spectrum of issues: from law preparation, territory planning and management to earth's crust movement in the Baltic Sea region and geodetic problems in the nearby area of the Ignalina Nuclear Plant.

I am glad to inform you that the participants of the symposium found time to prepare their presentations also in written form and these texts are published in the 40th volume of series of monographs Social Strategies.

This provides possibility to introduce the ideas and work of the symposium to a greater audience of scientists and professionals.

I am also glad that Vilnius Gediminas Technical University from now on will be one of the venues for various events of the FESF. I hope that the management of the Faculty will invite representatives of Lithuania to participate in the forthcoming activities of the Faculty, and that later, after Lithuania joins the EU, it will decide to arrange a symposium in the capital of Lithuania. I hope that our University will be granted a possibility to give our colleagues and research friends a warm welcome once again.

I would like to ask the FESF participants to express their ideas in various magazines published at our University:

"International Journal of Strategic Property Management" (Editors E. K. Zavadskas, A. Kaklauskas, VGTU, B. Sloan, Napier University) ISSN 1648-715X;
"Journal of Civil Engineering and Management" (Journal of Vilnius Gediminas Technical University and Lithuanian Academy of Sciences) Editor-in-Chief E. K. Zavadskas ISSN 1392-3730;
"Technological and economic development of economy" (Editor-in-Chief E. K. Zavadskas) ISSN 1392-8619;
"Business: theory and practice" (Editor-in-Chief R.Ginevičius) ISSN 1648-0627;
"Town Planning and Architecture (Editor-in-Chief J. Vanagas) ISSN 1392-1630;
"Geodesy and Cartography" (Editor-in-Chief J. Skeivalas) ISSN 1392-1541;
"Transport Engineering" (Editor-in-Chief A. Baublys) ISSN 1648-4142;
"Journal of Environmental Engineering and Landscape Management" (Editor-in-Chief P. Baltrėnas) ISSN 1648-6897.

I hope that in the future we will maintain our constructive cooperation and productive joint activities with the FESF participants.

Edmundas Kazimieras Zavadskas

Teil I
Beiträge zu Rahmenbedingungen

WILLY KRAUS

Globale Umweltproblematik und umweltpolitischer Föderalismus

I. Bemerkungen zur globalen Umweltproblematik

1. Nutzung der natürlichen Umwelt und Umweltschäden während der Anfänge der Industrialisierung

Wir können wohl nicht in Abrede stellen, dass durch den im 19. Jahrhundert einsetzenden weltweiten Übergang zur industriellen Produktionsweise, durch die Intensivierung der Handelsströme, durch den tiefgreifenden technologischen Wandel, durch die gewaltige Zunahme des Verkehrs usw. die Nutzung der natürlichen Umwelt stark angeregt und beansprucht worden ist. Hierbei traten auch bereits Umweltschäden auf, die aber meist lokal begrenzt und auf bestimmte konkrete Ursachen zurückzuführen waren. Indessen waren weite Bereiche des Ökosystems nicht betroffen.[1] Von „Umweltschutz" oder „Umweltschäden" war expressis verbis noch nicht die Rede, auch wenn es sich um solche handelte.

Diese lokal ausgelösten und auch wirksamen Schäden waren natürlich als Begleiterscheinungen eines grundlegenden Wandels erkennbar und spürbar. Es bestand allerdings kein Grund, hierbei schon von Globalisierungsfolgen zu sprechen, obwohl von der Sache her bereits jene Entwicklung eingesetzt hatte, die heute als Globalisierung freudig begrüßt oder gebrandmarkt bzw. gar verdammt wird. Im Mai 1997 hat der „Internationale Währungsfonds" in seinem „World Economic Outlook" das Phänomen der Globalisierung zutreffend als Wiederaufnahme von Entwicklungen in der Weltwirtschaft bezeichnet, die seit mehr als hundert Jahren zu beobachten sind. Erst in den 80er Jahren des vorigen Jahrhunderts tauchte weithin der Begriff „Globalisierung" auf, dem dann auch bestimme, ganz konkrete Sachverhalte zugeordnet wurden. Es handelt sich hierbei nicht nur um eine intensive Liberalisierung und eine weltweite Marktöffnung. Entscheidende Impulse hat dieser moderne Globalisierungsprozess durch das Sinken der Transportkosten und durch die Revolution der Entwicklung wie die Verbreitung von modernen Informations- und Kommunikationstechniken erhalten. Globalisierung kann heute kurz gefasst als eine umfassende, grenzüberschreitende Vernetzung des weltweiten Güter- und Dienstleistungsflusses beschrieben werden.

---

1 Artikel Umweltschäden. In: Öko Lexikon, München 2003, S. 220.

Der Gedanke an eine globalisierte Welt, die ohne jede Grenze allen offen steht, hat natürlich Zustimmung, Beifall und Aufbruchstimmung ausgelöst. Das lang ersehnte kosmopolitische Zeitalter vom ewigen Frieden scheint für die Befürworter in greifbare Nähe zu rücken und verspricht die Durchsetzung noch ausstehender Freiheitsrechte. Aber neben den zustimmenden Urteilen zur Globalisierung tauchen andererseits auch Befürchtungen, unheilvolle Ahnungen oder ablehnende apodiktische Behauptungen insbesondere im Hinblick auf ökologische, soziale, kulturelle und politische Begleiterscheinungen dieses Phänomens auf.

2. Entstehung einer neuen globalen Umweltproblematik

Zahlreiche Gegner der Globalisierung verweisen nun mit Nachdruck auf eine *neue*, auf eine neu entstandene globale Umweltproblematik, insbesondere ausgelöst durch die Ausweitung grenzüberschreitender ökonomischer Transaktionen. Ihre ständige Ausweitung führe zu weiteren beachtlichen Umweltschäden. Die Globalisierung sei auch nicht in der Lage, solche Umweltschäden wie z. B. Festlegung von angemessenen Umweltstandards zu verhindern, bzw. zu beseitigen. Zudem würden multinationale Konzerne in ihrer finanziellen Gier die Umwelt nicht beachten und sich über Umweltgesetzte einfach hinwegsetzen. Wir können wohl nicht in Abrede stellen, dass immer wieder gewaltige Umweltschäden flächendeckend auftreten wie z. B.

Treibhauseffekte,
Abbau der Ozonschicht,
Wüstenausdehnung,
Bodenerosion,
Waldsterben,
Bildung von Ölteppichen usw.

Bei zunehmendem globalen Wettbewerbsdruck werden ohne Zweifel neue ressourcensparende Technologien verwendet werden. Auch ist damit zu rechnen, dass ein verstärkter Transfer von Technologie und Wissen in Regionen mit bedrohter Umwelt erfolgen wird, wie z. B. von Abluftfiltern, Rauchgasentschweflungsanlagen, Abgaskatalysatoren usw. Gleichwohl scheint auch weiterhin eine zusätzliche Inanspruchnahme von Elementen der natürlichen Umwelt zu erfolgen. Das bedeutet aber, dass in diesen Fällen mit einer technologischen Absorptionswirkung nicht ohne weiteres gerechnet werden kann.

3. Einige Globalisierungserfahrungen

Wir erleben, dass häufig Auslandsinvestitionen dort vorgenommen werden, wo die Umweltstandards relativ niedrig sind. In anderem Zusammenhang habe ich darauf hingewiesen, dass z. B. taiwanesische Investoren mit Vorliebe auf dem chinesischen Festland investiert haben, um dort die relativ niedrigen Umweltstandards zu nutzen. Unter diesen Umständen hatte die Volksrepublik China lange Zeit auch wenig Interesse, energisch den Umweltschutz auszubauen. Man gewinnt jedoch nunmehr den Eindruck, dass sich inzwischen auch in China das Umweltbewusstsein ausbreitet. Heute erkennt man wohl viel klarer als zuvor: „Die Umweltzerstörung droht dem Wirtschaftswachstum buchstäblich die Luft und das Wasser zu nehmen.“[2]

Über den konkreten Taiwan-China-Fall hinaus wird man auch nicht ausschließen können, dass einzelne Nationalstaaten den Versuch unternehmen, durch faktische Herabsetzung ihrer Umweltstandards, sich im Globalisierungsprozess Wettbewerbsvorteile zu verschaffen, um möglichst große Investitionsströme ins Land zu locken.

Wir stellen außerdem fest, dass insbesondere die Großmächte dazu neigen, eigene Wege in der Umweltpolitik zugehen, obwohl weltweit oder regional ein allgemeiner Konsens über Umwelt-Schutzmaßnahmen erzielt worden ist. Russland hat im Juni 03 auf der internationalen Meeresschutz-Konferenz in Bremen sein Veto eingelegt, die gesamte Ostsee als besonders empfindliches Seegebiet auszuweisen. Das Übereinkommen der Vereinten Nationen über Klimaänderungen – 1992 in Rio de Janeiro verabschiedet – ist inzwischen von 186 Staaten und der Europäischen Union unterzeichnet worden. Indessen hat die derzeitige US-Administration erklärt, das Umsetzungsprotokoll (Kyoto-Protokoll) nicht ratifizieren zu wollen. Möglichen Sanktionen versteht man unter Berufung auf das traditionelle Souveränitätsrecht auszuweisen.[3]

Christoph Knill hat in seinem Buch über die europäische Umweltpolitik darauf hingewiesen, dass die formale und praktische Umsetzung europäischer Umweltpolitik zu beachtlichen Implementationsdefiziten geführt hat. „So sind im Jahre 2001 knapp 600 Beschwerden gegen umweltpolitische Rechtsverstöße bei der Kommission eingegangen.“[4] Wir stellen auch immer wieder fest, dass Trinkwassergrenzwerte nicht eingehalten werden, dass Stickstoffemissionen des Verkehrswesens nicht reduziert werden, dass die Energiewirtschaft weiterhin mit viel zu hohen Kohlendioxidemissionen arbeitet.

---

2 Konrad Seitz, China im 21. Jahrhundert. Alfred Herrhausen Gesellschaft für internationalen Dialog, Frankfurt 2000, S. 46.

3 Udo Simonis, Klimarahmenkonvention. In: Öko-Lexikon, a. a. O. S. 109. Vgl. hierzu auch Dietrich Brockhagen. Kyoto Protokoll. In: Öko-Lexikon a. a. O. S. 117 ff.

4 Christoph Knill, Europäische Umweltpolitik. Steuerungsprobleme und Regulierungsmuster im Mehr-Ebenen-System. Opladen 2003, insbesondere S. 174.

Mit zunehmendem globalen Denken und Verständnis wird man auch den globalen Umweltfragen Rechnung tragen und entsprechende handeln müssen. Globale Umweltgüter wie Klima, Erdatmosphäre usw. sind schließlich Ressourcen mit kollektiven Eigentumsrechten. Seien wir uns doch bewusst, dass die souveränen Rechte der Nationalstaaten dort ihre notwendigen natürlichen Grenzen finden sollten, wo die kollektiven globalen Eigentumsrechte verletzt werden können. Folgt man dem Schlussbericht der deutschen Enquete-Kommission „Globalisierung der Wirtschaft – Herausforderungen und Antworten" vom 13. Mai 2002 so haben „internationale völkerrechtliche Vereinbarungen zum Umweltschutz…Regierungen zur Erarbeitung von Umweltschutzgesetzen und zur Umsetzung von Umweltschutzmaßnahmen veranlasst, welche ohne Außenöffnung der Nationalstaaten nicht oder erheblich später erfolgt wären." Auf jeden Fall ist doch unverkennbar, dass sich nach und nach ein internationaler Ordnungsrahmen herausbildet. Inzwischen „existieren mehr als 170 multilaterale völkerrechtlich verbindliche Verträge zum Umweltschutz."[5] Besonders zu erwähnen sind die Konvention über weitreichende grenzüberschreitende Luftverschmutzung (1987, in Kraft 1989), das Montrealer Protokoll zum Schutze der Ozonschicht (1987, in Kraft 1989) und die Klimarahmenkonvention (1992, in Kraft 1993). Es wir allerdings darauf ankommen, dass diese Verträge auch umgesetzt werden.

Im Zusammenhang mit der globalen Umweltproblematik hat auch der Begriff der „nachhaltigen Entwicklung" (sustainable development) einen festen Platz in der politischen Diskussion eingenommen. Folgt man dem Brundtland-Report, so wird er in folgender Weise definiert: „Dauerhafte Entwicklung ist Entwicklung, die Bedürfnisse der Gegenwart befriedigt, ohne zu riskieren, dass künftige Generationen ihre eigenen Bedürfnisse nicht befriedigen können."[6]

4. Global Governance

Die globale Umweltproblematik hat schließlich zum Konzept „Global Governance" beigetragen, worunter man vor allem den Prozess der politischen Gestaltung der Globalisierung versteht.[7] Das Ziel besteht offenbar darin, ein Institutionen- und Regelsystem sowie neue Mechanismen internationaler Kooperation zu entwickeln, „welche die kontinuierliche Problembearbeitung

5 Ulrich Petschow / Susanne Dröge, Globalisierung und Umweltschutz. Die Rolle des Nationalstaates. In: Aus Politik und Zeitgeschichte, 4. Juni 1999. S. 23 ff.

6 Weltkommission für Umwelt und Entwicklung. Unsere gemeinsame Zukunft (Brundtland-Report). Hrsg. Von Volker Haufff, Greven 1987, 1.146. Hierzu auch Rudi Kurz, Innovationen für eine zukunftsfähige Entwicklung. In: Aus Politik und Zeitgeschichte. 9. Februar. 1996, S. 14 ff.

7 Artikel Global Governance. In: Udo Simonis (Hrsg.), Öko-Lexikon, München 2003, S. 90 ff.

globaler Herausforderungen und grenzüberschreitender Probleme erlauben."[8] Diese globalen Intentionen beziehen sich natürlich weitgehend auch auf Umweltfragen. Es handelt sich aber nicht um die Bildung einer Weltregierung. Indessen sollen internationale Organisationen die Koordinationsfunktion übernehmen und zur Herausbildung globaler Sichtweisen beitragen. „Staaten verpflichten sich vielmehr durch vertragliche Vereinbarungen zu Entscheidungen über gemeinsame Probleme."[9] Damit würde sich natürlich das traditionelle Verständnis von Souveränität verändern, falls sich das Konzept des „Global Governance" durchsetzt und beachtet wird. Denn mit diesem Konzept sind schließlich Souveränitätsverzichte verbunden.

Die Anhänger der „Global Governance" vertreten zudem die Auffassung, dass mit diesem Konzept eine Neuverteilung der weltwirtschaftlichen und weltpolitischen Gewichte in Richtung einer multipolaren Welt eingeleitet wird, begleitet von einem Prozess der Regionalisierung. In allen Weltregionen würden sich neue Kooperations- oder Integrationszonen formieren, wobei die Europäische Union als das am weitesten entwickelte Modell gelte.[10] Offenbar gibt es allerdings bisher nur einen einzigen transnationalen Staatenbund – die EU – in dem eine gemeinsame Umweltpolitik gemacht wird. Auch sehe ich vorerst kaum Ansätze zu einem effektiven Staatenbund außerhalb und neben der EU. Im Rahmen der Diskussion „Global Governance" ist auch eine institutionelle Integration der großen internationalen Organisationen auf der globalen Ebene vorgeschlagen wurde. Vereinte Nationen – UN, Internationaler Währungsfonds – IMF und die Weltbank sollen hierbei eine führende Rolle übernehmen. Der UN würde zudem die Aufgabe zufallen, ihre eigene Zersplitterung von Zuständigkeiten im Umweltbereich, verbunden mit überschneidenden Sachgebieten, aufzugeben.[11] Der deutsche Umweltforscher Udo E. Simonis hat daher die Gründung einer einheitlichen Organisation, etwa in Form einer Weltorganisation für Umwelt und Entwicklung – Global Environment and Development Organization – vorgeschlagen. Eine bessere Vernetzung des umweltpolitischen Bereiches sollte einen effizienten Beitrag zur Politikoptimierung leisten.[12] Jedenfalls stellt sich auch hier die entscheidende Frage:

8 Dirk Messner, Global Governance. In: Öko-Lexikon, a.a.O., S. 90.

9 Derselbe, ebenda, S. 91.

10 Derselbe, ebenda.

11 Hier sind zu nennen: Das UN-Umweltprogramm (UNEP), das UN-Entwicklungsprogramm (UNDP), die UN-Kommission für nachhaltige Entwicklung (CSD), de Weltorganisation für Meteorologie (WMO), die Internationale Ozeanographische Organisation (IOC) usw. Es kommen noch zahlreiche etablierte selbstständige Sekretariate für die inzwischen abgeschlossenen internationalen Umweltverträge hinzu. Vgl. hierzu Wolfgang Engelhardt, Globale Umweltpolitik. In: MUT-Forum für Kultur, Politik und Geschichte, Nr. 413, Januar 2002, S. 36 ff.

12 Udo E. Simonis, Frank Biermann, Politikinnovationen auf der globalen Ebene. In: Jahrbuch Ökologie 2000, München 2000, S. 26 f.

Wie soll „Global Governance“ in reale, aktive Politik umgesetzt werden? Bei der vorgeschlagenen institutionellen Integration der großen internationalen Organisationen wird ausdrücklich von der „Weltlegislative“[13] gesprochen. Jede Legislative, die aber mehr als eine leere Attrappe sein will, bedarf der exekutiven Ergänzung. Die Umweltpolitik kann sich nicht darauf beschränken, ein umfassendes Regelwerk zu entwerfen, es muss auch gehandelt werden, und zwar auf den angemessenen Handlungsebenen.

II. Umweltpolitischer Föderalismus

1. Zentralismus und Dezentralismus im Umweltschutz

Man kann den Eindruck gewinnen, dass der unaufhaltsamen Globalisierung unserer Zeit notwendig eine Zentralisation aller relevanten Entscheidungen – dies auch im Umweltbereich – folgen wird. Auch im Umweltschutz würden sich „Global Players“ bald kräftig breit machen. Zu diesen Befürchtungen treten dann noch die Ambitionen der Globalisierungsgegner. Auch von dieser Seite geht der Ruf nach Steuerung, Lenkung und Intervention auf zahlreichen Lebensgebieten aus. Vielfach fordern sie Kontrolle im Weltmaßstab.

Ich möchte glauben, dass ich wohl ausreichend darauf hingewiesen habe, dass das Aktionsfeld der Umweltpolitik auf globaler Ebene wesentlich erweitert worden ist und in Zukunft auch notwendig erweitert werden muss, wenn Umweltpolitik auf nationalstaatlicher Ebene nur in einer politisch begrenzten Reichweite agieren kann.

Aber man sollte allerdings auch beachten, dass Umweltprobleme sehr unterschiedliche räumliche Dimensionen aufweisen. Sie können auf lokaler Ebene auftauchen, – und dies in sehr vielen Fällen. Sie können auf der nationalstaatlichen Ebene ihren Einfluss geltend machen. Aber sie können auch einen globalen Umfang einnehmen. Das heißt aber im Prinzip, dass jede Ebene in der Lage sein sollte, die Umweltprobleme mit kompetenten Akteuren und Mitteln zu lösen.

Dem wird nun entgegen gehalten, dass im Verlaufe der Globalisierung die Souveränität von Nationalstaaten ausgehöhlt werde. Der globale Systemwettbewerb führe schließlich zum Souveränitätsverlust, Verlust an Demokratie, und zur Entmachtung der Politik. Die aus der Globalisierung resultierenden „Entgrenzungsprozesse“ (Ulrich Menzel) würden auch die Steuerungskapazitäten der klassischen Nationalstaaten in Frage stellen.

---

13 Dieselben, Politikinnovationen auf der globalen Ebene. In: Aus Politik und Zeitgeschichte. 26.11.1999.

Globalisierung bedeutet aber nicht, dass die Nationalstaaten nun mehr alle Aufgaben verlieren, also keine Beschlüsse mehr fassen und auch an der demokratischen Politikgestaltung nicht mehr teilnehmen. Es besteht auch nicht der geringste Grund, das nationalstaatliche Gewaltmonopol zur Einklagbarkeit und Erzwingung von Bürgerrechten irgendwie anzutasten. Erhöhte Anforderungen an Flexibilität und Anpassung sowie der unumgängliche Strukturwandel werden den Staat dazu zwingen, sich auf seine ureigenen Hoheitsbereiche zurückzuziehen und all das dem einzelnen und seinen Kommunen überlassen, was diese weit besser, gründlicher, und praxisnäher vollziehen können als eine ferne, aufgeblähte Zentrale. Der Schwerpunkt er Verwaltungstätigkeit – so hoffe ich – würde sich auf die lokalen Einheiten verlagern müssen, – zum großen Vorteil der nicht globalen Umweltschutzerfordernisse.

2. Subsidiaritätsprinzip im Rahmen der EU

Die globale Umweltpolitik der EU steht immer wieder im Kreuzfeuer heftiger Kritik. Insbesondere wird auf die „Regulierungswut“[14] dieser Großorganisation verwiesen, der als nimmersatter Moloch „die nationalen Handlungsspielräume mehr und mehr zum Opfer fallen“.[15] Auf der anderen Seite ist aber auch dem Sachverhalt Rechnung zu tragen, dass im Vertrag zur Gründung der Europäischen Gemeinschaft (EGV) im Artikel 174, Absatz 4, ausdrücklich dass umweltpolitische Subsidiaritätsprinzip aufgenommen worden ist. Es besagt, dass die Gemeinschaft nur dann im Umweltbereich tätig werden darf, wenn die „Ziele besser auf Gemeinschaftsebene erreicht werden können, als auf der Ebene der einzelnen Mitgliedsländer.“ Diese Orientierung kommt auch „im Prinzip der angemessenen Handlungsebene zum Ausdruck, welches bereits im ersten umweltpolitischen Aktionsprogramm formuliert wurde. Gemäß diesem Prinzip ist für jede Kategorie umweltpolitischer Probleme diejenige Handlungsebene zu wählen, die in Anbetracht der Reichweite und Natur des Problems am besten geeignet ist, die erforderlichen Maßnahmen zu beschließen und durchzuführen.“[16] Das bedeutet nun auch für die Umweltpolitik, dass man dem einzelnen Staatsbürger und seinen Kommunen all das überlässt, was diese besser, gründlicher und praxisnäher vollziehen können, als eine ferne, aufgeblähte Zentrale. Es ist zu hoffen, dass stets auch danach gehandelt wird. Im übrigen wissen wir seit Buchanans[17] berühmtem Aufsatz „Federalism and Fiscal Equity“, dass die Übertragung von Kompetenzen von der dezentralen Ebene auf eine zentrale Politikebene unter bestimm-

14 Christoph Knill, Europäische Umweltpolitik a. a. O., S. 11.

15 Ebenda.

16 Ebenda.

17 Buchanan, J. M., Federalism and fiscal equity. In: The American economic review, Vol. 40 (1950), S. 583 ff.

ten Bedingungen mit einer Verringerung der Effizienz öffentlicher Leistungen verbunden sein kann. Es steht wohl nicht im Wege, die Folgerungen aus dem Umkehrschluss zu ziehen, indem Kompetenzen von der zentralen Ebene auf eine dezentral Politikebene verlagert werden.

Natürlich können bei der Einhaltung dieses Prinzips Kompetenzschwierigkeiten entstehen. In manchen Fällen wird man sich die strittige Frage vorlegen müssen, auf welcher Ebene man am besten, am effizientesten die erforderlichen Umweltschutzmaßnahmen durchführt. Auch ist die Befürchtung aufgekommen, dass Mitgliedsstaaten, die „um eine Beschneidung ihrer Kompetenz fürchten oder eine bestimmte EU-Regelung aus ökonomischen Gründen ablehnen“[18], unter Berufung auf das Subsidiaritätsprinzip versuchen, „ein Tätigwerden der EU zu verhindern.“[19] Ich möchte aber glauben, dass diese Gefahr gegenüber den dezentralen Vorteilen des Subsidiaritätsprinzips und seinen nicht-hierarchischen Steuerungsformen doch relativ gering eingeschätzt werden sollten. Zur optimalen Kompetenzverteilung ist es wohl angebracht, an das aus der Neoklassik bekannte Externalitätenproblem anzuknüpfen.[20] Hieraus ist als Grundbedingung für ein effizientes „Angebot“ umweltpolitischer Maßnahmen abgeleitet worden, „dass die Auswirkungen der Umweltpolitik einer Gebietskörperschaft auf deren räumlichen Zuständigkeitsbereich begrenzt sein sollten.“[21] Generell wird man wohl sagen können, dass jene Ebene Umweltschaftaufgaben erfüllen sollte, die dazu am kostengünstigsten in der Lage ist.

3. Umsetzung der Umweltpolitik

Wir sollten bei der globalen Umweltproblematik auch nicht die Augen vor den notwendigen Implementierungen an Ort und Stelle verschließen. Wir werden bei der globalen Umweltpolitik doch zur Hauptsache mit dem globalen Maßnahmenkatalog, mit den globalen Aktionsprogrammen, mit dem Zustandekommen der zahlreichen Direktiven, Regeln, Bestimmungen, mit der Entwicklung politischer Programme vertraut gemacht. In der sehr eindrucksvollen Studie von Christoph Knill[22] erfahren wir hierüber eine ganze Menge, zum Beispiel Details der Grundlagen und Prinzipien, Entwicklungslinien der europäischen Umweltpolitik, Entscheidungsfindung, Politik des kleinsten

18 Christoph Knill, a. a. O., S. 46.

19 Ebenda.

20 Burkhard Huckestein, Umweltpolitik und Föderalismus – Ökonomische Kriterien für umweltpolitische Kompetenzen der Europäischen Gemeinschaft. Zeitschrift für angewandte Umweltforschung, Jg. 6 (1993), H. 3, S. 331.

21 Ebenda.

22 Christoph Knill, Europäische Umweltpolitik, a. a. O.

gemeinsamen Nenners, Interessenkollisionen, Interaktionsmuster, Politikformulierung usw. Kommt es dann aber zur Umsetzung, dann tritt schließlich naturgemäß der Einzel- oder Nationalstaat auf und dies mit seinen Gliederungen – mit den Provinzen, Regionen, Kommunen bis hin zu den Betrieben und Haushalten. Hier wird nun reale, praktische Umweltpolitik gemacht. Hier an Ort und Stelle wird nicht nur angeordnet und verordnet, sondern gehandelt, also umweltpolitisch umgesetzt.

Nach den europäischen Vertragsunterlagen obliegt die Durchführung gemeinschaftlicher Rechtsakte grundsätzlich den Mitgliedsstaaten. Die Kommission kann zur Durchsetzung korrekter Implementierungen ein Vertragsverletzungsverfahren gegen ein Mitgliedsland einleiten und schließlich auch den Europäischen Gerichtshof anrufen.[23] Aber die Kommission „verfügt nicht über die notwendigen personellen und finanziellen Ressourcen, um eine systematische Überwachung durchzuführen."[24] Anteilmäßig ist daher die Zahl der von Amtswegen ermittelten Vertragsverletzungen sehr gering. Hauptinstrument der Anwendungskontrolle sind deshalb die Beschwerden einzelner Bürger und Umweltverbände in den Mitgliedsstaaten. Im Jahre 2001 liefen 1300 solcher Beschwerden ein.[25]

Die *globale Umweltproblematik* konzentriert sich auf Ziele, Programme, Regeln, Verordnungen zur Einschränkung grenzüberschreitender Schadstoffemissionen.

Der *Umweltföderalismus* bis hin zur einzelwirtschaftlichen Umweltproblematik hat indessen die Hauptaufgabe, sie zu realisieren, wobei wir immer berücksichtigen sollten, dass es auch eine ganze Reihe nationaler, regionaler oder kommunaler Umweltbeeinträchtigungen gibt, die mit grenzüberschreitenden Vorgängen nichts zu tun haben und die deshalb ganz aus der globalen Umweltproblematik herausfallen, wie z.B. lokale Freisetzung gefährlicher Chemikalien, regionaler Gewässerschutz, binnenwirtschaftliche Abfalltransporte, örtlicher Naturschutz usw. In den genannten Fällen wird die Umwelt als nationales, regionales oder kommunales Gut aufgefasst.

Mit Recht wird heute von der Zunahme des Umweltbewusstseins gesprochen. Aber das kann doch nur bedeuten, dass primär Haushalte und Unternehmen gute, gesunde und vernünftige Umweltverhältnisse anstreben. Lebensqualität ist in allen Bevölkerungsschichten – und nicht nur bei Bürokraten – hoch gepriesen. Und die Unternehmen haben weithin längst erkannt, dass vernünftige Umweltverständnisse die Einstellung qualifizierter Arbeitkräfte erleichtern und gegebenenfalls auch zur höheren Produktqualität beitragen.

---

23 Derselbe, S. 162 f.
24 Ebenda, S. 165.
25 Ebenda, S. 166.

4. Trends neuer Umweltmaßnahmen

Mir scheint, dass die föderale Umweltschutzstruktur auch die erforderliche Nähe zu erkennbaren Trends neuer Umweltschutzmaßnahmen und Forschungsfelder eröffnet.

Warum sollten wir insbesondere bei der globalen Umweltproblematik nur auf Anordnungen, Direktiven, Verordnungen usw. „reagieren"? Warum wollen wir nicht stärker selbstständig „agieren"? Warum sind hierzu in der Regel noch Anreizmittel notwendig? Sie sollten jedenfalls weit mehr ins Spiel gebracht werden, als dies heute der Fall ist.

Wenden wir uns einzelnen föderativen Ebenen zu, den Forschungsinstituten und Unternehmen. Hier wird es doch vor allen Dingen darauf ankommen, einen geringeren Ressourceneinsatz je Produkteinheit (spezifischer Umweltverbrauch) und geringere Schadstoffemissionen und Abfallmengen je Produkteinheit (spezifische Umweltbelastung) zu ermöglichen.[26] Ferner wird es zunehmend notwendig sein, technologische Entwicklungen nicht nur auf nachsorgende (kurative) sondern auch auf vorsorgende (präventive) Möglichkeiten zu richten.

Es handelt sich hierbei nicht nur um nachsorgende Produktionsanlagen wie Filter-, Wasch- und Sorptionsanlagen zur Entstaubung und Abgasreinigung, Abwasserreinigungsanlagen usw. sondern auch um produktionsintegrierte Umweltschutztechnologien.[27] Es sind also nicht nur betriebliche Anstrengungen vorzunehmen um die gesetzlichen Vorgaben zum Schutz der Umwelt durch additive Maßnahmen umzusetzen. Jedenfalls hat die unternehmerische Seite in einem dezentralen, föderativen System zahlreiche Möglichkeiten, um aktiv und innovativ zur Lösung der großen Umweltprobleme unserer Zeit beizutragen.

Ich verweise ferner auf die Einrichtung ökologischer Informations- und Steuerungssystem in Unternehmen, auf den Aufbau von Recyclingnetzwerken, auf private und öffentliche Umweltpartnerschaften, auf das Umweltmanagement und schließlich auf den sich ausbreitenden Umweltschutzmarkt.

Am 13. Juni 2001 haben sich auf einer gemeinsamen Konferenz in Potsdam die Agrar- und Umweltminister der Länder für eine Agrarwende ausgesprochen. Mit diesem Beschluss sollte die Agrar- und Ernährungspolitik verstärkt in Richtung artgerechte Tierhaltung, Umwelt- und Naturschutz eingeschlagen werden. Leitbild bildet der „ökologische Landbau", ein Bundespro-

---

26 Udo E. Simonis, Ökologische Umorientierung er Industriewirtschaft. In: Aus Politik und Zeitgeschichte. 9.2.1996, S. 8.

27 Eberhard Böhm und Harald Hiessl, Trends in der Umweltschutztechnologie. In: Handbuch des integrierten Umweltmanagements. München Wien 1997, S. 234 ff.

gramm, das für 2002 und 2003 mit jeweils 33 Mio. € ausgestattet ist.[28] Die Verwendung, der Konsum der so erzeugten Produkte hängt aber nicht nur von dem zugenommenen Umweltbewusstsein der Bevölkerung ab, sondern auch vom Umweltverhalten. Es muss schließlich die Bereitschaft bestehen, einen möglichen Mehrpreis für umweltgerecht Produktvarianten gegenüber den traditionellen Produkten zu zahlen.[29] Natürlich muss die Gewähr bestehen, dass sog. Bioprodukte auch wirklich dem entsprechen, was die potenziellen Käufer sich gewünscht haben. Dazu gehören in ausreichendem Maße Auskünfte und Informationen.

Neben der Landwirtschaft bestehen zahlreiche weitere Politikbereiche, die dem innovativen Umweltschutz offen stehen.

III. Einige Schlussbemerkungen

Die globale Umweltproblematik hat offenbar lange Zeit einen besonders hohen Rang beansprucht. Dabei ist sie meist nur legislativer Art gewesen. Beschlüsse werden gefasst, Verordnungen erlassen, und Regelungen angeordnet. Umso mehr besteht doch Veranlassung, jene Akteure aufzutreiben, die faktisch Umweltpolitik betreiben, – die Einzelstaaten, bei einem föderativen Aufbau die nachgeordneten Ebenen bis zu den Kommunen, Unternehmen und Haushalten.

All das ist doch durch einen breiten Umweltfundamentalismus etwas verdrängt worden, dem die unterschiedlichsten Gruppen gehuldigt haben, – also nicht nur Naturfreunde. Mitunter wurde „das Biotop vor der eigenen Haustür zum Mittelpunkt der Welt stilisiert.[30] Mag sein, dass die extremem Positionen zur Ausweitung des Umweltbewusstseins beigetragen haben. Auf der anderen Seite hat er zur Umsetzung umweltpolitischer Maßnahmen relativ wenig beigetragen. Wir müssen vielmehr den Gesamtzusammenhang von Wirtschaft, Gesellschaft und Ökologie zu erfassen suchen. „Wir müssen die ökologischen, sozialen und wirtschaftlichen Interessen der Menschen zusammen führen, weil sie zusammen gehören“ – schrieb vor einiger Zeit die

---

28 Thomas Wardenbach / Frank Waskow, Der Förderpreis ökologischer Landbau. Ein wichtiger Baustein der Agrarwende. In: Jahrbuch Ökologie. 2003. München 2003, S. 237ff.

29 Herbert Meffert und Manfred Kirchgeorg, Ökologieorientertes Konsumentenverhalten als markt- und wettbewerbsstrategische Herausforderung für das Umweltmanagement. In: Ulrich Steger (Hrsg.), Handbuch des integrierten Umweltmanagements. München Wien 1997, S. 234 ff.

30 Heide Simonis, Kein Blatt vorm Mund. Für eine aktive Bürgergesellschaft. Hamburg 1997, S. 98.

Ministerpräsidentin von Schleswig-Holstein Heide Simonis.[31] „Nicht jede einzelne ökologische Maßnahme ist auch eine Überlebensfrage, sondern jeweils im Gesamtzusammenhang der ökonomischen und gesellschaftlichen Entwicklung zu betrachten."[32] Um so mehr sehen wir uns doch veranlasst, die Umweltpolitik nicht mehr als einen völlig selbstständigen, autonomen Politikbereich aufzufassen, sondern sie in ihrer Vielfältigkeit volksnah zu sehen, – global und föderal.

31 Dieselbe, ebenda, S. 100 f.
32 Ebenda, S. 97.

MAGUELONNE DÉJEANT-PONS

La Convention européenne du paysage Florence, 20 octobre 2000

Introduction

> « Le paysage…
> … participe de manière importante à l'intérêt général, sur les plans culturel, écologique, environnemental et social, et … constitue une ressource favorable à l'activité économique, dont une protection, une gestion et un aménagement appropriés peuvent contribuer à la création d'emplois ;
> … concourt à l'élaboration des cultures locales et … représente une composante fondamentale du patrimoine culturel et naturel de l'Europe, contribuant à l'épanouissement des êtres humains et à la consolidation de l'identité européenne ;
> … est partout un élément important de la qualité de vie des populations : dans les milieux urbains et dans les campagnes, dans les territoires dégradés comme dans ceux de grande qualité, dans les espaces remarquables comme dans ceux du quotidien ;
> … constitue un élément essentiel du bien-être individuel et social, et … sa protection, sa gestion et son aménagement impliquent des droits et des responsabilités pour chacun ».
>
> Préambule de la Convention européenne du paysage.

– *Le Conseil de l'Europe et l'environnement*

Organisation internationale intergouvernementale créée en 1949, le Conseil de l'Europe a son siège à Strasbourg, en France. Il comprend actuellement 45 États membres[1]. Ses principaux objectifs sont de promouvoir la démocratie, les droits de l'homme et la prééminence du droit ainsi que de rechercher des solutions communes aux grands problèmes de société de l'Europe d'aujourd'hui. L'Organisation s'engage ainsi en faveur de la protection de l'environnement et d'un développement territorial durable conforme aux « Principes directeurs pour le développement territorial durable du Continent européen » adoptés à Hanovre le 8 septembre 2000 par la Conférence européenne

1 Au 7 octobre 2003 : Albanie, Allemagne, Andorre, Arménie, Azerbaïdjan, Autriche, Belgique, Bosnie-Herzégovine, Bulgarie, Chypre, Croatie, Danemark, Espagne, Estonie, Fédération de Russie, Finlande, France, Géorgie, Grèce, Hongrie, Irlande, Islande, Italie, Lettonie, « l'ex-République yougoslave de Macédoine », Liechtenstein, Lituanie, Luxembourg, Malte, Moldova, Norvège, Pays-Bas, Pologne, Portugal, République tchèque, Roumanie, Royaume-Uni, Saint-Marin, Serbie-Monténégro, Slovaquie, Slovénie, Suède, Suisse, Turquie, Ukraine.

des Ministres responsables de l'aménagement du territoire des Etats membres du Conseil de l'Europe (CEMAT). Il s'agit de préserver la qualité de vie et le bien-être des Européens en prenant en compte les valeurs paysagères, naturelles et culturelles[2].

– *Les origines de la Convention*

Sur la base d'un premier projet élaboré par le Congrès des pouvoirs locaux et régionaux de l'Europe (CPLRE) dans le cadre de sa Recommandation 40 (1998), le Comité des Ministres du Conseil de l'Europe a décidé de créer en 1999 un groupe restreint d'experts chargé de la rédaction d'une Convention européenne du paysage, sous l'égide du Comité du Patrimoine culturel (CD-PAT) et du Comité pour les activités du Conseil de l'Europe en matière de diversité biologique et paysagère (CO-DBP). A la suite des travaux de ce groupe d'experts, avec participation des principales organisations internationales, gouvernementales et non gouvernementales, le texte final de la Convention a été adopté par le Comité des Ministres du Conseil de l'Europe, le 19 juillet 2000. La Convention a été ouverte à la signature à Florence, en Italie, le 20 octobre 2000 dans le cadre de la campagne du Conseil de l'Europe «L'Europe, un patrimoine commun». Au 7 octobre 2003, 27 États l'ont signée[3] et huit d'entre eux, la Norvège, la Moldova, la Roumanie, l'Irlande, la Lituanie, la Croatie, le Danemark et la Slovénie, l'ont approuvée ou ratifiée. La Convention entrera en vigueur le premier jour du mois qui suivra l'expiration d'une période de trois mois après la date à laquelle dix Etats membres du Conseil de l'Europe auront exprimé leur consentement à être liés par elle[4].

– *Pourquoi une convention sur le paysage?*

Élément essentiel du bien-être individuel et social et de la qualité de vie des populations, le paysage contribue à l'épanouissement des êtres humains ainsi qu'à la consolidation de l'identité européenne. Il participe de manière importante à l'intérêt général, sur les plans culturel, écologique, environnemental et social et constitue une ressource favorable à l'activité économique, avec le tourisme notamment.

---

2 En ce qui concerne le patrimoine naturel et culturel, voir les autres conventions du Conseil de l'Europe : Convention relative à la conservation de la vie sauvage et du milieu naturel de l'Europe (Berne, 19 septembre 1979), Convention pour la sauvegarde du patrimoine architectural de l'Europe (Grenade, 3 octobre 1985) et Convention européenne pour la protection du patrimoine archéologique (révisée) (La Valette, 16 janvier 1992).

3 Belgique, Bulgarie, Croatie, Danemark, Espagne, Finlande, France, Grèce, Italie, Lituanie, Luxembourg, Malte, Moldova, Norvège, Portugal, Roumanie, Saint-Marin, Slovénie, Suède, Suisse et Turquie (voir l'Annexe 1).

4 Par le dépôt d'un instrument de ratification, d'acceptation ou d'approbation.

Et pourtant, les évolutions des techniques de production agricole, sylvicole, industrielle et minière, ainsi que les pratiques en matière d'aménagement du territoire, d'urbanisme, de transport, de réseaux, de tourisme et de loisirs, et plus généralement les changements économiques mondiaux, ont très fréquemment conduit à une dégradation et à une banalisation des paysages.

Si chaque citoyen doit certes contribuer à préserver la qualité du paysage, les pouvoirs publics ont la responsabilité de définir le cadre général permettant d'assurer cette qualité. La Convention établit les principes juridiques généraux devant guider l'adoption de politiques nationales concernant le paysage ainsi que l'instauration d'une coopération internationale en la matière.

– *Structure de la Convention*

Le texte de la Convention comporte un préambule et quatre parties principales consacrées aux questions suivantes :

- définition des objectifs et du champ d'application de la convention, ainsi que des termes clés (chapitre I) ;
- énumération des mesures à prendre au niveau national (chapitre II) ;
- énoncé des fondements de la coopération européenne et des mesures à prendre au niveau international, ainsi que du rôle des Comités responsables du suivi de la mise en œuvre de la Convention (chapitre III) ;
- procédures d'adoption de la convention et questions connexes (chapitre IV).

I. La portée de la Convention

1. Le fondement de la Convention

1.1. Objectifs de la Convention

La Convention a pour objet de promouvoir la protection, la gestion et l'aménagement des paysages européens et d'organiser la coopération européenne dans ce domaine[5]. Il s'agit d'assurer la protection, la gestion et l'aménagement des paysages européens par l'adoption de mesures nationales et la mise en place d'une coopération européenne entre les Parties.

5 Article 3 de la Convention (Objectifs).

Le préambule de la Convention vise à souligner les enjeux qui la soustendent en mettant en avant les points suivants :

- la Convention s'inscrit dans le contexte des travaux du Conseil de l'Europe dans le domaine du patrimoine naturel et culturel, de l'aménagement du territoire, de l'environnement et de l'autonomie locale[6] ;
- le souci du développement durable énoncé à la Conférence de Rio en 1992 donne au paysage une place essentielle en tant que facteur d'équilibre entre un patrimoine naturel et culturel reflet de l'identité et de la diversité européenne, et ressource économique créatrice d'emplois et liée à l'essor d'un tourisme durable[7] ;
- le paysage joue un rôle important en tant qu'élément de l'environnement et du cadre de vie des populations aussi bien en zone urbaine que rurale et tant pour les paysages remarquables que pour ceux du quotidien. De ce fait, le public est invité à jouer un rôle actif dans sa gestion et son aménagement, et doit se sentir responsable de son devenir ;
- les Etats membres du Conseil de l'Europe, soucieux de promouvoir les idéaux qui sont leur patrimoine commun par des accords internationaux ont avec le paysage, un patrimoine précieux à entretenir et à gérer grâce à une coopération internationale effective et organisée autour d'un instrument juridique exclusivement consacré au paysage[8].

---

6 Le Préambule de la Convention constate que le but du Conseil de l'Europe est de réaliser une union plus étroite entre ses membres, afin de sauvegarder et de promouvoir les idéaux et les principes qui sont leur patrimoine commun, et que ce but est poursuivi en particulier par la conclusion d'accords dans les domaines économique et social.

7 Le Préambule de la Convention fait état du souci des Etats membres du Conseil de l'Europe de « parvenir à un développement durable fondé sur un équilibre harmonieux entre les besoins sociaux, l'économie et l'environnement ».

8 Le Préambule de la Convention reconnaît que la qualité et la diversité des paysages européens constituent une ressource commune pour la protection, la gestion et l'aménagement de laquelle il convient de coopérer. Les paysages d'Europe présentent un intérêt local, mais ont aussi une valeur pour l'ensemble de la population européenne. Ils sont appréciés au-delà du territoire qu'ils recouvrent et des frontières nationales. En outre, certains paysages présentent des caractéristiques communes de part et d'autre d'une frontière et des mesures transfrontalières sont alors nécessaires pour appliquer les principes d'action. Enfin, les paysages sont exposés aux influences, favorables ou défavorables, de processus qui peuvent se déclencher dans d'autres zones et faire sentir leurs effets par-delà les frontières. C'est pourquoi, il est légitime de s'occuper des paysages au niveau européen. La diversité et la qualité des valeurs culturelles et naturelles liées aux paysages européens forment un patrimoine commun des Etats européens, ce qui leur impose de prendre en charge ensemble les moyens propres à garantir de façon concertée la protection de ces valeurs. Seule une convention internationale à l'échelle du Conseil de l'Europe peut donc contribuer à atteindre cet objectif afin de fournir une référence juridique aux autres initiatives internationales œuvrant dans ce domaine.

1.2. Relation avec d'autres textes existants

Les Etats signataires de la Convention déclarent dans leur préambule «souhait[er] instituer un instrument nouveau consacré exclusivement à la protection, à la gestion et à l'aménagement de tous les paysages européens». La Convention représente effectivement aujourd'hui le premier traité international exclusivement consacré à la protection, à la gestion et à la mise en valeur du paysage européen. Quelques instruments juridiques internationaux concernent le paysage, soit directement, soit indirectement. Toutefois, aucun d'entre eux ne traite de manière directe, spécifique et complète des paysages européens et de leur préservation, malgré leur inestimable valeur culturelle et naturelle et les nombreuses menaces qui pèsent sur eux. La Convention est destinée à combler cette lacune: elle se distingue ainsi de la Convention concernant la protection du patrimoine mondial, culturel et naturel de l'Unesco du 16 novembre 1972, à la fois sur les plans formel et matériel. Les deux ont des vocations distinctes tout comme les organisations sous les auspices desquelles elles ont été élaborées. L'une a une vocation régionale, l'autre mondiale. La Convention européenne du paysage apparaît complémentaire de celle de l'Unesco. Sur le plan matériel, elle englobe tous les paysages, même ceux qui n'ont pas une valeur universelle exceptionnelle, mais ne comprend pas les monuments du patrimoine culturel, à la différence de la Convention de l'Unesco. De même, son principal objectif n'est pas d'établir une liste de biens d'intérêt exceptionnel et universel, mais d'établir un régime de protection, de gestion et d'aménagement de tous les paysages sur la base d'une série de principes. Chaque texte possède de ce fait sa spécificité.

Lors des travaux préparatoires ayant conduit à l'élaboration de la Convention, une référence a été constamment faite aux textes juridiques déjà existants aux niveaux international et national concernant le paysage[9]. La Convention

---

9 Parmi ces textes figurent, outre la Convention concernant la protection du patrimoine mondial, culturel et naturel de l'Unesco, la Convention pour la sauvegarde du patrimoine architectural de l'Europe, la Convention relative à la conservation de la vie sauvage et du milieu naturel de l'Europe, la Convention européenne pour la protection du patrimoine archéologique, la Recommandation 95 (9) du Comité des Ministres relative à la conservation des sites culturels intégrée aux politiques de paysage, la Recommandation (79) 9 du Comité des Ministres concernant la fiche d'identification et d'évaluation des paysages naturels en vue de leur protection, la Charte du paysage méditerranéen, le Règlement des Communautés européennes concernant les méthodes de production agricole compatibles avec les exigences de l'environnement ainsi que l'entretien de l'espace naturel, la directive des Communautés européennes concernant la conservation des habitats naturels ainsi que de la faune et de la flore sauvages, la directive des Communautés européennes concernant l'évaluation de l'impact sur l'environnement, ainsi que d'autres textes importants de droit national, communautaire et international.

souligne dans son préambule qu'elle garde à l'esprit les textes juridiques existant au niveau international dans les domaines de la protection et de la gestion du patrimoine naturel et culturel, de l'aménagement du territoire, de l'autonomie locale et de la coopération transfrontalière, notamment la Convention relative à la conservation de la vie sauvage et du milieu naturel de l'Europe (Berne, 19 septembre 1979), la Convention pour la sauvegarde du patrimoine architectural de l'Europe (Grenade, 3 octobre 1985), la Convention européenne pour la protection du patrimoine archéologique (révisée) (La Valette, 16 janvier 1992), la Convention-cadre européenne sur la coopération transfrontalière des collectivités ou autorités territoriales (Madrid, 21 mai 1980) et ses protocoles additionnels, la Charte européenne de l'autonomie locale (Strasbourg, 15 octobre 1985), la Convention sur la diversité biologique (Rio, 5 juin 1992), la Convention concernant la protection du patrimoine mondial, culturel et naturel (Paris, 16 novembre 1972), et la Convention sur l'accès à l'information, la participation du public au processus décisionnel et l'accès à la justice en matière d'environnement (Aarhus, 25 juin 1998).

Afin de prévenir des difficultés éventuelles avec d'autres instruments juridiques internationaux, la Convention indique ne pas porter atteinte aux dispositions plus strictes en matière de protection, de gestion ou d'aménagement des paysages contenues dans d'autres instruments nationaux ou internationaux contraignants qui sont ou entreront en vigueur[10].

2. Le champ d'application

2.1. Champ d'application matériel

La Convention prévoit que sous réserve des dispositions de l'article 15[11], elle s'applique à tout le territoire des Parties et porte sur les espaces naturels, ruraux, urbains et périurbains. Elle inclut les espaces terrestres, les eaux intérieures et maritimes. Elle concerne tant les paysages pouvant être considérés comme remarquables, que les paysages du quotidien et les paysages dégradés.

Il est intéressant de noter que le champ d'action des politiques et des mesures mentionnées dans la Convention doit se référer à la totalité de la dimension paysagère du territoire des Etats. A cet égard, la Convention s'applique à l'ensemble du territoire européen et porte sur les espaces naturels, urbains et

10 Article 12 de la Convention (Relations avec d'autres instruments). La rédaction de cet article s'inspire des dispositions types retenues déjà dans d'autres conventions internationales afin de régler le problème de l'articulation de conventions intervenant dans des domaines similaires.

11 Article 2 de la Convention (Champ d'application). Voir ci-après B.3, Champ d'application territorial.

périurbains, qu'ils soient terrestres, aquatiques (lacs, étangs) ou maritimes (littoral, mer territoriale). Elle ne concerne donc pas uniquement les paysages remarquables, mais aussi les paysages ordinaires du quotidien et les espaces dégradés. Le paysage est désormais reconnu indépendamment de sa valeur exceptionnelle puisque toutes les formes de paysages conditionnent la qualité du cadre de vie des citoyens et méritent d'être prises en compte dans les politiques paysagères. De nombreuses zones rurales et périurbaines notamment, connaissent des transformations profondes et doivent faire l'objet d'une plus grande attention de la part des autorités et du public.

L'originalité de la Convention réside dans cette application aussi bien aux paysages ordinaires qu'aux paysages remarquables. Ce vaste domaine d'application est justifié par les raisons suivantes : tout paysage constitue un cadre de vie pour la population concernée ; il existe des interconnections complexes entre paysages urbains et ruraux ; la plupart des Européens vivent dans des villes (grandes ou petites), dont la qualité paysagère influe énormément sur leur existence ; enfin, les paysages ruraux occupent une place importante dans la sensibilité européenne.

L'extension du champ d'application de l'action des pouvoirs publics en matière de paysage à la totalité de la dimension paysagère de leur territoire national ne signifie cependant pas qu'il faille appliquer les mêmes mesures et politiques à l'ensemble des paysages. Ces mesures et ces politiques devront pouvoir se référer à des paysages qui, selon leurs caractéristiques, nécessiteront des interventions locales diversifiées qui vont de la plus stricte conservation à la véritable création en passant par la protection, la gestion et l'aménagement. Ces interventions peuvent permettre un développement socio-économique important des territoires concernés.

La Convention n'est par ailleurs pas limitée aux seuls éléments culturels ou artificiels, ou aux seuls éléments naturels du paysage : elle se réfère à l'ensemble de ces éléments et aux relations entre eux.

2.2. Champ d'application géographique

La Convention est ouverte à la signature des Etats membres du Conseil de l'Europe[12]. Il est prévu qu'après son entrée en vigueur, le Comité des Ministres du Conseil de l'Europe pourra inviter la Communauté européenne et tout

12 Article 13 de la Convention (Signature, ratification, entrée en vigueur). La Convention est soumise à ratification, acceptation ou approbation. Les instruments de ratification, d'acceptation ou d'approbation sont déposés près du Secrétaire Général du Conseil de l'Europe. Pour tout signataire qui exprimera son consentement à être lié par la Convention après son entrée en vigueur, celle-ci entrera en vigueur le premier jour du mois qui suit l'expiration d'une période de trois mois après la date de dépôt de l'instrument de ratification, d'acceptation ou d'approbation.

Etat européen non membre du Conseil de l'Europe à y adhérer, par une décision prise à la majorité prévue à l'article 20.*d* du Statut du Conseil de l'Europe, et à l'unanimité des Etats Parties ayant le droit de siéger au Comité des Ministres[13].

2.3. Champ d'application territorial

La Convention prévoit dans un article intitulé «Application territoriale» que tout Etat ou la Communauté européenne peuvent, au moment de la signature ou au moment du dépôt de leur instrument de ratification, d'acceptation, d'approbation ou d'adhésion, désigner le ou les territoires auxquels elle s'applique. Par ailleurs, toute Partie peut, à tout moment par la suite, par une déclaration adressée au Secrétaire Général du Conseil de l'Europe, étendre son application à tout autre territoire désigné dans la déclaration[14]. Il est également prévu que toute déclaration faite en vertu des dispositions précédentes pourra être retirée en ce qui concerne tout territoire désigné dans cette déclaration, par notification adressée au Secrétaire Général[15].

Ces dispositions concernent la faculté laissée à certains Etats, pour des raisons constitutionnelles, de ne pas appliquer automatiquement un traité international ratifié à certains territoires, notamment d'outre-mer. Sont uniquement visés des territoires au statut particulier tels que les territoires d'outremer ou les îles Féroé et Groenland pour le Danemark ou Gibraltar, l'île de Man et les bailliages de Jersey et de Guernesey pour le Royaume-Uni. Il a cependant été entendu qu'il serait contraire à l'objet et au but de la Convention qu'une Partie exclue de l'application de cet instrument des parties de son territoire métropolitain et qu'il n'était pas nécessaire de préciser ce point explicitement dans la Convention.

13 Pour tout Etat adhérent ou pour la Communauté européenne en cas d'adhésion, la Convention entrera en vigueur le premier jour du mois qui suit l'expiration d'une période de trois mois après la date de dépôt de l'instrument d'adhésion près du Secrétaire Général du Conseil de l'Europe (article 14 de la Convention – Adhésion).

14 La Convention entrera en vigueur à l'égard de ce territoire le premier jour du mois qui suit l'expiration d'une période de trois mois après la date de réception de la déclaration par le Secrétaire Général (article 15 de la Convention – Application territoriale).

15 Le retrait prendra effet le premier jour du mois qui suit l'expiration d'une période de trois mois après la date de réception de la notification par le Secrétaire Général (article 15 de la Convention – Application territoriale).

2.4. Champ d'application temporel

La Convention présente l'avantage de s'appliquer pendant une durée indéterminée et d'être mise en œuvre sous les auspices d'une organisation internationale, en l'espèce le Conseil de l'Europe[16].

Comme toute convention internationale, la Convention constitue malgré tout un instrument juridique vivant qui évolue avec l'objet de ses dispositions. La prise en compte des valeurs et intérêts paysagers doit ainsi pouvoir évoluer avec le caractère variable de ces valeurs et de ces intérêts. Il est pour cela prévu que toute Partie ou les Comités d'experts visés à la Convention peuvent proposer des amendements à Convention[17]. Les amendements peuvent en effet permettre de l'adapter ou de l'améliorer. Les Comités d'experts mentionnés à l'article 10 de la Convention peuvent préparer les amendements et examiner ceux qui ont été proposés par les Parties[18].

16 Toute Partie peut cependant, à tout moment, dénoncer la Convention en adressant une notification au Secrétaire Général du Conseil de l'Europe. La dénonciation prend effet le premier jour du mois qui suit l'expiration d'une période de trois mois après la date de réception de la notification par le Secrétaire Général (article 16 de la Convention – Dénonciation).

17 Toute proposition d'amendement est notifiée au Secrétaire Général du Conseil de l'Europe qui la communique aux Etats membres du Conseil de l'Europe, aux autres Parties et à chaque Etat européen non membre qui a été invité à adhérer à la Convention. Toute proposition d'amendement est examinée par l'organe chargé de suivre la mise en œuvre de la Convention qui soumet le texte adopté à la majorité des trois quarts des représentants des Parties au Comité des Ministres pour adoption. Après son adoption par le Comité des Ministres à la majorité prévue à l'article 20.*d* du Statut du Conseil de l'Europe et à l'unanimité des représentants des Etats Parties ayant le droit de siéger au Comité des Ministres, le texte est transmis aux Parties pour acceptation. Tout amendement entre en vigueur à l'égard des Parties qui l'ont accepté le premier jour du mois qui suit l'expiration d'une période de trois mois après la date à laquelle trois Parties membres du Conseil de l'Europe auront informé le Secrétaire Général qu'elles l'ont accepté. Pour toute autre Partie qui l'aura accepté ultérieurement, l'amendement entrera en vigueur le premier jour du mois qui suit l'expiration d'une période de trois mois après la date à laquelle ladite Partie aura informé le Secrétaire Général de son acceptation (article 17 de la Convention – Amendements).

18 Voir ci-après B.1 : l'organe chargé de suivre la mise en œuvre de la Convention.

II. Le contenu de la Convention

1. Les obligations juridiques

1.1. Les définitions

Les termes employés dans la Convention sont définis à l'article 1er afin d'en garantir une interprétation uniforme par tous ceux qui entendent œuvrer pour le bon état des paysages européens:

- «paysage» désigne une partie de territoire telle que perçue par les populations, dont le caractère résulte de l'action de facteurs naturels et/ou humains et de leurs interrelations[19];
- «politique du paysage» désigne la formulation par les autorités publiques compétentes des principes généraux, des stratégies et des orientations permettant l'adoption de mesures particulières en vue de la protection, la gestion et l'aménagement du paysage;
- «objectif de qualité paysagère» désigne la formulation par les autorités publiques compétentes, pour un paysage donné, des aspirations des populations en ce qui concerne les caractéristiques paysagères de leur cadre de vie[20];
- «protection des paysages» comprend les actions de conservation et de maintien des aspects significatifs ou caractéristiques d'un paysage, justifiées par sa valeur patrimoniale émanant de sa configuration naturelle et/ou de l'intervention humaine;
- «gestion des paysages» comprend les actions visant, dans une perspective de développement durable, à entretenir le paysage afin de guider et d'har-

---

19 Le terme paysage est donc défini comme une zone ou un espace, tel que perçu par les habitants du lieu ou les visiteurs, dont l'aspect et le caractère résultent de l'action de facteurs naturels et/ou culturels (c'est-à-dire humains). Cette définition tient compte de l'idée que les paysages évoluent dans le temps, sous l'effet des forces naturelles et de l'action des êtres humains. Elle souligne également l'idée que le paysage forme un tout dont les éléments naturels et culturels sont considérés simultanément.

20 Les expressions «politiques du paysage» et «objectifs de qualité paysagère» visent les phases de la stratégie que les Etats doivent mettre au point en deux étapes:
- la politique du paysage est l'expression de la prise de conscience par les pouvoirs publics de la nécessité d'énoncer une politique publique en matière de paysage. Elle consiste à formuler des orientations fondamentales, des principes généraux et des choix stratégiques qui doivent servir de guide aux décisions de protection, de gestion et d'aménagement du paysage;
- un objectif de qualité paysagère consiste, pour un paysage particulier, après qu'il a été identifié et qualifié, à énoncer précisément les caractéristiques que les populations locales concernées souhaitent voir reconnaître pour leur cadre de vie.

moniser les transformations induites par les évolutions sociales, économiques et environnementales;

– «aménagement des paysages» comprend les actions présentant un caractère prospectif particulièrement affirmé visant la mise en valeur, la restauration ou la création de paysages[21].

Dans chaque zone paysagère, l'équilibre entre ces trois types d'activités dépend du caractère de la zone et des objectifs définis. Certaines zones peuvent mériter une protection très rigoureuse. A l'opposé, il peut y avoir des zones dont le paysage extrêmement abîmé demande à être entièrement remodelé. La plupart des paysages ont besoin d'une combinaison des trois modes d'action, et certains d'entre eux ont besoin d'un certain degré d'intervention.

Dans la recherche d'un juste équilibre entre protection, gestion et aménagement d'un paysage, la Convention ne cherche pas à préserver ou à «geler» des paysages à un stade donné de leur longue évolution. Les paysages ont toujours changé et continueront à changer, tant sous l'effet de processus naturels que de celui de l'action humaine. En réalité, l'objectif est d'accompagner les changements à venir en reconnaissant la grande diversité et la qualité des paysages hérités du passé et de s'efforcer de préserver, voire enrichir, cette diversité et cette qualité au lieu de les laisser péricliter.

---

21 L'article 1 de la Convention contient les définitions se rapportant à trois expressions qui reviennent fréquemment dans la Convention, à savoir « protection », « gestion » et « aménagement » des paysages, qui sont des principes d'action sur le paysage envisagés de manière dynamique et prospective :

– protection des paysages, concerne les mesures prises dans le but de préserver le caractère et la qualité existants d'un paysage auquel les populations attachent une grande valeur du fait de sa forme naturelle ou culturelle particulière. Cette protection doit être active et s'accompagner de mesures d'entretien pour maintenir les aspects significatifs d'un paysage;
– gestion des paysages, concerne les mesures prises conformément au principe de développement durable pour accompagner les transformations induites par les nécessités économiques, sociales ou environnementales. Les mesures pourront concerner l'organisation de ces paysages ou les éléments de leur composition. Elles visent à assurer l'entretien régulier d'un paysage et à veiller à ce qu'il évolue harmonieusement et de manière à satisfaire les besoins économiques et sociaux. La gestion doit être dynamique et tendre à améliorer la qualité des paysages en fonction des aspirations des populations ;
– aménagement des paysages, concerne le processus formel d'étude, de conception et de construction par lequel de nouveaux paysages sont créés de manière à répondre aux aspirations de la population concernée. Il s'agit d'élaborer de véritables projets d'aménagement, notamment dans les espaces les plus touchés par le changement et fortement détériorés (banlieues, zones périurbaines et industrielles, littoraux). Ces projets d'aménagement visent à restructurer profondément des paysages dégradés.

1.2. Les obligations

1.2.1. Concernant l'ordre national

Les Parties contractantes s'engagent à protéger, gérer et/ou aménager leurs paysages par l'adoption de toute une série de mesures, générales et particulières au niveau national, et ce, dans le respect du principe de subsidiarité. Chaque Partie met en œuvre la Convention, en particulier ses articles concernant les mesures à prendre dans l'ordre national, selon la répartition des compétences qui lui est propre, conformément à ses principes constitutionnels et à son organisation administrative, en tenant compte de la Charte européenne de l'autonomie locale. La Convention doit donc être mise en œuvre à l'échelon administratif le plus approprié pour l'adoption de mesures relatives au paysage et le cas échéant, les collectivités locales et régionales ainsi que leurs groupements doivent être assurés d'une participation officielle à ce processus.

Dans le cas où les autorités locales et régionales disposent des compétences requises, la protection, la gestion et l'aménagement des paysages sont en effet plus efficaces si la responsabilité de leur mise en œuvre est confiée – dans le cadre constitutionnel législatif posé au niveau national – aux autorités les plus proches de la population concernée. Chaque Etat doit définir précisément les tâches et les mesures attribuées à chaque niveau (national, régional ou local), et énoncer des règles pour la coordination de ces mesures entre les différents niveaux, notamment en ce qui concerne les instruments relatifs à l'urbanisme et à l'aménagement du territoire[22].

Les Parties contractantes s'engagent à mettre en œuvre quatre mesures générales :

- la reconnaissance juridique du paysage en tant que composante essentielle du cadre de vie des populations, expression de la diversité de leur patrimoine commun culturel et naturel, et fondement de leur identité ;
- la définition et la mise en œuvre des politiques du paysage visant la protection, la gestion et l'aménagement des paysages ;
- des procédures de participation du public, des autorités locales et régionales et des acteurs concernés par la conception et la réalisation des politiques du paysage ;

---

22 La Convention laisse aux Parties le choix des moyens à utiliser dans leurs ordres juridiques internes pour s'acquitter des obligations découlant de celle-ci. Les moyens juridiques, administratifs, fiscaux et financiers mis en place dans chaque pays aux fins de la mise en œuvre de la Convention doivent s'inscrire le plus harmonieusement possible dans les traditions nationales. De plus, en vertu du principe de subsidiarité, la responsabilité des mesures en faveur du paysage incombe aux pouvoirs publics de niveaux local et régional, et pas seulement de niveau national et international.

– l'intégration du paysage dans les politiques d'aménagement du territoire, d'urbanisme et dans les politiques culturelle, environnementale, agricole, sociale et économique, ainsi que dans les autres politiques pouvant avoir un effet direct ou indirect sur le paysage[23].

Les Parties contractantes s'engagent par ailleurs à mettre en œuvre, de manière consécutive, cinq mesures particulières:

– la sensibilisation: il s'agit d'accroître la sensibilisation de la société civile, des organisations privées et des autorités publiques à la valeur des paysages, à leur rôle et à leur transformation;
– la formation et l'éducation: il s'agit de promouvoir: la formation de spécialistes de la connaissance et de l'intervention sur les paysages; des programmes pluridisciplinaires de formation sur la politique, la protection, la gestion et l'aménagement du paysage, destinés aux professionnels du secteur privé et public et aux associations concernées; et des enseignements scolaires et universitaires abordant, dans les disciplines intéressées, les valeurs attachées au paysage et les questions relatives à sa protection, à sa gestion et à son aménagement;
– l'identification et la qualification: il s'agit de mobiliser les acteurs concernés en vue d'une meilleure connaissance des paysages et de guider les travaux d'identification et de qualification des paysages par des échanges d'expériences et de méthodologies, organisés entre les Parties à l'échelle européenne[24];
– la formulation d'objectifs de qualité paysagère: il s'agit formuler des objectifs de qualité paysagère pour les paysages identifiés et qualifiés, après consultation du public;
– la mise en œuvre des politiques du paysage: il s'agit de mettre en place des moyens d'intervention visant la protection, la gestion et/ou l'aménagement des paysages[25].

1.2.2. Concernant l'ordre international

Les Parties contractantes s'engagent à coopérer au niveau international sur un plan européen lors de la prise en compte de la dimension paysagère dans les politiques et programmes internationaux, et à recommander, le cas échéant,

23 Article 5 de la Convention (Mesures générales). Cet article détermine les mesures nécessaires à la mise en œuvre de la Convention dans chaque Partie.

24 Chaque Partie s'engage ainsi: à identifier ses propres paysages, sur l'ensemble de son territoire; à analyser leurs caractéristiques ainsi que les dynamiques et les pressions qui les modifient; à en suivre les transformations; et à qualifier les paysages identifiés en tenant compte des valeurs particulières qui leur sont attribuées par les acteurs et les populations concernés (Article 6 de la Convention – Mesures particulières).

25 Article 6 de la Convention (Mesures particulières).

que les considérations concernant le paysage y soient incorporées[26]. Elles s'engagent par ailleurs à coopérer pour renforcer l'efficacité des mesures prises conformément aux dispositions de la Convention, et en particulier : à offrir une assistance technique et scientifique mutuelle par la collecte et l'échange d'expériences et de travaux de recherche en matière de paysage ; à favoriser les échanges de spécialistes du paysage, notamment pour la formation et l'information ; et à échanger des informations sur toutes les questions visées par les dispositions de la Convention[27].

Les paysages transfrontaliers font l'objet d'une disposition spécifique : les Parties s'engagent à encourager la coopération transfrontalière au niveau local et régional et, au besoin, à élaborer et mettre en œuvre des programmes communs de mise en valeur du paysage[28].

---

26 Article 7 de la Convention (Politiques et programmes internationaux). La Convention doit faciliter une prise en compte accrue du paysage dans les diverses instances internationales ainsi que dans les programmes internationaux. Pour cela, les Parties spécialement sensibilisées au problème du paysage doivent jouer un rôle actif, en coordonnant leurs réflexions et leurs propositions au sein de l'organe chargé de suivre la mise en œuvre de la Convention. Il est par ailleurs prévu que le Conseil de l'Europe devrait organiser une coopération particulière sur le thème du paysage avec d'autres organisations internationales, gouvernementales, en particulier l'Unesco, l'Union européenne et l'IUCN, et d'autres organisations non gouvernementales.

27 Article 8 de la Convention (Assistance mutuelle et échange d'informations). Ces dernières années, on a constaté une extension considérable de l'intérêt – politique, professionnel et universitaire – que suscitent les questions paysagères, d'où le développement d'un ensemble d'expériences et de compétences techniques dont les Etats membres, les collectivités locales et régionales, et les autres acteurs peuvent s'inspirer pour la mise en œuvre de la Convention. Dans le même temps, les moyens disponibles pour cet échange d'idées – et pour les aspects techniques de l'étude paysagère – ont été améliorés radicalement par le progrès des communications électroniques et l'arrivée d'Internet. Grâce à cette évolution, l'échange d'idées et l'assistance mutuelle peuvent s'opérer sur une base bien plus large qu'il y a ne serait-ce qu'une décennie ; c'est ainsi que, dans toute l'Europe, les acteurs locaux peuvent participer à cet échange qui permet d'instaurer une véritable « démocratie paysagère ».

28 Article 9 de la Convention (Paysages transfrontaliers). Cet article engage les Parties à mettre au point des programmes transfrontaliers pour l'identification, la qualification, la protection, la gestion et l'aménagement des paysages transfrontaliers. Dans l'élaboration de ces programmes, tout en respectant le principe de subsidiarité défini par la Charte européenne de l'autonomie locale, les Parties sont invitées à s'appuyer sur les collectivités locales et régionales en utilisant comme instruments de mise en œuvre ceux préconisés par la Convention-cadre européenne du 21 mai 1980 sur la coopération transfrontalière des collectivités ou autorités territoriales, et ses protocoles additionnels.

1.3. Le Prix du paysage du Conseil de l'Europe

La Convention prévoit l'attribution d'un Prix du paysage du Conseil de l'Europe. Celui-ci constitue une reconnaissance de la politique ou des mesures prises par des collectivités locales et régionales ou des organisations non gouvernementales en matière de protection, de gestion et/ou d'aménagement durable de leurs paysages, faisant preuve d'une efficacité durable et pouvant aussi servir d'exemple aux autres collectivités territoriales européennes.

Ce prix contribue par conséquent à stimuler les acteurs locaux pour encourager et reconnaître une gestion exemplaire des paysages. Il est décerné par le Comité des Ministres, sur proposition de l'organe chargé du suivi de la mise en œuvre de la Convention.

Peuvent ainsi se voir attribuer le Prix du paysage du Conseil de l'Europe les collectivités locales et régionales et leurs groupements qui, dans le cadre de la politique de paysage d'une Partie à la Convention, ont mis en œuvre une politique ou des mesures visant la protection, la gestion et/ou l'aménagement durable de leurs paysages, faisant la preuve d'une efficacité durable et pouvant ainsi servir d'exemple aux autres collectivités territoriales européennes. La distinction peut également être attribuée aux organisations non gouvernementales qui ont fait preuve d'une contribution particulièrement remarquable à la protection, à la gestion ou à l'aménagement du paysage.

Les candidatures au Prix du paysage seront transmises par les Parties à l'organe chargé de suivre la mise en œuvre de la Convention. Les collectivités locales et régionales transfrontalières et les regroupements de collectivités locales ou régionales concernés (au sein d'un seul pays ou sur une base transfrontalière) peuvent être candidats, à la condition qu'ils gèrent ensemble le paysage en question. Sur proposition de l'organe chargé de suivre la mise en œuvre de la Convention, le Comité des Ministres définit et publie les critères d'attribution du Prix du paysage, adopte son règlement et décerne le prix.

L'attribution du Prix du paysage doit conduire les sujets qui en sont titulaires à veiller à la protection, à la gestion et/ou à l'aménagement durables des paysages concernés[29]. Le Prix vise donc à stimuler un processus que pourraient lancer les Etats dans toute l'Europe pour encourager et reconnaître une gestion exemplaire des paysages. Il pourrait ainsi couronner un processus géré au niveau national comportant éventuellement l'organisation de concours

---

29 Article 11 de la Convention (Prix du paysage du Conseil de l'Europe). Le Rapport explicatif précise que les collectivités locales et régionales, leurs groupements, et des organisations non gouvernementales pourront concourir pour le prix du paysage par le biais de leur Etat membre. L'Etat partie pourra ainsi évaluer les candidatures, éventuellement dans le cadre d'un concours national pouvant donner lieu à des prix ou récompenses, et présenter à l'organe chargé de suivre la mise en œuvre de la Convention, le gagnant national ou un nombre limité de candidats en vue de l'attribution du prix.

nationaux identiques et un soutien financier aux collectivités locales et régionales concernées.

2. Le dispositif institutionnel

2.1. L'organe chargé de suivre la mise en œuvre de la Convention

Le Conseil de l'Europe fournit le Secrétariat de la Convention et dispose de structures où toutes les Parties à la Convention peuvent se faire représenter[30].

La Convention prévoit que les Comités d'experts compétents existants, établis en vertu de l'article 17 du Statut du Conseil de l'Europe, sont chargés par le Comité des Ministres du Conseil de l'Europe, du suivi de sa mise en œuvre[31]. Il est en effet apparu que les objectifs de la Convention seraient plus aisément atteints si les représentants des Parties ont la possibilité de se rencontrer régulièrement pour mettre au point des programmes communs et coordonnés et assurer de façon conjointe le suivi de la mise en œuvre de la Convention.

Compte tenu de la pluridisciplinarité de la notion et des activités liées au paysage, il a été décidé que le suivi de cette mise en œuvre pourrait ainsi être confié au Comité pour les activités du Conseil de l'Europe en matière de diversité biologique et paysagère (CO-DBP) et au Comité du patrimoine culturel (CC-PAT) qui, au sein du Conseil de l'Europe, œuvrent dans le domaine traité par la Convention et qui ont un accès direct au Comité des Ministres. Il a aussi été considéré qu'afin d'accomplir cette tâche, ces comités peuvent se réunir conjointement dans le cadre d'une conférence, afin que la Convention profite d'un forum de discussion approprié. La Déclaration de la Deuxième Conférence des Etats contractants et signataires de la Convention européenne

---

30 Le texte de la Convention tel qu'il a été fait à Florence le 20 octobre 2000, est notamment déposé en français et en anglais dans les archives du Conseil de l'Europe. Il est prévu que le Secrétaire Général du Conseil de l'Europe en communique copie certifiée conforme à chacun des Etats membres du Conseil de l'Europe ainsi qu'à tout Etat ou à la Communauté européenne invités à adhérer à la présente Convention. Par ailleurs, le Secrétaire Général du Conseil de l'Europe notifie en particulier aux Etats membres du Conseil de l'Europe, à tout Etat ou la Communauté européenne, ayant adhéré à la présente Convention : toute signature ; le dépôt de tout instrument de ratification, d'acceptation, d'approbation ou d'adhésion ; toute date d'entrée en vigueur de la Convention conformément aux articles 13, 14 et 15 ; toute déclaration faite en vertu de l'article 15 ; toute dénonciation faite en vertu de l'article 16 ; toute proposition d'amendement, ainsi que tout amendement adopté conformément à l'article 17 et la date à laquelle cet amendement entre en vigueur ; tout autre acte, notification, information ou communication ayant trait à la Convention (Article 18 de la Convention – Notifications).

31 Article 10 de la Convention (Suivi de la mise en œuvre de la Convention).

du paysage, adoptée à Strasbourg le 29 novembre 2002 et dont le Comité des Ministres a pris note le 28 mai 2003, a par ailleurs demandé d'associer le Comité des hauts fonctionnaires de la Conférence européenne des ministres responsables de l'aménagement du territoire (CEMAT) aux Comités d'experts compétents, chargés en vertu de l'article 10 de la Convention du suivi de sa mise œuvre[32].

L'Assemblée parlementaire et le Congrès des pouvoirs locaux et régionaux de l'Europe (CPLRE) doivent être associés aux travaux de ces Comités liés à la Convention. Etant donné les responsabilités croissantes des autorités locales et régionales en ce qui concerne la protection, la gestion et l'aménagement des paysages, le CPLRE pourra adresser des avis au Comité des Ministres sur les rapports préparés par l'organe chargé du suivi de la mise en œuvre de la Convention, sur la base de l'article 2, alinéa 2 de la Résolution statutaire (2000) 1.

La Convention prévoit qu'après chacune des réunions des Comités d'experts, le Secrétaire Général du Conseil de l'Europe transmet un rapport sur ses travaux et son fonctionnement au Comité des Ministres et que l'organe chargé de son suivi propose au Comité des Ministres les critères d'attribution et le règlement du Prix du paysage[33].

2.2. Les Conférences des Etats contractants et signataires de la Convention

Deux Conférences des Etats contractants et signataires à la Convention européenne du paysage ont été organisées à Strasbourg les 22 et 23 novembre 2001 et 28 et 29 novembre 2002[34]. Celles-ci ont eu pour objet de réunir, en vue de l'entrée en vigueur de la Convention, les Etats contractants et signataires, les Etats invités à la signer ainsi que des observateurs.

Ces Conférences ont pour objet:

- de promouvoir la signature et/ou la ratification de la Convention afin qu'elle entre rapidement en vigueur;
- d'apporter une assistance juridique aux Etats signataires et aux Etats membres du Conseil de l'Europe invités à signer la Convention;
- de préparer la mise en œuvre effective de la Convention après son entrée en vigueur.

---

32 Décision du Comité des Ministres du Conseil de l'Europe (840e réunion, 28 mai 2003) de prendre note de la Déclaration de la deuxième Conférence des Etats signataires et contractants de la Convention européenne du paysage (28-29 novembre 2002, Strasbourg), telle qu'elle figure à l'Annexe 22 du document CM (2003) 66.

33 Article 10 de la Convention (Suivi de la mise en œuvre de la Convention). Le Rapport explicatif de la Convention prévoit que le CPLRE est appelé à participer activement aux activités de suivi et à exprimer son avis sur les critères d'attribution du Prix du paysage du Conseil de l'Europe.

34 Voir les rapports de ces Conférences: T-FLOR 1 (2001) 19 et T-FLOR 2 (2002) 27.

A l'occasion de la deuxième Conférence, une Déclaration a été adoptée[35].

Des travaux préparatoires tendant à faciliter l'entrée en vigueur de la Convention ont été réalisés sur les cinq thèmes suivants[36]:

- politiques du paysage: contribution au bien-être des citoyens européens et au développement durable (approches sociale, économique, culturelle et écologique);
- identification, qualification du paysage et objectifs de qualité paysagère, en tirant parti des ressources culturelles et naturelles;
- sensibilisation, éducation et formation;
- instruments innovateurs en vue de la protection, de la gestion et de l'aménagement du paysage;
- prix du paysage.

Une première réunion des Ateliers pour la mise en œuvre de la Convention européenne du paysage s'est par ailleurs tenue à Strasbourg les 23 et 24 mai 2002[37] et une deuxième réunion des Ateliers se tiendra à nouveau à Strasbourg les 27 et 28 novembre 2003.

Trois thèmes sont à présent à l'étude en vue de la tenue de la Troisième Conférences des Etats contractants et signataires à la Convention européenne du paysage qui se tiendra en 2004:

- l'intégration du paysage dans les politiques et programmes internationaux (article 7 de la Convention) et les paysages transfrontaliers (article 9 de la Convention);
- le paysage et le bien-être individuel et social (préambule de la Convention);
- le paysage et l'aménagement du territoire (article 5, d de la Convention).

Un document rassemblant des informations reçues concernant une fiche synthétique de présentation des politiques de paysage menées dans les Etats membres du Conseil de l'Europe a été réalisé[38] et un site Internet de la Convention mis sur place[39]. Il est prévu que la Convention entrera très prochainement en vigueur.

---

35 Voir l'Annexe 2.

36 Voir les documents et rapports de ces Conférences: T-FLOR 2 (2002) 20, 21, 22, 23 et 24.

37 Voir le document: T-FLOR 2 (2002) 18 et 18 addendum.

38 Voir le document: T-FLOR 2 (2002) 11.

39 *ww.coe.int/Conventioneuropéennedupaysage; www.coe.int/EuropeanLandscapeConvention*

Conclusion

Les modes de vie contemporains font que les citoyens aspirent de plus en plus à retrouver un cadre de vie non défiguré et à préserver leur patrimoine, tant naturel que culturel. De par cette demande sociale croissante, le paysage trouve – ou retrouve – ses lettres de noblesse et commence à être perçu comme une composante majeure des politiques environnementales. Il représente en outre un atout majeur pour le développement régional sur le plan touristique.

La Convention suscite de grands espoirs : il s'agit de reconnaître l'importance et la valeur des paysages et de réconcilier le droit à la rentabilité et le droit au bien-être, à la santé, à l'esthétique et à la beauté.

Annexe 1

COUNCIL OF EUROPE CONSEIL DE L'EUROPE

Convention européenne du paysage
(European Landscape Convention)
STE n° : 176
Traité ouvert à la signature des Etats membres du Conseil de l'Europe et à l'adhésion de la Communauté européenne et des Etats européens non membres

Situation au 07/10/03

Ouverture à la signature :
Lieu : Florence
Date : 20/10/00

Entrée en vigueur :
Conditions : 10 Ratifications.
Date :

Etats membres du Conseil de l'Europe :

Etats	Date signature	Date ratification	Date entrée en vigueur	Renv.	R.	D.	A.	T.	C.	O.
Albanie										
Andorre										
Arménie	14/05/03									
Autriche										
Azerbaïdjan										
Belgique	20/10/00									
Bosnie-Herzégovine										
Bulgarie	20/10/00									
Croatie	20/10/00	15/01/03								
Chypre	21/11/01									
République tchèque	28/11/02									
Danemark	20/10/00	20/03/03						X		
Estonie										
Finlande	20/10/00									

France	20/10/00									
Géorgie										
Allemagne										
Grèce	13/12/00									
Hongrie										
Islande										
Irlande	22/03/02	22/03/02								
Italie	20/10/00									
Lettonie										
Liechtenstein										
Lituanie	20/10/00	13/11/02								
Luxembourg	20/10/00									
Malte	20/10/00									
Moldova	20/10/00	14/03/02								
Pays-Bas										
Norvège	20/10/00	23/10/01								
Pologne	21/12/01									
Portugal	20/10/00									
Roumanie	20/10/00	07/11/02								
Russie										
Saint-Marin	20/10/00									
Serbie-Monténégro										
Slovaquie										
Slovénie	07/03/01	25/09/03								
Espagne	20/10/00									
Suède	22/02/01									
Suisse	20/10/00									
l'ex-République yougoslave de Macédoine	15/01/03									
Turquie	20/10/00									
Ukraine										
Royaume-Uni										

Etats non membres du Conseil de l'Europe :

Etats	Date signature	Date ratification	Date entrée en vigueur	Renv.	R.	D.	A.	T.	C.	O.

Nombre total de signatures non suivies de ratifications :	19
Nombre total de ratifications/adhésions :	8

Renvois :
a: Adhésion – s: Signature sans réserve de ratification – su: Succession – r: Signature « ad referendum ».
R.: Réserves – D.: Déclarations – A.: Autorités – T.: Application territoriale – C.: Communication – O.: Objection.

Source : Bureau des Traités sur http://conventions.coe.int

Annexe 2

Déclaration de la Deuxième Conférence des États contractants et signataires de la Convention européenne du paysage, réunie à Strasbourg les 28 et 29 novembre 2002

La Deuxième Conférence des États contractants et signataires de la Convention européenne du paysage, réunie à Strasbourg les 28 et 29 novembre 2002, par la voie des Comités d'experts compétents chargés par le Comité des Ministres du Conseil de l'Europe du suivi de sa mise en œuvre: le Comité Directeur du Patrimoine Culturel (CDPAT) et le Comité pour les activités du Conseil de l'Europe en matière de diversité biologique et paysagère (CO-DBP),

Considère :
- l'importance de la Convention européenne du paysage pour le respect des valeurs promues par le Conseil de l'Europe: défense des droits de l'homme, garantie de la démocratie et recherche de solutions aux grands problèmes de société;
- la contribution fondamentale de la Convention pour la mise en œuvre des résultats de la Conférence des Nations Unies sur le développement durable tenue à Johannesburg en août-septembre 2002;
- la nécessité de favoriser l'entrée en vigueur rapide de la Convention,

Tient à souligner :
- que la Convention européenne du paysage, prenant acte de la diversité des paysages, invite à définir et à mettre en place des politiques du paysage, ce qui écarte toute prétention d'imposer une politique uniforme du paysage ;
- que le paysage est un facteur de bien-être de l'individu dans sa singularité ainsi qu'en tant que membre de la société civile, qui bénéficie non seulement à ceux qui y résident en tant qu'élément familier de leur vie quotidienne mais aussi à ceux qui le traversent dans le cadre d'activités touristiques ou professionnelles;
- que le paysage constitue un élément clé du développement territorial durable du continent européen dans la mesure ou il concilie les besoins sociaux, environnementaux et culturels, tout en contribuant au développement économique et en particulier à la création d'emplois ;
- que le paysage est façonné par les processus naturels et culturels et que, pour le ménager, il est nécessaire d'avoir conscience de son histoire et de ses caractéristiques particulières ;
- que le principe de l'intégration du paysage dans tous les types de territoires et dans toutes les politiques sectorielles est une condition fondamentale pour ne pas limiter la politique du paysage aux seuls paysages déjà protégés ;
- qu'il convient de maîtriser l'évolution des paysages par une protection, une gestion et un aménagement à échelle humaine ;
- qu'une prise en compte optimale du paysage, notamment dans les zones périurbaines, contribue de manière importante au bien-être des populations et permet par-là de maintenir la cohésion sociale territoriale ;
- que les pouvoirs locaux et régionaux sont particulièrement concernés par la mise en œuvre efficace et cohérente de la Convention européenne du paysage avec l'appui du Congrès des pouvoirs locaux et régionaux du Conseil de l'Europe ;

– que la Convention d'Aarhus sur l'accès à l'information, la participation du public au processus décisionnel et l'accès à la justice en matière d'environnement contribue à faciliter la mise en œuvre de la Convention européenne du paysage,

Invite les États contractants et signataires :

1. à mettre en place, au plus tôt et au niveau approprié, les institutions propres à assurer à travers des organes permanents la coordination et la consultation en matière de paysage, en tenant compte du principe de subsidiarité ;
2. à développer et à mettre à disposition des outils facilitant le respect du paysage dans les politiques sectorielles et des instruments incitatifs ;
3. à être attentifs au respect du principe de cohérence afin d'éviter les contradictions entre les politiques paysagères et les autres politiques sectorielles ;
4. à veiller à la formulation des objectifs de qualité paysagère et à la mise en œuvre de politiques de protection, de gestion et d'aménagement, de manière appropriée ;
5. à réaliser, au cas où ils ne l'auraient pas déjà fait, des travaux d'identification et de qualification des paysages en collaboration active entre les différents acteurs, qu'ils soient des représentants élus, des scientifiques, des professionnels ou des habitants ;
6. à développer au mieux les actions de sensibilisation, de formation et d'éducation prévues par la Convention, notamment pour développer les compétences professionnelles nécessaires et pour faciliter le dialogue entre les citoyens, les élus et les autres acteurs ;
7. à favoriser la recherche relative à la connaissance quantitative et qualitative des valeurs que les populations attribuent aux paysages, en articulant les divers champs de signification du paysage ;
8. à promouvoir la connaissance des savoirs traditionnels et une mutuelle reconnaissance de ces savoirs et des connaissances scientifiques ;
9. à favoriser l'accès aux résultats de la recherche scientifique et leur compréhension non seulement par le monde scientifique, mais également par l'ensemble de la société ;
10. à promouvoir la coopération sur des projets traitant des systèmes de valeurs attribuées aux paysages par les sociétés concernées,

Demande au Comité des Ministres du Conseil de l'Europe d'associer le Comité des hauts fonctionnaires de la Conférence européenne des ministres responsables de l'aménagement du territoire (CEMAT) aux Comités d'experts compétents, chargés en vertu de l'article 10 de la Convention du suivi de sa mise en œuvre,

***Décide* de transmettre la présente Déclaration au Comité des Ministres du Conseil de l'Europe.**

Martin Lendi

Ethische Grundorientierungen zu Fragen der Raumplanung/Raumordnung – eine Skizze

Hinweise auf ethische Anhaltspunkte für das Planen und Entscheiden wie auch das Handeln in der räumlichen Planung – Etappenbericht auf dem Weg einer offenen Diskussion

In zahlreichen Wissenschafts-, Politik- und Lebensbereichen häufen sich gegenwärtig Ethikfragen. An Thesen und Debatten mangelt es nicht. Überraschenderweise haben sich die räumliche Planung als öffentliche Aufgabe und die Wissenschaften, die ihr zur Seite stehen, während langer Zeit der Auseinandersetzung entzogen. Die Umweltwissenschaften hielten es anders. Sie sind auf die ethischen Aspekte zugegangen. So wird es für die räumliche Planung, die sich in Tuchfühlung mit dem Schutz der Umwelt sieht, Zeit, sich Rechenschaft über ihr Verhältnis zur Ethik zu geben und sich zu fragen, ob sich für sie resp. ihre Träger ethische Anhaltspunkte anbieten oder gar aufdrängen. Mit den nachstehenden Bemerkungen wird versucht, die Probleme anzusprechen und eine Skizze möglicher Grundorientierungen für die Alltagsarbeit und die Lehre zu vermitteln – nicht als abschliessende Aussagen, sondern als Einladung, hier und jetzt mitzudenken, im Wissen, dass Grundorientierungen Hilfen und Stolpersteine zugleich sind.

Ethische Grundorientierungen zu vermitteln, ist eine heikle Aufgabe.[1]

Dies gilt für die Lehre von der Ethik. Und dies trifft überall dort zu, wo ethische Erwägungen anzustellen sind, so vor dem Hintergrund täglichen Entscheidens und Handelns, des Vordringens der Naturwissenschaften, der Gesetzgebung, beim Strafen, nicht minder aber gegenüber den Doktrinen zur räumlichen Planung und vis à vis der öffentlichen Aufgabe der Raumpla-

---

1 Diese Skizze – ein Zwischenergebnis auf einem langen Weg – stellt sich Problemen und wirft Fragen auf, denen wir nicht ausweichen können. Voraussichtlich werden auch vertiefte Überlegungen nicht zu einer Klärung des Verhältnisses von Ethik und Planung führen. Dies hindert nicht, den roten Faden immer wieder zu suchen. Die Diskussion soll angeregt und gefördert werden. Im Gesamtkontext der Ethik der Raumplanung bildet allerdings die hier im Umrissen dargelegte Frage, ob Grundorientierungen oder ob gar ein Ethikkodex für Raumplaner zu erarbeiten und zu vermitteln seien und, wenn ja, in welcher Art, nur eine von vielen. Ihr voraus müssen grundsätzliche Erörterungen gehen. Im vorliegenden Text können die vorangehenden und die nachfolgenden Fragen nur gestreift werden.

nung / Raumordnung.[2] Ihr Zukunftsanspruch ruft nach Zielen, Werten, Vorgaben. Aber: Die Ethik ist nicht unangefochten. Sachzwänge und Normativitätsansprüche, objektive und subjektive, basierte und spekulative, bedrängen sie. Die Klippen ethischer Grundorientierungen sind damit angesprochen. Sie liegen bei der räumlichen Planung, bei den Planern sowie den von der Planung Betroffenen, sie liegen aber auch bei den Fakten politischer, wirtschaftlicher, gesellschaftlicher und natürlicher Entwicklungsstränge, den dortigen Engpässen, sogar in der Ethik als solcher und in den um ethische Grundorientierungen ringenden Lehren. Die Ethik wie die Planung sind nicht problemfrei.

Die *Ethik* versteht sich unter divergierenden philosophischen und theologischen wie auch methodischen Ansätzen sehr unterschiedlich. Auch dringt sie in variationsreichen Intensitätsgraden in die Alltagprobleme vor, teilweise nimmt sie davon Abstand, teilweise übernimmt sie sich an ihnen. Sie ist jedenfalls nicht leicht fassbar. Und doch meldet sie sich zu Wort. Selbst im Gewissen des Einzelnen ist sie, wenn auch unterschiedlich gewichtet, gegenwärtig. Der Verdacht, moralisch voreingenommen zu sein, lastet ihr an. Der implizierte Vorwurf gereicht ihr aber zur Mahnung. Als letzte Ressource muss sie geistige Freiheit vorausschicken und dennoch Unordnung vermeiden, also die Gratwanderung von verpflichtenden und doch Freiheit voraussetzenden Hinweisen wagen, indem sie zum Besinnen anhält. Ihr letzter Sinn liegt nicht in verordneten Ordnungen , auch nicht im guten Leben, sondern in der Beunruhigung, in der zum Bedenken anstiftenden Unruhe.

Die *Planung*, die sich bemüht, Gefahren für den Lebensraum abzuwehren und dessen Entwicklung hin zur Zukunftsfähigkeit von Politik, Wirtschaft und Gesellschaft in Raum und Zeit sachte zu steuern, steht unter dem Vorurteil, einseitig zweckrationalen Gesichtspunkten zu folgen. Auch leidet sie unter der wiederkehrenden Vorhalt, auf finalisierte Endzustände hin zu wirken, die Zukunft vorwegzunehmen und dabei die Entscheidungskraft zu lähmen. Solche Einwände haften an ihren Fersen, obwohl sie sich dagegenstellt. Den-

---

2 Die dt. Akademie für Raumforschung und Landesplanung (ARL) mit Sitz in Hannover hat vor wenigen Jahren einen Arbeitskreis beauftragt, Gedanken zur Ethik in der Raumplanung zu entwickeln. Die Niederschrift der Reflexionen dieser Gruppe erscheint voraussichtlich im Jahre 2004 unter dem Titel „Ethik in der Raumplanung“. Dort wird die vorliegende Gedankenführung nicht im vollen Wortlaut zu finden sein, ist aber in den Beiträgen des Autors sinngemäss enthalten.

Im Arbeitskreis der ARL wurde die Frage, ob es angezeigt sei, für die in der Raumplanung Tätigen ethische Grundaussagen zu formulieren, kontrovers diskutiert. Als Leiter des Arbeitskreises hielt sich der Verfasser dieser Skizze zurück, übernahm es aber, einen Entwurf möglicher Grundorientierungen vorzulegen. Daraus gingen – über mehrere Etappen entwickelt, teilweise publiziert, teilweise manuskriptmässig verbreitet – die hier ausgebreiteten Erwägungen hervor. Sie verpflichten den Autor, nicht aber den Arbeitskreis und die Akademie, denen ich allerdings grössten Dank schulde, weil sie mich inspirierten. Es wird abzuwarten sein, wie sich die weitere Diskussion entwickelt. Ich bin ihr gegenüber offen.

noch kommen weder Politik, Wirtschaft noch Gesellschaft ohne Auseinandersetzung mit der Zukunft aus. Planung steht für das Befassen mit der Zukunft, Raumplanung für die zukunftsfähige Erhaltung und Gestaltung des Lebensraumes mit Chancen der individuellen, politischen, wirtschaftlichen und sozialen Lebensentfaltung.

Orientierungsbedürfnisse

Eine der Hauptsorgen der räumlichen Planung gilt der Orientierung. Sie erwartet von ihr Eckwerte des Selbstverständnisses und bewertete Vorgaben des lenkenden Einwirkens. Worauf soll sie gründen, woran soll sie ihre Aussagen, ihre Ziele und Massnahmen messen? Reicht es hin, auf praktikable Methoden zu vertrauen und sie anzuwenden? Genügt es, sich dem Recht zu stellen und sich von ihm ausrichten zu lassen? Weisen ökonomische Faktoren den Weg? Taugt es, rational Zielsysteme zu entwerfen?

Probleme, sich zu orientieren, stellen sich in der heutigen Zeit nicht nur der räumlichen Planung. Sie sind allenthalben präsent – im Alltag genau so wie im ausholenden Nachdenken.[3] Mitten in der überbordenden Informationsflut und sich ausweitenden räumlich/sachlichen Komplexitäten wird es anhaltend schwieriger, sich über die Grundlagen des Denkens, Entscheidens und Handelns Rechenschaft zu geben. Häufiger denn je werden Abkürzungen – nicht selten in Form von Rezepten – angeboten. Beispielsweise fragt sich, ob Prinzipien wie jenes der Nachhaltigkeit, Postulate wie jenes der Würde der Kreatur, Gewissheiten – mitten unter der Last der Ungewissheiten – verschaffen. Lassen sich, ein weiteres Beispiel, mit den Methoden des effizienten Sichtens der Wissensfülle oder gar der Komplexitätsreduktion gesichertes Wissen und Wege des Problemlösens erarbeiten? Gibt es Meta-Wissenschaften, die Kapitulationen im Dickicht der Undurchsichtigkeiten abwenden und Durchbrüche erlauben? Ist gar die Ethik, auf deren Spur wir uns hier bewegen, eine qualifizierte Hilfe, wenn nicht gar *die* Hilfe?

---

3 Orientierungsbedürfnisse drücken sich mindestens zweifach aus. Einerseits gilt die Suche der Überwindung des Ungewissen und des Nichtwissens durch Wissenserarbeitung resp. dem Umgang mit dem Nichtwissen, anderseits ruft sie nach Eckwerten, von denen aus Wegweisung im Nichtwissen und Ungewissen möglich wird. Zu diesen zwei Aspekten ist die Literatur unendlich. Sie berührt die Philosophie, die Ethik, die Theologie, die Psychologie und die Planungswissenschaften. Als Vertreter der beiden Ansätze können Karl Popper und Hans Jonas angesprochen werden. Siehe unter anderem *Popper Karl*, Über Wissen und Nichtwissen, Vortrag gehalten in Frankfurt am Main anlässlich der Verleihung der Ehrendoktorwürde am 8. Juni 1976, in: Auf der Suche nach einer besseren Welt, Vorträge und Aufsätze aus dreissig Jahren, München / Zürich 1987; *Jonas Hans*, Das Prinzip Verantwortung, Frankfurt am Main 1979.

Unter dem Gewicht der Fragen und offenen Antworten drohen anerkannte Methoden der Problemerkenntnis, der Problemannahme und des bewussten Problemlösens an Bedeutung einzubüssen. Die Planer, die planenden Behörden und die partizipierenden Bürgerinnen und Bürger werden vermehrt im Stich gelassen. Sie beginnen, an ihrer Kompetenz, an ihrem Auftrag, an den Methoden des Planens zu zweifeln – sie resignieren, oder erfinden die Raumplanung in unendlichen vielen Variationen neu – immer wieder, wiederkehrend. Andere insistieren mit Hartnäckigkeit auf dem Vertrauten, beklagen die Ignoranz Dritter und umgeben sich mit schützender Wissenschaftlichkeit.

Gesucht ist die Fähigkeit, sich zu orientieren und zu besinnen. Gefragt sind also nicht Gewissheiten, sondern die Neugierde und der Wille, der Frage nach dem, was wir – mitten im Ungewissen, bedrängt von Informationsströmen – tun müssen, nachzugehen. Keine einfache Aufgabe – weder für die Lehre, noch für die Praxis, schon gar nicht für die räumliche Planung, die sich schwer tut, sich zu erklären und zu begründen, so elementar die Aufgabe ist, den Lebensraum zu schützen, zu gestalten und anhaltend die Voraussetzungen für das Leben in Freiheit und Würde zu stärken. Gesucht sind Orientierungen, die, wie bereits gesagt, nicht beruhigen, sondern helfen, jene Unruhe zu stiften, die offene Fragen als solche erkennt und sich mit ihnen auseinandersetzt.

Zum Verständnis räumlicher Planung

Was räumliche Planung, was Raumplanung ist, dies mag schwer zu verstehen sein.[4] Ihre Leistungen sind aber beeindruckend. Ihr gelang es in der zweiten Hälfte des 20. Jahrhunderts bei wachsenden Ansprüchen des Staates, der Wirtschaft, der Gesellschaft und der einzelnen Menschen, den Lebensraum

---

4 Zum Verständnis der räumlichen Planung vgl. u. a.: *Akademie für Raumforschung und Landesplanung (Hrsg.),* Handwörterbuch der Raumordnung, Hannover 1995; *eadem,* Methoden und Instrumente räumlicher Planung, Hannover 1998; *eadem*, Grundriss der Landes- und Regionalplanung, Hannover 1999. Grundlegend sind vor allem die Aufsätze von *Ritter Ernst-Hasso*, Stellenwert der Planung in Staat und Gesellschaft, und von *Wolf Klaus*, Theoretische Aspekte der räumlichen Planung, in: ARL, Methoden und Instrumente räumlicher Planung, a. a. O., S. 6 ff. resp. S. 39 ff. Siehe sodann *ARL/VLP*, Deutsch-Schweizerisches Handbuch der Planungsbegriffe, Hannover/Bern 1999, und *Jenkins Helmut W. (Hrsg.)*, Raumordnung und Raumordnungspolitik, München/Wien 1996; *Lendi Martin,* Grundriss einer Theorie der Raumplanung, 3. A., Zürich 1996; *Lendi Martin/Elsasser Hans*, Raumplanung in der Schweiz, eine Einführung, 3. A., Zürich 1991; *Lendi Martin*, Raumplanung im Umbruch. Auf dem Weg zu einer politisch, sachlich und ethisch indizierten Raumplanung, Schweizerisches Zentralblatt für Staats- und Verwaltungsrecht, Zürich 1999, S. 193 ff.; *Spitzer Hartwig,* Einführung in die räumliche Planung, Stuttgart 1995.

zu erhalten und Gestaltungsfreiräume offen zu halten, sicherlich unzulänglich, aber gemessen am Druck gesellschaftlicher Prozesse doch beachtenswert: Infrastrukturen tragen, Landschafts-, Siedlungs- und Transport-/Versorgungsstrukturen sind aufeinander abgestimmt, Dörfer, Städte, Agglomerationen charakterisieren sich, offene Räume bleiben weitgehend unüberbaut, Wälder und Gewässer werden geschützt, – Landschaften prägen den Lebensraum, Bauten beleben die Sinne. Ob als Stadt-, als Orts-, Regional-, Landes- oder gar überstaatliche Planung – sie bewahrt und gestaltet, sie weiss mit Knappheiten (vom Boden über die Finanzen bis zum Konsens) umzugehen, sie haushaltet und begünstigt gleichzeitig ein leistungsfähige Wirtschaft, eine solidarische Gesellschaft und das ökologische Gleichgewicht.

Die räumliche Planung bleibt auch weiterhin nötig. Das Leben spielt sich eben in Raum und Zeit ab, also in Räumen und über die Zeiten hinweg, hier und dort, aus der Geschichte durch die Gegenwart hindurch in die Zukunft hinein, nicht frei von Gegebenheiten, Schranken, Randbedingungen und Grenzen. Stadt und Land, Regionen, Staaten, supranationale Organisationen, ja sogar die werdende Weltordnung – sie alle basieren auf Lebensräumen. Diese sind gegeben, nun aber nicht im natürlichen Urzustand, sondern als von den Menschen samt ihrer Kultur durchdrungene, teilweise tangierte, wenn nicht gar verletzte. Knappheiten gehen mit einher. Grenzen der Belastbarkeit werden sichtbar. Das Befassen mit den Lebensvoraussetzungen sowie den personalen und institutionellen Entfaltungsmöglichkeiten ist unter diesen Umständen zwingend – bei unvermeidlichen Engpässen. Raumplanung konfrontiert – was immer sie im Detail in Plänen, Rechtssätzen und Verträgen anordnet – Politik, Wirtschaft und Gesellschaft mit der Zukunftsfähigkeit. Sie ist in diesem Sinne bewahrende und gestaltende Auseinandersetzung mit dem politischen, wirtschaftlichen, sozialen und ökologischen Geschehen im Raum als dem Gefäss der Lebensvoraussetzungen und des vielgestaltigen Lebens – über die Zeiten hinweg in die Zukunft hinein. Erwehren muss sie sich des Rückzuges gestaltender Politik, widersetzen muss sie sich vermeintlichen Sachzwängen, den Fluchtwegen in die Unverbindlichkeit und ins Beliebige muss sie Hürden entgegenstellen, den Versuchungen umfassender Ordnungsmacht hat sie zu widerstehen, und dort, wo noch in Restanzen vorhanden, abzusagen.

Räumliche Planung optiert für Schritte von Menschen für Menschen – ein Ansatz, der aus sich heraus unvollkommen, aber um der Menschen und ihrer Sache willen zu akzeptieren ist. Raumplanung ist ein Weg begrenzter Macht, oft unzureichend sachlicher Ausleuchtung, ungenügend bedachter Normativität, bisweilen knapper demokratischer Legitimation, aber doch belebt vom Willen, Zukunftsfähigkeit mit Zukunftsoffenheit – in Verantwortung – zu verbinden.

Zum Verständnis ethischen Fragens

Die Raumplanung, ein Charakteristikum, ist normativ ausgerichtet, wendet sie sich doch mit Verhaltensanforderungen an die Menschen. Sie sagt, was getan, was unterlassen werden soll. Häufig verlangt sie sogar nach verbindlichen Normen, nach Rechtssätzen. Dahinter stehen in jedem Fall ethische Entscheidungen. In besonderem Masse gebieten die planerischen Ermessensspielräume – sie sind in zentralen Belangen grösser als in andern Rechtsgebieten – und die mit der modernen Planung einhergehende Wissensmacht innehaltendes Reflektieren und gewichtetes Handeln. Die ethische Frage hält also an, sich auf die Grundlagen, die Randbedingungen und vorauseilenden Denkbezüge zu besinnen. Sie gebietet bedachtes Entscheiden, besonnenes Handeln – unter Umständen sogar Verzicht auf nicht gebotene Massnahmen. Gedankenlosigkeit ist der Feind der Ethik.[5] Gedankenzuwendung aber setzt geistige Lebhaftigkeit, ja sogar Unruhe voraus.[6]

Allerdings – dies mag überraschen – steht bei erster Annäherung nicht die Ethik, sondern die sachlich-fachliche Kompetenz im Vordergrund. Und dies nicht ohne Grund. Auf sie kann nicht verzichtet werden. Auch bewährte Doktrinen haben ihren eigenen Stellenwert. Sie verhelfen zu durchdachten Methoden und Lösungen – oft hinreichender Bestimmungsgrund für das tägliche

---

5 Zur Ethik in der Raumplanung siehe vor allem die angekündigte Publikation der dt. Akademie für Raumforschung und Landesplanung zur Ethik in der Raumplanung (Fn.2). Dort werden sich umfangreiche Literaturverzeichnisse finden. Als Autoren zeichnen: Corinna Clemens, Evelyn Gustedt, Karl-Hermann Hübler, Martin Lendi, Petra Michel-Fabian, Bernd Streich, Markus Vogt und Gerlind Weber.

Jenseits dieser Veröffentlichung siehe beispielsweise die Publikationen von Hans Ruh, der sich zwar in seinen grösseren Abhandlungen nicht direkt zur Raumplanung äussert, der aber sinnverwandt Fragen der Boden- und der Umweltethik ausleuchtet, so *Ruh Hans*, Argument Ethik, Zürich 1991; *idem*, Störfall Mensch, Gütersloh 1995. Direkte Aussagen finden sich in einer kleinren Abhandlung in: *idem*, Ethik und Raumplanung. Informationshefte RP, Bundesamt für Raumplanung, 2/1988. Ferner: *Lendi Martin*, Ethik in der Raumplanung, in: Goppel / Schaffer (Hrsg.) Raumplanung in den 90er Jahren, Festschrift für Karl Ruppert, Augsburg 1991, S. 571 ff.; *idem,* Ethik, in: ARL, Handwörterbuch der Raumordnung, Hannover 1995, S. 232ff.; *idem,* Politisch, sachlich und ethisch indizierte Raumplanung – am Beispiel der Schweiz, Wien 1998. Die Umweltethik aus der Sicht des Rechts wird neuerdings angesprochen von *Petersen Volkert*, Umweltethik – zur Bedeutung eines im Umweltrecht bisher eher vernachlässigten Begriffs, UPR 6/2003, S. 201 ff. Ferner: *Lendi Martin*, Rechtsethische Anforderungen an Ingenieure, in: Straube M. / Weimar R. (Hrsg.), Jurist und Technik zwischen Wissenschaft und Praxis, Festschrift für Josef Kühne, Wien 1984, S. 19ff.

6 Siehe dazu *Leist Anton*, Unordnung durch Ethik, Eine letzte Ressource richtig verstehen, in: Schweizerische Monatshefte 10/2003, S. 39 ff. – eine knappe, sehr kritische und souveräne Auseinandersetzung mit dem Ethik-Boom.

Entscheiden und Handeln. Weder als Raumplaner noch als einzelne Menschen stehen wir gleichsam in jedem Moment unseres Lebens vor der Wahl zwischen dem was (qualifiziert) Geboten oder Verboten ist: Alles hat seine Zeit, das Sachliche und das ethische Gebotene / Verbotene – beide Dimensionen. Nicht auszuschliessen und sogar wahrscheinlich, dass sie sich begegnen oder bisweilen überschneiden. Unethisch wäre es, das Können – das Handeln lege artis – zu übergehen, auch wenn es besonnen zu handhaben, also ethisch einzubinden ist.

Für viele Tatbestände ist das Sachliche durch das Recht erfasst. Dieses ist Grundlage und Schranke des Handelns, und deshalb, weil es ethische Urteile einschliesst, der Ethik nahe. Es verknüpft sachlich Angezeigtes, ethisch Gebotenes und äusserlich Verbindliches. Das Raumplanungsrecht bringt dies deutlich zum Ausdruck. Es handelt festschreibend von Zielen, Trägern, Instrumenten, Massnahmen, vor allem auch von Verfahren, in denen das Geschehen gebündelt und raum / zeit-kompatibel mit ausholenden Horizonten – nachhaltig – umgesetzt wird.

Die ethischen Reflexionen beschlagen ein weites Feld. Mitten im Nachdenken über die räumliche Planung und das raumplanerische Tun kann es in einer ersten Annäherung hinreichen, auf die unmittelbar greifbaren, elementaren Aussagen der Lehren von der Ethik zurückzugreifen. Im Vordergrund stehen dafür die persönliche Freiheit, die Hochachtung vor der menschlichen Würde, der Respekt vor dem Leben, die bedachte Sorgfalt und die Verantwortung für das Leben der kommenden Generationen. Die Aufzählung will weder Vollständigkeit noch eine Prioritätenordnung vortäuschen.[7] Wichtig sodann: Solche Massstäbe determinieren das konkrete Entscheiden und Handeln nicht; denn Ethik ist nicht eine Retorte der Herstellung des jederzeit ethisch handelnden Menschen, wohl aber Grund zum Innehalten, zum Gewissenhaften, sicherlich dann, wenn Störungen auftreten oder verfängliche Routine droht. In diesem Sinne begleitet die Ethik die Raumplanung, die Raumplaner sowie die von ihr Berührten und Betroffenen, die mitzuwirken bereit sind, also jene, die sich bewegen lassen, die von der Ethik umgetrieben werden.

Ethische Phänomene sind, ungeachtet der Gegenwärtigkeit, für die einzelnen Menschen und Raumplaner vorweg als Grenzereignisse erfahrbar. In ihnen gilt es sich zu bewähren. Sie stellen sich allerdings nicht nur in konkreten, existenziellen Situationen, sondern auch dort, wo grundsätzlich gedacht,

---

7 Hier müsste näher ausgeführt werden, welche Werte resp. Postulate usw. näher zu betrachten wären. Im Rahmen dieser kurzen Abhandlung ist dies nicht möglich und sinnvoll. Vorausgesetzt sind auf jeden Fall die Freiheit des Menschen und seine Würde – nicht zu vergessen die Gerechtigkeit mit ihren Weiterungen hin zur Gleichheit und zur Rechtssicherheit. Für die räumliche Planung ist der Respekt vor dem Leben von besonderer Relevanz, steht doch für sie der Lebensraum im Vordergrund, allerdings nicht ohne Rückbezüge zur Welt der Werte, die das Verhalten der Menschen in Raum und Zeit prägen oder doch prägen sollten.

wo Doktrinen ent- resp. verworfen werden und wo Dispositionen für Lehrbücher zur Raumplanung formuliert werden. Ein Beispiel: Ob ich die ethische Komponente des Prinzips der Nachhaltigkeit – intergenerationelle Verantwortung – für das Entwickeln von Planungstheorien und -methoden ernst nehme oder verwerfe, stellt eine Entscheidung dar, die sich mitten in abstrakten, vermeintlich für das tätige Leben irrelevanten Gedankengängen als ethisch zu reflektierende Grenzentscheidung entpuppen kann, so die Konsequenzen bedacht werden.

Versuch einer Annäherung an Grundorientierungen

Der Schritt zu normativen Grundorientierungen liegt nahe. Allerdings ersetzen sie ethisches Fragen und Bedenken nicht. Reife zeigt sich im Fragen und Nachdenken. Den endformulierten Raumplaner-Kodex kann es deshalb nicht geben. Es geht einzig um gedankliche Hilfestellungen, nicht als Rezepte, wohl aber als mögliche Anhaltspunkte, die das kritische Reflektieren erleichtern, ohne es zu erübrigen.

Die Grundorientierungen sind alles andere als widerspruchsfrei. Sogar in ihrer Abfolge sind sie nicht zwingend. Sie sind auch nicht vollständig in der Summe und in ihren einzelnen Aussagen. Obwohl sie abstrakt, zeit- und ortlos tönen – sie beanspruchen nicht Allgemeingültigkeit und schon gar nicht durchsetzbare Verbindlichkeit. Sie repräsentieren weder das Ethos im Sinne einer gefestigten sittlichen Grundhaltung noch spiegeln sie den raumplanerisch relevanten Niederschlag der Lehre der Ethik. Wären sie logisch perfekt gegliedert, austariert strukturiert und auf Vollständigkeit angelegt, so würden sie ethisches Bedenken in kritischen Situationen ausblenden. Und gerade dies wollen sie nicht.

Ethik als „Geschenk der Freiheit", als „imperativer Ausdruck der im Menschen vorausgesetzten Freiheit" oder als „Weg verantwortungsvollen Handelns" kann keine Regeln zu subsumierbaren, ablesbaren, determinierbaren Entscheidungen liefern. Sie akzeptiert die Freiheit der bedachten, besonnenen Entscheidung. Allerdings mahnt sie, wirklich besonnen zu entscheiden, insbesondere Sachzwänge zu befragen, (vermeintlich) planerische Folgerichtigkeiten kritisch zu bedenken und sich Entscheidungen mit Konsequenzen, die als fragwürdig aufhorchen lassen, mit begründenden Erwägungen in den Weg zu stellen. Die persönlichen ethischen Ausgangspositionen, subjektiv geprägt, oft selbst-rechtfertigend angelegt, gilt es in jedem Fall gegenüber den ethischen Grundorientierungen zu exponieren.

In diesem Sinne sind die nachstehenden Aussagen nicht als Lehrsätze oder Verhaltensregeln zu verstehen. Weder nach einem Planer Grundkonsens noch nach einem Schwur auf ethische Minima resp. Maxima wird gefragt. Es geht um nichts anderes als Hinweise, aber nicht auf Nebensächlichkeiten, sondern auf Aussagen, die umgetrieben sind von dem, was getan werden muss. Sie zeugen zudem von Verknüpfungen mit elementaren Rückbezügen zu Postulaten und Werten, die zu reflektieren sich lohnen könnte, um, wir unterstreichen es nochmals, in konkreten Situationen besonnen entscheiden und beim Entwickeln von allgemeinen Lehren Tiefe und Weite planerischer Normativität bedenken zu können. Grundorientierungen vermitteln nicht Gewissheiten, sondern stiften – der rote Faden klingt auch hier an – zum Nachdenken an.

a) Grundorientierungen in der Alltagsarbeit

1. – Die räumliche Planung, vertreten durch die Beauftragten und die Involvierten, ist und bleibt sich ihres ausholenden Gegenstandes – Schutz des Lebens wie auch der Lebensvoraussetzungen und der politischen, wirtschaftlichen, gesellschaftlichen wie auch individuellen Lebensentfaltung in Raum und Zeit, in den Spannungsfeldern von sachlichen Zwängen und moralischen Anforderungen einerseits sowie von Entscheidungsfreiheit und Normativität anderseits – stets kritisch bewusst.

2. – Die räumliche Planung erkennt in der Ethik einen – für Tiefen und Weiten der Rückbezüge, so auf die Freiheit und die Würde des Menschen – offenen Reflexionsraum, der Distanz und bedachtes Sinnen und Besinnen für das konkrete Entscheiden und Handeln schafft.

3. – Die räumliche Planung fördert Kreativität und Innovation der in der Raumplanung Tätigen. Sie festigt fachliche Kompetenz, würdigt das methodische Können und unterstützt das Tun mit fachlicher und ethischer Motivation.

4. – Die räumliche Planung bemüht sich um den sorgfältigen, begründenden und Schranken akzeptierenden Umgang mit dem Mächtigen und der Macht, insbesondere der staatlichen, der wirtschaftlichen und der gesellschaftlichen Erscheinungsformen. Sie begrenzt und kontrolliert selbstredend auch die Planungsmacht.

5. – Die räumliche Planung strebt nach Akzeptanz und Konsens in der Entscheidfindung, sei es im Rahmen der Gemeinwesen, sei es in Kooperation mit Berührten und Betroffenen – stets insistierend auf den Argumenten für das Langfristige. Insbesondere verweist sie auf die Rücksichtnahme zugunsten der kommenden Generationen.

6. – Die räumliche Planung berücksichtigt die Anliegen jener Menschen und gesellschaftlichen Minderheiten, die sich ungenügend artikulieren und ihre Interessen nicht ausreichend vertreten können.

7. – Die räumliche Planung drängt mitten im sich dynamisch entwickelnden Leben in Politik, Wirtschaft und Gesellschaft auf stabile Lebens-/Lebensraumverhältnisse über die Zeiten hinweg, d. h. sie zielt auf eine nachhaltige Ent-

wicklung in Wirtschaft, Gesellschaft und Umwelt – je in sich und insgesamt in gegenseitiger Abstimmung.

8. – Die Planung widmet sich der haushälterischen Raumnutzung – mit Einschluss der Boden- und Ressourcennutzung – und bedenkt die Belastbarkeitsgrenzen. Sie setzt Prioritäten, ausgehend vom Schutz der Lebensvoraussetzungen und der Zukunftsfähigkeit der Gesellschaft. Sie gewährleistet die Versorgung und Entsorgung. Sie schützt insbesondere Wasservorkommen nach Qualität und Quantität.

9. – Die räumliche Planung achtet – vor dem Hintergrund unumgänglicher menschlicher Eingriffe in die Natur – auf Biodiversität, Ausgleichsräume und die Qualität der Landschafts- und Siedlungsgestaltung, auch in ästhetischer Hinsicht. Die Städte und Agglomerationen sowie die offene Landschaft werden in ihrem Zusammenhang ganzheitlich und je für sich in ihren spezifischen Anforderungen betrachtet.

10. – Die räumliche Planung begünstigt jene Mobilität, die auf Dauer tragbar ist. Sie achtet auf kurze Wege, verweist Arbeitsteilungen in die Regionen und fördert die Substituierung des materiellen durch den immateriellen Verkehr.

11. – Die räumliche Planung verzichtet auf Massnahmen, deren Auswirkungen nicht oder noch nicht ausreichend bekannt sind, sofern das Risiko für das Leben in Raum und Zeit unabsehbar oder erheblich sein könnte. Irreversible Prozesse, die lebenszerstörende Wirkungen zeitigen könnten, trachtet sie zu vermeiden. Planerische Vorgaben für künftige Vorgänge, die sich als schwer handhabbar erweisen, hält sie zurück, es sei denn, sie könne die laufende Beobachtung, die Steuerbarkeit und gefahrenabwehrende Korrekturen gewährleisten.

12. – Die räumliche Planung insistiert bei den privaten und öffentlichen Unternehmungen auf das Bedenken der Gegebenheit des Raumes, der Knappheit räumlicher Nutzungsmöglichkeiten und der optimalen Ausrichtung der Aktivitäten auf die Anforderungen des Lebensraumes; die massgebenden Faktoren zu Standort-, Verkehrs-, Immissions-, Ressourcenbelangen usw. sind in die Entscheidfindungen einzubeziehen.

13. – Die räumliche Planung spricht mit der Öffentlichkeit über die Begrenztheit des Lebensraumes und die globalen, regionalen wie auch örtlichen Gefahren der Übernutzung der Ressourcen und der Überlastung von empfindlichen Räumen. Sie wirkt stufengerecht an der öffentlichen Problembewusstseinsbildung mit. Sie bezieht die Planungsberührten in die Planungen ein.

14. – Die räumliche Planung nimmt Interessenkonflikte und Wertantinomien auf und setzt sich mit ihnen auseinander. Sie trifft abgewogene, zieladäquate, in sich stimmige Entscheidungen und steht zu diesen, insbesondere gegenüber nicht sachbezogenen Interessen. Sie vermittelt Rechtssicherheit und veranlasst Novellierungen, wenn stichhaltige Gründe diese gebieten.

15. – Die räumliche Planung wehrt jeglicher Art von Diskriminierung und missbilligt sie. Sie achtet in ihren Verfahren auf Fairness. Sie bewahrt vor Willkür, achtet auf Treu und Glauben und schützt insbesondere sensitive Daten. Sie verwahrt sich gegen Spekulation und Korruption.

16. – Die räumliche Planung ordnet die von ihr befolgte planerische Zweckrationalität Recht und Gerechtigkeit unter. Sie ist dem demokratischen Rechtsstaat und insbesondere den Grundrechten verpflichtet.

b) Grundorientierungen in der Lehre

17. – Die Lehre von der räumlichen Planung unterstellt die rationale Problemdurchdringung der Suche nach der Wahrheit, d.h. sie exponiert sich der Konkurrenz unterschiedlicher Auffassungen zu Staat, Wirtschaft und Gesellschaft. Sie fördert vorweg und vor allem die sachlich-fachliche Kompetenz. Darüber hinaus bemüht sie sich, das eigene planerische Befassen wiederkehrend in Frage zu stellen.

18. – Die Lehre von der räumlichen Planung achtet auf kritisches Hinterfragen und Begründungsnotwendigkeiten. Sie bezieht ethische Problemstellungen ein und ermuntert zur ethischen Argumentation.

19. – Die Lehre von der räumlichen Planung strebt in ihrer Grundlegung an, die notwendigen normativen Dimensionen durch kritische Auseinandersetzung mit den Theorien zum politischen, wirtschaftlichen, gesellschaftlichen und ökologischen Geschehen zu gewinnen sowie ihre Wertungen argumentierend zu begründen.

20. – Die Lehre von der räumliche Planung widmet sich im besonderen Masse den Grundanforderungen des sich über Generationen hinziehenden Lebens in Raum und Zeit. Sie besteht auf dem langfristig Relevanten auch dann, wenn dieses durch Zwänge der Tagesaktualitäten droht verdrängt zu werden. Das Standhalten zugunsten des Intergenerationellen als dem Kern des Prinzips der Nachhaltigkeit ist ihr in besonderem Masse aufgetragen.

21. – Die Lehre von der räumlichen Planung betont die Stellung des Menschen im Planungsgeschehen, weil er zur Verantwortung fähig ist. Ihn einzubinden und mit ihm eben diese Verantwortung zu diskutieren, ist eine ihrer wesentlichen Grundfunktionen. Die Verfahren hoheitlicher Planungen, demokratischer Planungsabläufe, partizipativer Planungen und andere Formen der Kooperation entbinden nicht davon.

22. – Die Lehre von der räumlichen Planung vermeidet den Rückgriff auf irgendwelche Ideologien und zu sich selbst genügenden Rationalitäten. Sie ist sich bewusst, dass die räumliche Planung vom natürlichen und gesellschaftlichen Leben samt seinen Widersprüchlichkeiten nicht abgekoppelt werden kann. Sie sieht sich mit einer Vielfalt von Gegebenheiten und konkurrierender Interessen konfrontiert. Darum bemüht sich die Lehre, die Fähigkeit zu schulen, mit den Fakten umzugehen sowie Interessenkonflikte und Wertantinomien zu ertragen und auszugleichen – unter Ausrichtung auf Grundüberlegungen zu Freiheit, Menschenwürde und Schutz des Lebens.

Kritische Reflexion

Die hier entworfenen und vorgelegten Grundorientierungen liessen sich hinlänglich kommentieren. Allerdings würde die Gefahr mitlaufen, sie als moralische Verbindlichkeiten zu verstehen. Es handelt sich aber gerade nicht um systematisierte auf Widerspruchsfreiheit angelegte Regeln, die gleich Rechtssätzen anzuwenden wären. Sie sind, wenn sie in Betracht gezogen werden, vom Sinn her zu erwägen sowie abzuwägen und zwar bevor Entscheidungen in der Sache getroffen werden. Sie werden im selbständigen Fragen und Antworten lebendig. Gegenüber den Grundorientierungen ist eben aus philosophischen, theologischen und raumplanerischen Gründen sorgfältiges Hinterfragen angezeigt. Weder die Ethik noch die räumliche Planung ertragen finalisierte, gültige Aussagen. Sie leben von Konflikten, von Horizonten, von Bewegtem und Bewegendem, von einer Ethik, die nicht verkürzt, sondern auftut, die Klärung, wenn nicht sogar gedanklich Ordnung dadurch bewirkt, dass sie das Unklare und das Unordentliche bedenkt.

Um den Sinn für das Ethische zu schärfen, nicht um es vorzuzeichnen, wagten und wagen wir die hier vorgelegte Skizze.[8] Die Orientierungen sind letztlich zweiseitig: Sie zeigen Brüche und Bruchstücke, Unvollendetes, Unvollkommenes, Unordentliches genau so auf, wie sie Ansätze zu Ordentlichem andeuten. Die Skizze zu diskutieren, ist Aufgabe der Leser. Sie lebhaft zu debattieren, ist für die Raumplaner und die Lehre von der Raumplanung hilfreich. Sie zu verwerfen, kann nötig werden. Das Fragen aber nicht aufzugeben, das ist Pflicht.

Ziel ist eine fachkundige, bedachte, besonnene – eine gewissenhafte Raumplanung, die sich den sachlichen Komplexitäten, den Interessen- und Wertekonflikten stellt und welche die innere Unruhe zulässt, diese (ethisch) vertieft anzugehen.

---

8 Es bleibt bei einer Skizze. „Zehn Gebote“, wie sie oft katechismusartig nachgezeichnet werden, stehen nicht an. Die neueste Version eines Kodex legt die Schweizerische Akademie der Technischen Wissenschaften „Ethik im technischen Handeln“ vor (Zürich 2003). Darnach trägt der Ingenieur u. a. die persönliche, ethische Verantwortung für sein Handeln wie er auch zur Wahrhaftigkeit verpflichtet ist. Solche und weitere Postulate hören sich positiv an. Letztlich helfen sie aber nicht weiter, weil sie zu wenig konkret sind und den Konflikten (Gewissens-, Interessen-, Wertekonflikte usw.) nicht Rechnung tragen.

Paul Trappe

Le système – un point de vue sous l'angle des sciences appliquées

Je suis sociologue; par conséquent les théories et les hypothèses que je formule doivent se vérifier en regard de la pratique. Je tire mes exemples de projets de recherche réalisés par l'Institut de sociologie de l'Université de Bâle, projets qui, comme on peut le constater aisément, s'inscrivent pleinement dans la pratique et la réalité sociétales. Ceux-ci ne visent en effet pas à expliquer le système, ils représentent au contraire, chacun en sol un système; ils évoluent de fait dans une multitude de systèmes, tout en étant interprétés respectivement, et ça d'une manière particulière comme il sera à démontrer, comme formant un système (du moins dans un cadre modeste). S'il s'agissait là d'un procédé scientifique illégitime, nous ne pourrions plus alors analyser de manière sensée les divers problèmes de notre société, ainsi que ceux relevant du domaine des sciences naturelles ou de la médecine. Il s'agit donc ici du sens donné au concept de système par les scientifiques qui, de leur propre initiative ou en réponse à un mandat public, doivent trouver des solutions à des problèmes.

Ceux-ci n'ont pas affaire au système global ou système du monde, ni à une interprétation du concept de système, largement répandue aujourd'hui, en particulier chez les sociologues germanophones, qui déborde de son cadre, en raison de sa prétention notamment heuristique. Un des risques au demeurant est que ce concept de «théorie du système» entraîne des conséquences inattendues tant au plan social que politique ou au niveau de la théorie scientifique, influençant notamment le processus de connaissance.

Les thèmes faisant l'objet d'études réalisées par l'Institut que j'ai nommé auparavant (mais aussi par d'autres instituts) concernent notamment: la fonction des normes sociales, outre celle du droit, les nouveaux pauvres et les mesures de politique sociale dans différents milieux sociaux suisses, l'évolution du système social suisse, plusieurs travaux sur les régions ou pays en voie de développement, principalement des pays méditerranéens et africains au sud du Sahara. Dans ces territoires géographiques, plus particulièrement des études consacrées à des réformes agraires très concrètes et délimitées, à des organismes de soutien et à des actions d'entraide dans le domaine de l'infrastructure sociale et économique, ou abordant encore avec le temps de plus en plus souvent la question relativement négligée jusqu'ici de savoir qui doit en fait assumer, dans le cadre de projets d'irrigation ou de boisage devenus indispensables, la gestion autonome, l'entretien ainsi que les mesures de protection nécessaires.

Ensuite, nous avons affaire à des systèmes sociaux, qui dans leur contexte spécifique évidemment, sont inscrits dans un ensemble assurément plus vaste et qu'il faut représenter sous forme de système. Lors de l'application de connaissances sociologiques à la pratique, j'aimerais toutefois souligner dès le départ, que la «théorie du système», qui est dans toutes les bouches, ne m'aide guère. Sur la base de mon travail quotidien, mais aussi en tant qu'individu politiquement engagé, je ne peux que formuler ce jugement sévère, à savoir que dans la pratique la «théorie du système» ne m'est pas utile.

J'aimerais commencer par donner la définition de quelques notions, qui dans le cadre d'une telle théorie sont considérées comme fondamentales, mais que l'on retrouve cependant aussi dans d'autres théories, voire probablement depuis toujours dans chaque processus de connaissance scientifique, à savoir – ici brèvement caractériser – les concepts de:

- Système: en tant qu'ensemble organisé de facteurs; cet ordre structuré doit mis en évidence pour le processus de connaissance. Les facteurs sont quant à eux variables, en ce sens les systèmes ont leur propre identité. Ainsi, deux systèmes qui accusent une similitude considérable ne peuvent jamais être identiques – même pas dans «la théorie» –, ne serait-ce qu'en vertu du simple principe qu'une chose ne peut être identique qu'à elle-même.
- Fonctions: autrement dit, les actions, les processus qui à l'intérieur d'un système remplissent un rôle spécifique, nécessaire au maintien et à l'évolution du système.
- Structure: c.-à-d. l'agencement des éléments, les rapports relativement stables et contraignants pour tous les éléments du système. Ce qui ne veut pas dire que chaque fonction respecte effectivement la structure dans tous ses enchaînements.
- Théorie: assertions vérifiées et réfutables (simples propositions, hypothèses) quant à des relations prouvées, des interdépendances ou de simples corrélations entre les différents éléments du système.
- Milieux environnants: en d'autres termes, les systèmes environnants, par rapport auxquels le système étudié doit affirmer son identité. Ce dernier subit soit directement, mais souvent aussi seulement de manière indirecte, l'influence du milieu environnant, dont les effets se ressentent ponctuellement ou continuellement et concernent certains éléments du système ou le système dans son ensemble. Etant donné que les faits environnants peuvent appartenir aux différentes disciplines de recherche une interdisciplinarité de recherche est indispensable

Cependant, je veux bien admettre que l'actuel débat autour de la théorie du système peut être également stimulant. Je crains toutefois son idéologisation – parce que le débat passe à côté de la réalité – parce que les formulations pleines d'esprit de cette théorie ne peuvent être appliquées telles quelles à la réalité, et enfin, parce que, en dépit du plaisir de noter une différenciation croissante de cette théorie, il y a un élément qui n'est pas suffisamment pris en compte et qui selon moi est au centre de toute recherche scientifique: à savoir l'individu et la société. L'individu agissant est, pour citer librement Max Weber, l'unité fondamentale et irréductible, le point de départ du système.

L'idéologisation est un danger immanent à cette théorie – Idéologie entendue ici comme une assertion se voulant scientifique, fondée sur une interprétation erronée et qui prétend être de la «théorie» (Theodor Geiger: «L'idéologie comme a-théorique qui se veut théorique», en original: «theoretisch gemeintes a-Theoretisches»). En plus de l'idéologisation, la supposition que les propositions de cette théorie fourniraient les principes pour expliquer une réalité complexe, pour autant que l'on réduise suffisamment cette complexité. Autrement dit, cela implique que je ne saisis plus la réalité dans son ensemble, mais au mieux une réalité réduite, amenuisée – or cela n'est pas la réalité, mais seulement une partie de celle-ci, peut-être essentielle, mais peut-être aussi tout à fait insignifiante.

L'intention de la théorie du système est de fournir aux travaux de recherche scientifique un cadre qui serait en outre applicable à tous les domaines scientifiques – donc pas seulement aux sciences sociales. Un tel objectif ne peut par sa nature même qu'être très général. Le système peut être entendu ici comme système du monde ou système global. Il peut s'agir aussi de sous-systèmes à l'intérieur desquels les fonctions et les structures d'une part, les limites par rapport au systèmes environnants d'autre part, déterminent un système. En anticipant toutefois quelque peu, nous avons à proprement parler plutôt affaire à des réseaux de systèmes et à des recoupements considérables de systèmes et c'est sur ce problème que notre réflexion doit porter. Les visées de la théorie du système sont donc extrêmement amples; or, plus elles s'élargissent, plus elles s'éloignent de la réalité humaine et sociale. Leur contenu informatif est limité. Elles sont dans une grande mesure une construction de l'esprit. En effet, la multitude infinie de données réelles et effectives peut non seulement dissoudre tout système, mais aussi prouver l'absurdité d'un système. Il n'est pas rare de voir que face à une question complexe, tirée de la réalité, un théoricien du système est plus que dépassé.

Un exemple: celui qui limite le système du monde à l'actuel processus de «mondialisation» et représente ce dernier comme un processus de croissance économique, dispose rapidement de données relatives à des possibilités de croissance jamais vues et dont on peut effectivement tirer parti. Mais comme il n'aborde pas d'autres processus, il néglige notamment les processus de marginalisation, par exemple de pays et de groupes d'individus, qui non seulement relativisent l'importance du processus de croissance économique, mais peuvent également l'influencer négativement, voire même, dans certaines circonstances, provoquer un effondrement.

Les sciences appliquées doivent donc travailler avec leurs propres outils, elles n'obtiennent pas d'aide directe de la théorie du système et étudient leurs champs de problèmes selon des méthodes éprouvées et devant continuellement faire leurs preuves dans la pratique – ou alors échouer. J'aimerais évoquer ce qui ressort déjà de cette affirmation, à savoir que je dois beaucoup à la théorie de la connaissance de Karl Raimund Popper et que je tiens celle-ci pour étant la plus plausible de toutes les théories de la connaissance qui ont été formulées. Les méthodes appliquées jusqu'ici, même si elles débouchent parfois sur des contrevérités, des erreurs d'appréciation ou de fausses théories, ont toujours fait leurs preuves dans la pratique, tout en prenant également en compte que la complexité des milieux environnants ne pourrait jamais être appréhendée totalement. Ces méthodes, en se fondant sur la connaissance, possédaient toujours une ouverture essentielle pour corriger les résultats obtenus, considéraient et considèrent toujours la falsification comme élément inhérent à toute science. L'assertion, selon laquelle la complexité doit être réduite à des fins d'analyse et d'élaboration de théories, est à mes yeux triviale, même prétentieuse.

A mon avis, la théorie du système est un principe au développement très différencié, voire un ensemble complexe de principes. Or, l'existence humaine est dans une grande mesure une confrontation permanente avec la violation de principes; les principes ne pourront pas être reconnus dans leur totalité, du moins pas par tous les individus de la même manière, voire même peut-être pas du tout. Les principes portent donc toujours en eux le risque de ne pas être observés.

Qui pis est: le rapport entre les fonctions et les structures d'un système n'est pas coercitif; à l'intérieur du système, elles peuvent, selon le libre choix, être observées ou non (également en cas de lacunes du système ou de clauses générales). Ceci est particulièrement vrai pour ces fonctions que, par exemple, Robert King Merton a désignées, il y a quelques dizaines d'années déjà, par le terme de «fonctions latentes». Cet arbitraire peut cependant aussi concerner les fonctions manifestes, lorsque celles-ci peuvent ou doivent être interprétées en fonction du contexte situationnel et comportent de ce fait le risque d'une interprétation divergente selon le cercle d'individus considéré, et des conséquences variables.

Les fonctions manifestes peuvent littéralement être supplantées par des dysfonctionnement: par exemple, lorsque des facteurs de pouvoir entrent en jeu de manière inattendue. La contingence au sens philosophique est et sera toujours l'élément d'incertitude inhérent à tout système. Nous sommes confrontés à une pluralité de systèmes et donc à une pluralité de milieux environnants; au vu de cela, je pense que nous avons forcément aussi affaire à une pluralité d'erreurs d'interprétation dans le contexte social. Ceci pour des raisons structurelles, car les individus (pour la plupart du moins) ne créent pas leurs propres structures, mais reprennent celles qui existent déjà, les trouvent à leur disposition; rares sont les personnes qui peuvent exercer une influence réelle sur les structures et encore moins les modifier.

Dès sa naissance, l'être humain s'inscrit dans un grand nombre de systèmes (système de parenté, système de normes sociales, système juridique, système politique, système de stratification sociale, etc. – sans oublier les milieux naturels). Au cours du processus de socialisation, l'individu apprend à s'orienter à l'intérieur des divers systèmes et acquiert les aptitudes pour agir surtout conformément aux exigences du système ou selon les possibilités offertes par celui-ci. Le processus de socialisation est, en particulier à notre époque, axé de manière on ne peut plus évidente sur toujours plus de systèmes identifiables. Parler ici d'identité de système serait trompeur – compte tenu des faits, il serait plus approprié de parler de réseaux de systèmes à l'allure presque chaotique, à l'intérieur desquels un individu évolue et au sein desquels il apprend à concrétiser ses choix, voire même à développer des priorités.

Il faut ensuite se demander si chacun de nos actes sociaux peut être réduit à des fonctions spécifiques au système, dont les effets devraient en outre assurer le maintien du système. Les possibilités de choix et les espaces, dans lesquels la libre interprétation est donnée, rendent précisément les dysfonctionnements possibles. J'aimerais rappeler que dans le cadre de notre système de droit, et plus particulièrement dans le cadre du processus de globalisation, nous trouvons une multitude d'espaces de liberté, régies par, aucune loi, qui incitent littéralement aux agissements illégaux. Lorsque cela se produit de manière importante, on peut tout à fait parler de «droit de l'imitation» (Gabriel de Tarde) ou encore de «loi de la série». Le citoyen pense peut-être qu'il serait idiot de ne pas profiter de cette liberté d'action, alors qu'il la connaît peut-être et que son «entourage» en tire pleinement parti.

Je ne connais au fond qu'un seul système dont le développement social en Europe s'étend sur plusieurs millénaires et dont l'importance tant pour les individus, les groupes, les peuples que pour les rapports entre les nations a été plus que marquante: je veux parler du système juridique. Dans notre milieu culturel, ce système a été élaboré pendant plusieurs millénaires et il constitue en fait le système qui nous a édicté les structures auxquelles certains des processus sociaux doivent impérativement obéir. Il existe aussi d'autres systèmes, pour la plupart normatifs, comme les normes sociales qui, en plus des lois, s'appliquent à des milieux aussi bien restreints qu'étendus, ou encore celui des normes et des institutions des communautés religieuses.

Le système juridique détermine aujourd'hui encore notre culture, alors que sous l'angle de la théorie du système, il n'est considéré que comme un «secteur» dans la relation système-milieu environnant. Il représente à mes yeux le système fondamental. En effet, les troubles, les conflits sociaux ou les catastrophes de notre temps s'expliquent dans une large mesure sur la base de structures juridiques – ou peuvent en être déduits.

Nous avons depuis longtemps conscience de cela. Permettez-moi de vous citer une phrase tirée d'un fragment et attribuée à Thalès de Milet: il dit à Pittakos «L'autorité suprême est celle du mot écrit», autrement dit celle des lois écrites. Je remplace la notion d'«autorité» par celle de «système», ce qui ne trahit certainement pas la pensée de Thalès, et je dis que parmi tous les systèmes empiriquement vérifiables, le système du droit constitue dans la vie sociale le système du droit écrit, qui est le plus efficace et le plus utile parmi les différents systèmes de normes. D'où la nécessité de préserver la stabilité de la «règle de loi» ou «rule of law», mais aussi sa souplesse.

De tous les systèmes sociaux, le système juridique est le plus proche de la réalité et a atteint au cours des millénaires un dégré de complexité élevé ainsi qu'une stabilité structurelle; il possède en outre, ce qui ne surprendra aucun juriste, une faculté d'adaptation remarquable. C'est un droit ouvert – comme j'espère dans une société ouverte (Popper). Heureusement, il n'a jamais été revendiqué sérieusement de réduire la grande complexité du système juridique.

Le sociologue allemand de Bielefeld, Richard Grathoff, a publié un article, dont le titre est révélateur: «L'uniformité des systèmes et la diversité du monde vivant» (en allemand: «Ueber die Einfalt der Systeme und die Vielfalt der Lebenswelt»), dans lequel il voulait en fait rendre attentif à l'efficience et à la fonction relativement mineures des systèmes, en face de la complexité du «Lebenswelt». Ce faisant, il renvoyait, du moins implicitement, au fait que les champs de problèmes à analyser doivent également prendre en compte leurs variables spécifiques, les facteurs d'influence que le champ de problèmes met précisément en évidence. Or, pour réaliser cela, c'est-à-dire l'identification de problèmes, il faut véritablement maîtriser les détails spécifiques au problème, qui produisent de effets et font apparaître le problème hors d'une systématique présentée comme non problématique. Cela veut donc dire que l'analyse porte ici sur les détails qui déclenchent le problème et que la méthode utilisée est plutôt inductive qu'un procédé déductif, fondé sur une théorie du système.

Nous connaissons tous des exemples de situations conflictuelles qui de mémoire d'homme sont traitées et résolues sur la base de critères juridiques. Le juriste, chargé d'appliquer le droit, a affaire à des cas qui de par leur particularité structurelle ont souvent eu des répercussions sur la législation. Ainsi, les règles de droit pénal appliquées sur le continent européen définissent en tout cas très clairement ce qui se passe lorsque l'on parle de vol, de suicide, d'assassinat, de cambriolage, de diffamation, etc. En droit civil, il n'en est guère autrement. Je peux d'emblée identifier comme éléments de fait clairement définis par la loi, le régime juridique, des faits comme le trouble de la possession, la violation du droit de propriété ou du droit des contrats et ainsi de suite. A ce propos, je me demande dans quelle mesure il faut analyser ici le contexte d'un tel délit ou de la violation des normes, à moins que l'on ne veuille à tout prix trouver des circonstances atténuantes, ce qui n'est toutefois réalisable que si l'on dispose d'hypothèses concrètes et vérifiables s'inscrivant dans le contexte du délit proprement dit. Face à la complexité de la réalité sociale, les facteurs influençants considérés ont donc été réduits à un minimum relatif – qui s'avère suffisant pour la pratique du droit. Ces questions ne se posent pas seulement aux juristes, mais aussi dans d'autres disciplines, notamment en psychologie ou en sociologie. Ainsi, les thèmes que nous traitons dans le cadre de notre recherche et que j'ai évoqués auparavant, permettent assurément une délimitation de ce qui constitue le milieu environnant – et même une délimitation très claire:

- En psychologie, je pense que le fondement pour la détermination et donc la délimitation d'un champ de problèmes a été fourni par Kurt Lewin, à travers son ouvrage de référence, «Théorie du champ dans les sciences sociales» («Field Theory in the Social Sciences»). Il a explicitement pris comme point de départ de sa théorie la notion et le phénomène de «système», respectivement la «théorie du champ», reprise aux sciences naturelles et plus particulièrement à la physique.
- Burkhardt Holzner, quant à lui, a exposé des élements théoriques déterminants pour les sciences sociales, ne serait-ce par le bias de son essai déjà très remarqué lors de sa parution et traitant des «Ensembles dans les sciences sociales» («Wholes in the Social Sciences»). La volonté de circonscrire sans cesse des milieux environnants nouveaux est devenue obsolète.
- En philosophie, on peut renvoyer a la phénoménologie qui, à travers un réductionnisme continu au phénomène étudié, visait à découvrir l'essence de ce phénomène.
- Tout cela a donc ses prémisses: Je pense aux disciples de Emile Durkheim, en particulier è Marcel Mauss et à son concept du «phénomène social total», et plus récent, à Georges Balandier en traitant la «situation coloniale dans son ensemble». Il existe encore d'autres exemples qui ne font qu'attester qu'un système doit être appréhendé comme un ensemble certes délimitable, mais aussi complexe, aux connexions nombreuses et subissant les influences d'autre systèmes («milieux environnants»). Sans oublier qu'en dépit de cette complexité et d'un dynamisme constant, un système représente malgré tout un moyen d'orientation au sein d'une société complexe (voir E. Durkheim, P. Heintz, P. Trappe et d'autres).

Dans ce débat brèvement cité ici manque encore une source fondamentale, qui devrait être recouverte de nouveau: La «topique» comme structure relativement stable dans les discours humains – en effet sur les lignes posées par les grands précédents, comme Aristoteles, Cicero, Giambattista Vico, Theodor Viehweg et al.

Je n'aimerais pas développer plus avant ces références, mais me contenter de signaler encore un principe, qui a depuis longtemps fait ses preuves dans la théorie et la pratique juridiques et a anticipé – de mon point de vue – certains éléments de la «théorie du système». Les adeptes de la théorie du système, qui pour la plupart sont des juristes de formation, devraient connaître ce principe. A savoir que dans le code de procédure civile, mais aussi dans les codes de procédure pénale, voire même encore dans d'autres règles procédurières de l'application du droit, le juge exige expressément de prendre en compte dans l'appréciation du problème, du litige ou du délit soumis «toutes les circonstances du cas d'espèce». Or, je me demande comment un juriste parvient à déceler toutes ces circonstances. J'ose affirmer que dans sa recherche, ce dernier rencontre parfois des circonstances qui sont précisément atypiques au système défini dans le cadre de la «théorie du système». A mon avis, cela implique, sans vouloir attaquer la théorie du système dans ses principes de base exposés jusqu'à ce jour, qu'il faut en fait effectuer une inversion de la démarche, c'est-à-dire non pas partir des fonctions et des structures, mais des modes de comportement possibles, démarche qui met en évidence les fonctions atypiques ou les dysfonctionnements à l'intérieur d'un système apparemment délimité ainsi que la volonté de produire des effets susceptibles de modifier les structure du système.

Un système devient inutile si l'on n'y intègre pas les dysfonctionnements. Lorsqu'un système fonctionne parfaitement, ne présente aucun trouble, est même «autoréférentiel» ou «autopoiétique», il n'y a pas lieu de chercher à résoudre des problèmes sur la base d'une approche scientifique. La résolution de problèmes, l'appréciation de cas juridiques a à voir avec les facteurs perturbateurs à l'intérieur d'un système et leur traitement. Le système peut alors apparaître comme insuffisamment structuré ou «chargé» de dysfonctionnements – voire même chaotique. Face à une telle situation, un défenseur de la théorie du système apporte-t-il actuellement des solutions orientées vers la pratique? Je crains que non. Je souhaiterais que la théorie du système ne soit pas platonique – et ne le devient pas.

Littérature

Balandier, Georges, «Phénomènes sociaux totaux et dynamique sociale», in: Cahiers Internationaux de Sociologie, vol. XXX, Paris 1961, p. 23-34.

Capelle, Wilhelm, Die Vorsokratiker, Fragmente und Quellenberichte (Kröners Taschenausgabe, Bd. 119), Stuttgart 3. Aufl. 1940.

Carbonnier, Jean, Le flexible droit, 1969.

Durkheim, Emile, De la division du travail social, 2e éd. augmentée, Paris 1902.

Geiger, Theodor, Arbeiten zur Soziologie, Teil III B, (Soziologische Texte, Band 7), Neuwied-Berlin 1962.

Grathoff, Richard, «Ueber die Einfalt der Systeme und die Vielfalt der Lebenswelt, Eine Antwort auf Niklas Luhmann», in: ARSP LXXIII/2, Wiesbaden-Stuttgart 1987, S. 251-263.

Heintz, Peter, «Komplexe Gesellschaften», in: R. König ed., Soziologie, Das Fischer Lexikon, Frankfurt am Main 1958.

Holzner, Burkart, «Wholes in the social sciences: A Survey of concepts and their uses», in: ARSP XLIX/2-3, Neuwied-Berlin 1963, S. 237-254.

Lewin, Kurt, Field Theory in the Scoial Sciences, New York 1951 (dt. Bern 1963).

Luhmann, Niklas, Soziale Systeme, Grundriss einer allgemeinen Soziologie (in: Theorie der Gesellschaft, Lizenzausgabe 2002 für die Wissenschaftliche Buchgesellschaft, in 10 Bänden), Darmstadt 2002.

Mauss, Marcel, Sociologie et Anthropologie, Paris 1950.

Merton, Robert King, Social Theory and Social Structure, Third ed., New York-London 1968.

Popper, Karl Raimund, Logik der Forschung, 4. Aufl. Tübingen 1971.

Trappe, Paul, « Gemeinschaft – Gesellschaft – Person, Sozialer Wandel im 19. und 20. Jahrh. », Atti del XXIII Convegno Internazionale 1996, Meran 1997.

Trappe, Paul, « Gruppentheorie und Gesellschaftspolitik – Feststellungen zu sozialen intermediären Gewalten und Gruppenpluralismus », in: P. Trappe, ed., Politische und gesellschaftliche intermediäre Gewalten im sozialen Rechtsstaat (Social Strategies, vol. 22), Basel 1990.

Trappe, Paul, Die Entwicklungsfunktion des Genossenschaftswesens (Soziologische Texte, Bd. 31, Neuwied-Berlin 1966 (insbesondere über « phénomène social total », S. 42 ff.).

Viehweg, Theodor, Topik und Jurisprudenz, Ein Beitrag zur rechtswissenschaftlichen Grundlagenforschung, Fünfte, durchgesehene und erweiterte Auflage (Beck'sche Schwarze Reihe, Bd. 110), München 1974 (l. Aufl. 1954).

Weber, Max, Wirtschaft und Gesellschaft, Grundriss der verstehenden Soziologie, Tübingen 1921.

Willke, Helmut, Systemtheorie, Eine Einführung in die Grundprobleme der Theorie sozialer Systeme, 4., überarbeitete Aufl. (UTB 1161), Stuttgart-Jena 1993.

Willke, Helmut, Systemtheorie II: Interventionstheorie, Grundzüge einer Theorie der Intervention in komplexe Systeme (UTB 1800), Stuttgart-Jena 1994.

Willke, Helmut, Systemtheorie III: Steuerungstheorie, Grundzüge einer Theorie der Steuerung komplexer Sozialsysteme (UTB 1840), Stuttgart-Jena 1995.

GUIDO LEIDIG

Effizienz als Gestaltungsprinzip des Umweltrechts

A. Einleitung

Der sozioökonomisch-ökologische und kulturelle Wandel führt zu einer Zuspitzung des Verhältnisses von Gesellschaft, Zeit und Recht.[1] Die Frage nach der Effizienz[2] der Gesetzgebung – hier: Umweltgesetzgebung – stellt sich vor dem Hintergrund zunehmender Globalisierung einerseits und steigender Dynamik und Komplexität (= Dynaxität[3]) andererseits in allen Gesellschaftssystemen. Lässt sich vor diesem Hintergrund überhaupt noch ein Umweltrecht entwickeln, welches in der Lage ist, die natürlichen Ressourcen zu schützen? Denn: Die derzeit dominanten Problemfelder weisen – wenn überhaupt – nur noch eine schwache, insignifikante ökologische „Tönung" auf. Ökonomische Fragestellungen und Probleme bestimmen weitgehend die Inhalte der gesellschaftspolitischen Diskussion: Welche Rahmenbedingungen gilt es z.B. zu implementieren, um das Wirtschaftswachstum zu stimulieren? Ob diese Stimuli ökologieverträglich sind, ob sie mit bestehendem Umweltrecht kompatibel sind – wen interessiert dies im Falle eines Zielkonflikts wirklich?

Sollen Umweltgesetze künftig in einer dynamisch-komplexen Gesellschaft, Welt, überhaupt eine Chance haben, regulierend in bestimmte Prozesse, die in aller Regel nichtlinearen Charakter haben,[4] einzugreifen, und auch Wirkungen zu erzeugen, dann hat sich die Gestaltung des Umweltrechts am „Effizienzprinzip" zu orientieren.

Was immer der Zweck sein mag, den man mit dem Recht/Umweltrecht verfolgen möchte, die Verwirklichung hängt stets davon ab, ob Menschen oder Systeme auf Rechtsnormen so reagieren, wie man sich dies vorgestellt hat. Denn: Umweltrechtssetzung impliziert Folgenorientierung[5] – oder m. a. W.: Folgenprognose und Folgenbewertung in einer durch Dynaxität charakterisierten Gesellschaft. Um die Folgen von Umweltrechtsnormen abschätzen zu

---

1 Hierzu s. Weimar/Leidig 2002, S. 1 ff.

2 Statt vieler grundlegend Leidig 1986, S. 231 ff.; Leidig 1986, S. 15 ff.; Leidig 1988, S. 73 ff.; Leidig 1984 pass. sowie Eidenmüller 1995, pass; Gawel 2000, S. 9 ff.

3 Vgl. Leidig 2002, S. 56 f.

4 Dazu s. m. w. N. Leidig 1997, S. 127 ff.; Leidig 1999, S. 77 ff.; Leidig 1999, S. 43 ff.; Leidig 2002, S. 431 ff.; Leidig 2002, S. 121 ff.; Leidig 2000, S. 59 ff.

5 Zu diesem Ansatz s. Deckert 1995; Coles 1991; Eidenmüller 1995, S. 398 Anm. 4 m. w. N.

können,[6] benötigt man Verhaltensmodelle vom Menschen[7] und von Systemen.[8]

Ein Urteil über die Funktion der Effizienz des Umweltrechts – derzeit und in der Zukunft – ist naturgemäß signifikant davon beeinflusst – um nicht zu sagen abhängig – wie man diesen in verschiedenen Wissenschaften verwendeten Begriff definiert.

B. Effizienz-Begriffs-Facetten

Der Begriff Effizienz wird nicht nur in den Wirtschaftswissenschaften, geschweige in der Rechtswissenschaft einheitlich verwendet;[9] es erscheint deshalb angebracht, vorab der Frage nachzugehen, was unter „Effizienz" zu verstehen ist. Gibt es einen „allgemeinen" Bedeutungsinhalt des Begriffs „Effizienz", oder hängt der Begriffsinhalt davon ab, in welchem disziplinären Kontext dieser Terminus jeweils verwendet wird?

Die Ausdrücke „Effizienz" und „Effektivität" haben – trotz ihres lateinischen Ursprungs – erst aus dem angelsächsischen Bereich in die verschiedenen Fachsprachen Eingang gefunden. Etymologisch leitet sich der Begriff „Effizienz" von „effectus" ab, dem die Bedeutung von „Wirkung" bzw. „Erfolg" entspricht.[10] Unmittelbares Pendant zum lateinischen Ursprungswort ist der Ausdruck „Effektivität", für den wiederum Begriffe wie „Wirksamkeit" oder „nutzbare Leistung" synonym verwendet werden.[11]

Im verwaltungswissenschaftlichen bzw. betriebswirtschaftlichen Sprachgebrauch hat sich neben dem Ausdruck „Effizienz" der als äquivalent ver-

---

6 Rechtssetzung sollte nicht allein mit der Tätigkeit des Gesetzgebers gleichgesetzt werden; hierzu Weimar / Leidig 2002, S. 50 f. Denn auch Gerichte und Verwaltungsbehörden setzten im ökologischen Bereich Recht. Das Rechtssystem (und seine Akteure) gewinnt an Komplexität.

7 Die Wirtschaftswissenschaften bieten hier primär das Modell des „homo oeconomicus" (Rationalwahl-Modell) an; dazu statt vieler Eidenmüller 1995, S. 28 ff. m. w. N.; ergänzend Bosshardt 2001.

8 Zu dem Problemfeld der Gestaltung des Rechts in komplexen Gesellschaftssystemen s. Di Robilant 2002, S. 353 ff.

9 Zum Folgenden Leidig 1986, S. 231 ff. m. w. N.; Leidig 1986, S. 15 ff.; Leidig 1988, S. 73 ff.; Eidenmüller 1995, pass.

10 Vgl. Gzuk 1975, S. 12.

11 Die Termini „Effizienz" und „Effektivität" finden im umgangssprachlichen Gebrauch vielfach synonyme Anwendung. Übertragen auf rechtsökologische Problemstellungen spricht man von „Effektivität", wenn ein Entscheidungsträger zur Lösung eines Umweltproblems ein strategisch geeignetes Instrument einsetzen. Effizienz zielt hingegen auf die Umsetzung und Kontrolle des rechtsökologischen Instrumentariums ab.

standene Ausdruck „Leistungswirksamkeit" eingebürgert. Im angelsächsischen Sprachkreis sind die inhaltsgleichen Begriffe „effectiveness" und „efficiency", seltener auch Ausdrücke wie „success" und „worth" zu finden.[12] Sie werden oft in der Bedeutung von Zielerreichung bzw. Zielerreichungsgrad verwendet. In den Wirtschaftswissenschaften bedeutet Effizienz – zu unterscheiden von „Rationalität" – vielfach das Verhältnis von Ertrag zu Aufwand oder Nutzenzuwachs zu Nutzenentgang (sogen. Opportunitätskosten). Es finden sich auch differenzierende Definitionen der Effizienz: So unterscheidet Kyrer[13] zwischen „Produktionseffizienz" (technische Produktionseffizienz, ökonomische Produktionseffizienz) und „Planungseffizienz", wobei er unter Planungseffizienz das Ausmaß versteht, in dem ein Ziel oder Zielbündel durch eine Reihe von Operationen und Aktivitäten erreicht werden kann. Nach Kotarbiński[14] lassen sich drei Bedeutungen von Effizienz unterscheiden: universale, synthetische und manipulative. Im universalen Sinne ist Effizienz die allgemeine Bezeichnung einer jeden praktischen Valenz, also ist Genauigkeit ebenso eine Effizienz wie Ergiebigkeit, Einfachheit usw. Effizient im synthetischen Sinne ist die Gesamtheit dieser Valenzen. Man handelt nach Kotarbiñski umso effizienter, je mehr das Handeln die Valenzen der „guten Arbeit" in sich birgt. Diesen beiden relativ allgemeinen Bedeutungsgehalten von Effizienz stellt er andererseits eine ziemlich spezielle Bedeutung gegenüber, die er als manipulative Effizienz bezeichnet. Frohn[15] glaubt deshalb feststellen zu können, dass der Ausdruck Effizienz in den Sozialwissenschaften zur Zeit noch keine mit allgemein anerkannten Instrumenten messbare, sondern eher eine intuitiv zu erfassende Größe darstellt, deren Reichweite durch Synonyme wie „Wirksamkeit", „Leistungsfähigkeit", „Ergiebigkeit", „Sachlichkeit", „Vorteilhaftigkeit" abgesteckt wird.

Nicht weniger erstaunt es, dass der Begriff der „Effizienz" samt seiner Nuancen im Rahmen der Rechtswissenschaft nicht einheitlich verwendet wird.[16] So unterscheidet Leisner[17] drei Formen von Effizienz: „Zielerreichungseffizienz", Effizienz als „optimales Zweck-Mittel-Verhältnis" und „zielunabhängige" Effizienz. Ohne dass der Begriff der Effizienz bisher eine dogmatische Vertiefung erfahren hat, tritt er selbst terminologisch – wie in den klassischen Sozialwissenschaften – unter zahlreichen Synonymen auf: „Leistungsfähigkeit", „optimale Funktionswahrnehmung", „Rationalität", „Optimierung", „Wirkung", „soziale Wirksamkeit" „Wirksamkeit", „Effektivität" und in der angelsächsischen Literatur unter der Bezeichnung „efficieny" /„efficacy".[18]

12 Vgl. Gzuk 1975, S. 12 m. w. N.
13 1972, S. 125 ff.
14 Vgl. Kotarbiński 1965 zit. n. Leidig 1986, S. 251 m. w. N.
15 1977, S. 44.
16 Dazu s. Leisner 1971, S. 6; Eidenmüller 1995.
17 1971, S. 7.
18 Zusammenfassend hierzu Leidig 1986, S. 233 m. w. N.

Geiger[19] bestimmt die Verbindlichkeit einer Rechtsnorm als $v = e/s$, wobei: e = Effektivitätsquote und s = die Gesamtzahl der Fälle ist, in denen Normadressaten aktuell in normtypische Situationen geraten. Die Effektivitätsquote (e) berechnet sich dabei nach Geiger folgendermaßen: $e = (s \rightarrow bg) + ((s \rightarrow c\dot{g}) \rightarrow r)$. Dabei bezeichnet „b“ die Anzahl konformen Verhaltens (g), „c“ die Anzahl nonkonformen Verhaltens ($\dot{g}$). Nach Löhr[20] ist Effizienz kein Wert an sich, sondern immer nur im Hinblick auf ein bestimmtes Ziel, einen tatsächlichen und/oder rechtlichen, verwaltungsexternen oder auch -internen Erfolg zu sehen. Effizienz dient der Normverwirklichung. Ihr Ausbleiben bedeutet „soziale Unwirksamkeit von Gesetzen“ (Noll). Effizienz ist in dem Umfang verwirklicht, in dem bestimmte Aufgaben (als Ziele) relativ zum Ressourceneinsatz erfüllt werden, ohne dass bestimmte Funktionsvoraussetzungen vernachlässigt werden. Schon diese wenigen Beispiele verdeutlichen, dass im juristischen Schrifttum unterschiedliche Auffassungen zu den Bedeutungsinhalten des Begriffs „Effizienz“ anzutreffen sind. Gemeinsam ist diesen Ansätzen indes, dass sie „Effizienz“ durchaus (auch) als „Rechtsbegriff“ beanspruchen.

In den weiteren Darlegungen soll unter „Effizienz“ der Grad der Zielerreichung verstanden werden. Formal betrachtet stellt „Effizienz“ also einen Erfolgsindex dar, der Auskunft gibt über „Wirkungen“ von Gesetzen in dem für sie je relevanten Sektor. Voraussetzung hierfür ist, dass die Leistungseffekte in den Kategorien der operationalisierten Gesetzesziele formuliert werden. Die Aussage über die Effizienz einer rechtsökologischen Norm ist danach identisch mit der Aussage, ob und inwieweit die in der Norm formulierten Ziele erreicht werden.

C. Effizienz und Umweltgesetzesgestaltung

1. Funktion der Effizienz[21]

Effizienz von Gesetzen kann man von zwei funktionellen Aspekten her aufrollen. Zum einen kann sie a posteriori auf der Vollziehungsebene analysiert werden. Bedeutsam für die (Umwelt-)Gesetzgebung ist dieser Aspekt insofern, als Mängel bei der Zielerreichung an den Gesetzgeber rückgemeldet werden können. Diesen Aspekt der Effizienz kann man mit dem Ausdruck „Vollzugseffizienz“ wohl am besten charakterisieren. Die Überprüfung der Vollzugseffizienz erfolgt derzeit keineswegs planmäßig, sondern – wenn und

19 1964, S. 71.
20 1977, S. 78.
21 Dazu und zum Folgenden insbes. Leidig 1986, S. 234 m. w. N.

soweit überhaupt – mehr oder weniger zufällig – sowohl auf nationaler als auch internationaler Ebene. Erforderlich wäre deshalb eine institutionalisierte Kontrolle der Effizienz auf der Ebene der Vollziehung (Effizienzkontrolle). Neben diesem a posteriori-Aspekt erscheint es – z.B. im Rahmen des Umweltrechts – erforderlich, die intendierte Effizienz von rechtsökologischen Gesetzen bereits antizipatorisch bei deren Formulierung zu berücksichtigen (Effizienzprognose). Zu beachten ist in diesem Zusammenhang, dass die Kontrolle und Prognose von Effizienz im Rahmen der in diesem Beitrag zu erörternden Problematik deshalb mit besonderen Schwierigkeiten behaftet ist, weil es sich nicht um die Erfassung eindimensionaler, „universaler" Effizienz, sondern um das Erfassen mehrdimensionaler, „synthetischer" Effizienz in komplexen Systemen handelt. Hier ist nicht nur eine Valenz als Effizienzkriterium zu berücksichtigen, sondern eine Vielzahl.

Von diesen beiden Aspekten der Effizienz ist eine weitere Version zu unterscheiden: die Verfahrenseffizienz. Gemeint ist damit die auf Gesetzgebung als Prozess bezogene Effizienz, die die effiziente, also an vorgegebenen Zielen orientierte Abwicklung des Gesetzgebungsverfahrens selbst betrifft.

2. Effizienz-Kategorien und Umweltrecht[22]

Damit die Effizienz von Umweltgesetzen ermittelt werden kann, sind nicht nur operationale Ziele sowie entsprechende Instrumente bereitzustellen, es sind auch Effizienzkategorien zu formulieren und in einem Indikatorensystem zu vereinigen. Dabei ist generell Gleichheit bei den zu verwendenden Effizienzkriterien und hinsichtlich der Messinstrumente anzustreben; andernfalls sind die Ergebnisse nicht vergleichbar.

Zur Bewertung der Effizienz von Umweltgesetzen kommen verschiedene Effizienzkategorien in Betracht: ökologische, allokative, sozio-ökonomische, reduktive sowie temporale Effizienz.

Unter „ökologischer Effizienz" von Umweltgesetzen versteht man hier die unmittelbare bzw. schnelle Beseitigung/Reduzierung und Verhinderung von Umweltschäden. Operationalisieren könnte man diese Effizienzkategorie durch Subkriterien wie etwa Leistungsfähigkeit beim Gewässerschutz, Immissionsschutz, Strahlenschutz, bei der Abfallbeseitigung, in der Landespflege, im Erholungsbereich sowie beim Naturschutz/Bodenschutz.

Mit „allokativer Effizienz" ist das Ausmaß der Verringerung volkswirtschaftlicher Kosten und in Verbindung damit der Grad der Internalisierung externer Kosten gemeint. Die „sozio-ökonomische Effizienz" schließlich gibt an, in welchem Umfang andere potenzielle Zielvorstellungen durch Umweltgesetze berührt werden. Umweltziele sind immer nur einige von vielen gesellschaftlichen Zielen, die es zu realisieren gilt; mithin dürfen sie nicht ver-

---

22 Vgl. dazu Leidig 1986, S. 246 f.

absolutiert werden. Die „sozio-ökonomische Effizienz" von Umweltgesetzen ist daher stets umso höher, je weniger durch Gesetze dieser Provenienz die Verwirklichung anderer politischer Ziele negativ und je mehr sie positiv beeinflusst wird.

Was die „reduktive Effizienz" angeht, so gibt sie Auskunft über das Konfliktregulierungspotenzial von Umweltgesetzen. Anhand dieser Effizienzkategorie kann ermittelt werden, inwieweit ein Umweltgesetz in der Lage ist, ökonomisch-ökologische Konflikte aus sich selbst heraus zu bewältigen.

Bei der „temporalen Effizienz" wird ermittelt, ob das betreffende Gesetz die intendierten Umweltziele in dem vorgesehenen Zeitraum zu erreichen vermag oder nicht. Sie ist stets umso höher, je kürzer der Zeitraum ist, in dem die vorgegebenen Umweltziele realisiert werden (und umgekehrt).[23]

Um die Effizienz von Umweltgesetzen zu gewährleisten bzw. Effizienz als Gestaltungsprinzip[24] zu implementieren, ist es fernerhin notwendig, bereits bestehende Gesetze resp. im Gesetzgebungsprozess befindliche zu evaluieren. Ansonsten ist die Effizienzdiskussion „ineffizient".

3. Effizienz-Evaluation und Umweltrecht[25]

Der Begriff „Effizienz-Evaluation" bezieht sich auf die Ermittlung von Soll-Zuständen, die Feststellung von Ist-Zuständen und den Soll/Ist-Vergleich. Seine konstitutiven Begriffselemente lassen sich somit wie folgt zusammenfassen:

- Effizienzprognose
- Effizienzfeststellung
- Effizienzkontrolle

Allgemein lassen sich Prognosen als informationshaltige und empirisch begründete Aussagen über zukünftige Sachverhalte – hier: anzustrebende Soll-Zustände – auf der Basis des derzeitigen und vergangenheitsorientierten Wissens der am Prognoseprozess partizipierenden Personen / Institutionen interpretieren. Unter „Effizienzprognose" soll im Folgenden die logisch-rationale Durchdringung zukünftigen Geschehens bzw. künftiger Entwicklungen verstanden werden, die durch die Anwendung wissenschaftlicher Methoden geschieht und der Ermittlung von Informationen über den künftigen Effizienzgrad einzelner Umweltgesetze dient.[26]

23 Zur Zeitstruktur des Rechts s. Weimar / Leidig 2002, S. 8 ff. m. w. N.

24 Zum Begriff „Prinzipien" vgl. Eidenmüller 1995, S. 461 ff. m. w. N.

25 Vgl. Leidig 1988, S. 74 ff.

26 Wesentliche Fortschritte resultieren hier aus anderen Wissenschaftssystemen: Künstliche-Intelligenz-Forschung, Chaos- / Komplexitätstheorie, Artificial-Life-Science. Sie eröffnen neue Möglichkeiten, die Effizienz von Umweltgesetzen zu evaluieren; dazu s. a. Leidig 1995; Weimar / Leidig 2002, pass.; Leidig 2001.

„Effizienzfeststellung" hat den Zweck, Informationen über den zur Zeit realisierten Effizienzgrad von Umweltgesetzen zu liefern: Feststellung des Ist-Zustandes. Von der Güte und Vollständigkeit der zu eruierenden Fakten hängt in entscheidendem Maße die Qualität der weiteren Ergebnisse ab. Die Effizienzfeststellung schließt auch die Ordnung des Datenmaterials nach bestimmten – vorher festgelegten – Kriterien ein.

„Effizienzkontrollen" bestehen in der Gegenüberstellung von Soll- und Ist-Werten, verbunden mit einer eingehenden Analyse der ermittelten Abweichungen und ihrer Ursachen. Dem Effizienzkontrollprozess ist somit die Ermittlung von Soll- und Ist-Werten vorgelagert. Reine Ist/Ist-Vergleiche, also Vergleiche von derzeit erreichten mit vergangenen Ist-Zuständen, sind keine Effizienzkontrollen im oben formulierten Sinn.

Wenn man den mit der Effizienz-Evaluation verfolgten Zweck erreichen will, ist es notwendig, dass der Prozess der Effizienz-Evaluation sowohl geplant (Planung der Effizienz-Evaluation) als auch organisiert wird (Organisation der Effizienz-Evaluation). Aufgabe der Planung der Effizienz-Evaluation ist die Erarbeitung und Auswahl von Vorgehens- und Verhaltensweisen, mit denen die vorher festgelegten Ziele erreicht werden sollen. Dieser Prozess der Planung der Effizienz-Evaluation von Umweltgesetzen lässt sich in folgende Phasen aufteilen:

- Formulierung der mit der Effizienz-Evaluation verfolgten Ziele.
- Suche nach alternativen Instrumenten, mit denen die Ziele erreicht werden können.
- Bewertung der alternativen Instrumente im Hinblick auf ihre Zielwirksamkeit.
- Entscheidung darüber, mit welchen Instrumenten die Effizienzbewertung durchgeführt werden soll.

Des Weiteren ist es erforderlich, die Bewertung der Effizienz von Umweltgesetzen organisatorisch zu fundieren, und zwar durch die Einrichtung eines Effizienz-Evaluations-Systems, durch Bestimmung der mit der Effizienz-Evaluation betrauten Personen/Instanzen und durch die Festlegung der Berichterstattung.

Die Installierung eines solchen Systems soll bewirken, dass alle wesentlichen Daten beachtet werden. Einerseits sollen keine Informationen verloren gehen, andererseits soll die Effizienz-Evaluation möglichst reibungslos – unter Beachtung des ökonomischen Prinzips – durchgeführt werden. Deshalb liegt es nahe, die Bewertungsvorgänge zu systematisieren.

D. Effizienz und Umweltgesetzesauslegung

Die Effizienz von Umweltrecht bezieht sich nicht nur auf den Prozess der Gestaltung / Generierung von Rechtsnormen, um die natürlichen Ressourcen zu schützen, sondern auch die Gesetzesauslegung muss sich an dem Kriterium der Effizienz messen lassen.

Relevant sind in diesem Kontext zwei Problemfelder: Effizienz als

- Politik des Umweltgesetzes
- Gesetzeskonkretisierung.

1. Effizienz als Politik des Umweltgesetzes[27]

Unproblematisch erscheint eine Berücksichtigung des Effizienzkonzepts bei der Anwendung des Umweltrechts zunächst dort, wo Effizienz die Politik des rechtsökologischen Gesetzes ist. Überall dort, wo der Gesetzgeber klar zu erkennen gegeben hat, dass er mit dem Erlass einer bestimmten Regelung das Effizienzziel im ökologischen Sektor fördern möchte, ist es den Richtern nicht nur erlaubt, bei der Interpretation von Rechtsbegriffen ökologisch-ökonomische Erwägungen anzustellen; sie sind sogar dazu verpflichtet, dies zu tun. Wenn es der Wille des demokratisch legitimierten Parlamentes ist, durch ein bestimmtes Umweltgesetz das Effizienzziel im Umweltschutzbereich derzeit oder künftig zu fördern, dann sind die an Gesetz und Recht gebundenen Gerichte gehalten, diesen Willen auch möglichst in der Realität zur Geltung zu bringen. Dies können sie tun, indem sie bei der Gesetzesauslegung im Zweifel die Entscheidungsvariante wählen, die das gesetzgeberische Ziel möglichst optimal verwirklicht, die also unter Effizienzgesichtspunkten die beste ist. Eine Folgenanalyse und Folgenbewertung im Lichte der hier vorgestellten Effizienzkategorien ist zur Implementierung der legislativen Zielstruktur insoweit unerlässlich.

Das Beispiel des Umwelthaftungsgesetzes (BRD) steht hier stellvertretend für jene Fälle, in denen der Umweltgesetzgeber durch bestimmte rechtliche Normierungen die Allokationseffizienz im Zivilrecht verbessern möchte. „Lässt sich ein derartiger gesetzgeberischer Wille nachweisen, dann ist es legitim, wenn die Gerichte bei der Auslegung der entsprechenden Bestimmungen dieser Zielsetzung möglichst Rechnung tragen, indem sie die unter Effizienzgesichtspunkten beste Entscheidung treffen.“[28]

---

27 Allgemein zum Folgenden Eidenmüller 1995, S. 452 ff.
28 Eidenmüller 1995, S. 454.

2. Effizienz als Gesetzeskonkretisierung[29]

Wesentlich differenzierter und problematischer zu beurteilen ist jene Konstellation, in der Effizienz zwar mit Sicherheit nicht die „Politik des Umweltgesetzes" ist, in der man sich aber immerhin fragen kann, ob eine effizienzsensible Gesetzesinterpretation nicht wenigstens eine zulässige Gesetzeskonkretisierung darstellt.

Eine Berücksichtigung der Effizienzkonzepte anhand der verschiedenen Effizienzkategorien bei der Gesetzesauslegung kommt dann in Betracht, soweit Effizienz die Politik des Gesetzes ist oder soweit eine effizienzorientierte Rechtsanwendung zumindest als zulässige Gesetzeskonkretisierung angesehen werden kann. Unproblematisch ist eine Berücksichtigung dieses Effizienzkonzepts durch die Gerichte dort, wo der Gesetzgeber ihnen dazu ein klares Mandat erteilt (Effizienz als Politik des Gesetzes). Dies ist derzeit noch äußerst selten. Schwierigkeiten wirft vor allem die zweite Fallgruppe auf. „Eine effizienzorientierte Rechtsanwendung steht dabei immer unter dem Vorbehalt gegenläufiger juristischer Wertungen und sollte im Übrigen aus Gründen der institutionellen Legitimation und Kompetenz nur äußerst zurückhaltend erfolgen."[30]

Zusammenfassung und Ausblick

„Effizienz" als Thema in Wissenschaft und Praxis, hier im Hinblick auf den Schutz natürlicher Ressourcen und die konstruktive Fortentwicklung des Umweltrechts auf nationaler und internationaler Ebene fokussiert, ist zweifelsohne im Vordringen begriffen. Insbesondere die Propagierung des Effizienzgedankens durch die Ökonomie gilt vielen anderen Wissenschaftsdisziplinen, wie die der Umweltrechtswissenschaft, als Herausforderung und Chance. Effizienz als Gestaltungsprinzip ist Teil in einem allgemein zu beobachtenden Trend zur Ökonomisierung (aller) Lebensverhältnisse. Diesem Szenario darf sich auch das Umweltrecht nicht verschließen. Es ist eingebunden in einen globalen, an Dynamik und Komplexität gewinnenden Prozess, der zu gravierenden Veränderungen auf allen Ebenen des Gesellschaftssystems führt. Nicht nur Strukturen, sondern auch Einstellungen und Wertesysteme derzeitiger und künftiger Generationen dürften neu „gemischt" werden. Und die Effizienz des Umweltrechts hängt davon ab, ob die Menschen auf „Rechtsnormen" so reagieren, wie man sich dies vorgestellt hat. Angesprochen ist hiermit die „Evolution des (Umwelt-)Rechtssystems" in der Zeitstruktur.[31]

---

29 Dazu Eidenmüller 1995, S. 454 ff. m. w. N.

30 Eidenmüller 1995, S. 459.

31 Zum Folgenden Weimar / Leidig 2002, S. 5.

Bei der Analyse des Verhältnisses von Evolution und Rechtssystem lässt sich feststellen, dass das Recht kein unmittelbares „Abbild" der jeweiligen Zeit darstellt. Ein Wandel von Situationen und Anschauungen führt nicht automatisch zu einem Wandel des Rechtssystems und seiner (rechtsökologischen) Normen. Denn das Recht / Umweltrecht ist nicht eine bloße Kopie der (ökologisch-ökonomischen) Wirklichkeit. Als solche wäre es auch sicherlich kein Recht/Umweltrecht.

Vielmehr muss der Rechtsanwender auf Veränderungen in einer spezifischen Art und Weise reagieren. Gesellschaftliche Veränderungen wirken auf den Umgang mit dem Recht und seiner Gestaltung im Sinne einer entsprechenden Veränderung des (Umwelt-)Rechts ein. Die auf dem Glauben an die Statik rechtlicher Regelungen gründende – vermeintliche – Rechtssicherheit stellt – wie sich daraus ergibt – nur einen Gesichtspunkt unter mehreren dar, was hier nur angedeutet sei.

Die Zeitstruktur des (Umwelt-)Rechts wird – jedenfalls in der herkömmlichen Rechtswissenschaft – vor allem im Hinblick auf Veränderungen bei der Auslegung des Inhalts von Normen thematisiert. Dabei kann ein Wandel der Verhältnisse eine Vielzahl ganz unterschiedlicher Auswirkungen für Normen und Einzelentscheidungen hervorrufen. Auswirkungen auf die Auslegung bilden immer nur einen Teil der Möglichkeiten, wenn auch in der Rechts- und Verwaltungspraxis sicherlich den zahlenmäßig umfangreichsten. Ein Auslegungswandel ist nur die erste und unmittelbare Reaktion des Rechts und des Rechtsanwenders – des Richters, des Verwaltungsentscheiders, auch des Anwalts – auf einen relevanten Wandel der sozialen Verhältnisse.

Darüber hinaus gibt es auch Veränderungen, die mit Hilfe einer Normanpassung durch entsprechende Auslegung nicht mehr bewältigt werden können. In Fällen dieser Art sind grundsätzlich verschiedene „Lösungen" möglich, wobei von dieser Problematik im Einzelnen hier allerdings nicht die Rede sein soll. So sehr es einerseits vorstellbar ist, Normen auch dann noch – bis zur Aufhebung durch den Gesetzgeber – Geltung beizumessen, wenn sie aufgrund von Wandlungen der Wirklichkeit ihren Sinn verloren haben, so findet sich auf der anderen Seite die geradezu gegenteilige Position, die ihnen ohne förmliche Aufhebung die Geltung schon dann absprechen will, wenn die faktische Grundlage für ihr Vorhandensein entfallen ist.

Neuere Entwicklungen im Umweltrecht müssen also mindesten zwei basalen Aspekten Rechnung tragen: Dem Effizienzprinzip und der sich beschleunigenden Zeitstruktur des Rechts bzw. des Rechtssystems. Gesichtspunkte, die bislang in vielen Fällen eher eine marginale Rolle gespielt haben.

Literatur

Bosshardt, Ch. (Hrsg.), Problembereiche interdisziplinärer Forschung, Bern u. a. 1999.

Bosshardt, Ch., Homo Confidens. Eine Untersuchung des Vertrauensphänomens aus soziologischer und ökonomischer Perspektive, Bern u. a. 2001.

Coles, Ch., Folgenorientierung im richterlichen Entscheidungsprozeß: ein interdisziplinärer Ansatz, Frankfurt / M. u. a. 1991.

Deckert, M. R., Folgenorientierung in der Rechtsanwendung, München 1995.

Dixon-Gough, R. W. (Hrsg.), Land Reform and Sustainable Development, Aldershot u. a. 1999.

Eidenmüller, H., Effizienz als Rechtsprinzip, Tübingen 1995.

Frohn, H., Gesetzesbegriff und Gewaltenteilung, Diss. Köln 1977.

Gawel, E., Effizientes Umweltordnungsrecht – Anforderungen und Grenzen aus ökonomischer Sicht, in: Gawel / Lübbe-Wolff (Hrsg.), 2000, S. 9 ff.

Gawel, E. / Lübbe-Wolff, G. (Hrsg.), Effizientes Umweltordnungsrecht, Baden-Baden 2000.

Geiger, Th., Vorstudien zu einer Soziologie des Rechts, Neuwied – Berlin 1964.

Gzuk, R., Messung der Effizienz von Entscheidungen, Tübingen 1975.

Jäggi, V. / Mäder, U. / Windisch, K. (Hrsg.), Entwicklung, Recht, Sozialer Wandel. Festschrift für Paul Trappe zum 70. Geburtstag, Bern u. a. 2002.

Kotarbiński, R., Traktat über die gute Arbeit, Wroclaw u. a. 1965.

Krawietz, W. / Weimar, R. (Hrsg.), Die Ordnung des Bodens im Fortschritt der Wissenschaften, Frankfurt / M. u. a. 1986.

Kyrer, A., Effizienz und staatliche Aktivität, Wien 1972.

Leidig, G., Ökologisch-ökonomische Rechtswissenschaft, Frankfurt/ M. u. a. 1984.

Leidig, G., Gesetzgebung und Effizienz, in: Tammelo / Mock (Hrsg.), 1986, S. 231 ff.

Leidig, G., Effizienz der Umweltgesetzgebung, in: Krawietz / Weimar (Hrsg.), 1986, S. 15 ff.

Leidig, G., Effizienzevaluation von Umweltgesetzen, in: Lopez Calera / Seele (Hrsg.), 1988, S. 73 ff.

Leidig, G., Chaosforschung und Umweltschutz, Basel 1995.

Leidig, G., Rechtsökologische Forschung und Chaostheorie, in: ZÖR 52 (1997), H. 1, S. 127 ff.

Leidig, G., Chaostheorie und Zukunftsherausforderungen, in: Bosshardt (Hrsg.), 1999, S. 77 ff.

Leidig, G., Ecological land development and multidisciplinary research, in: Dixon-Gough (Hrsg.), 1999, S. 43 ff.

Leidig, G., Natural Environment, Natural Law and Natural Sciences – Aspects of a Multidisciplinary Approach, in: Vera Lex 2000, New Series Vol. 1, Nr. 1/2, S. 59 ff.

Leidig, G., Natural Sciences and Ecological Regional Planning: Chaos Theory, Artificial Intelligence Research, Artificial Life Research, Virtual Reality Research, in: http://www.paricenter.com/library/papers/leidig01.php, 2001.

Leidig, G., Dynaxibility – Modewort oder Zauberformel?, in: D & MM 2002, Nr. 21-22, S. 56 f.

Leidig, G., Komplexe Systeme und Unternehmensführung, in: Leidig / Mayer (Hrsg.), 2002, S. 121 ff.

Leidig, G., Raumplanung im Spannungsfeld von Emergenz, Imergenz und Komplexität, in: Weiß / Zangger (Hrsg.), 2002, S. 431 ff.

Leidig, G. / Mayer, Th. (Hrsg.), Betriebswirtschaft und Mediengesellschaft im Wandel. Festschrift für Diethelm Schmidt und Lorenz Rottland, Wiesbaden 2002.
Leisner, W., Effizienz als Rechtsprinzip, Tübingen 1971.
Löhr, R.-P., Die kommunale Flächennutzungsplanung, Siegburg 1977.
Lopez Calera, N. M. / Seele, W. (Hrsg.), Politisches System und Bodenordnung, Frankfurt / M. u. a. 1988.
Di Robilant, E., Probleme der Gestaltung des Rechts in der komplexen Gesellschaft, in: Jäggi / Mäder / Windisch (Hrsg.), 2002, S. 353 ff.
Tammelo, I. / Mock, E. (Hrsg.), Rechtstheorie und Gesetzgebung, Frankfurt / M. u. a. 1986.
Weimar, R. / Leidig, G., Evolution, Kultur und Rechtssystem. Beiträge zur New Political Ecology, Frankfurt / M. u. a. 2002.
Weiß, E. / Zangger, T. (Hrsg.), Aufgaben der Regionen im Hinblick auf Raumplanung, Bodenrecht und Umweltschutz, Bern u. a. 2002.

MARTIN LENDI

Steuerung der Siedlungsentwicklung – Übersicht zu Intentionen und Modalitäten der Lenkung[1]

Notwendigkeit und Art der Steuerung der Siedlungsentwicklung können unter planerischen, rechtlichen und ökonomischen, aber auch unter ökologischen Gesichtspunkten betrachtet werden. Über Interpretationen, Berechnungen und Bewertungen der relevanten Faktoren der Bemessung des Siedlungsbedarfs gehen die Meinungen oft erheblich auseinander. Im Kern aber sind zwei Fragen offen, a) welche Siedlungsentwicklung wir zu erwarten haben, und b) ob, in welchem Masse, wo und wann sie unter welchen Zielvorstellungen und mit welchen Qualitätsansprüchen mit hinreichender Wirkung gesteuert werden soll. Eine zusammenfassende Problemanalyse zu Sinn und Zweck der Siedlungssteuerung steht aus.[2] Darüber darf die Fülle an wissenschaftlichen Grundlegungen und Abhandlungen zur Raumplanung nicht hinwegtäuschen.[3]

---

1 Dieser Text entstand als Diskussionsbeitrag, gerichtet an den schweizerischen Rat für Raumordnung, zu offenen Fragen der Regelung der Siedlungsentwicklung.

2 Eine zusammenfassende Sicht vermittelt *ARL (Hrsg.)*, Flächenhaushaltpolitik, Feststellungen und Empfehlungen für eine zukunftsfähige Raum- und Siedlungsentwicklung, Hannover 1999. Für die faktische Entwicklung der Siedlungsräume werden – abgesehen von der Bevölkerungs- und den wirtschaftlichen Trends sowie den steigenden Ansprüchen an Wohn- und Arbeitsräume – vor allem die Verkehrsstrukturen verantwortlich gemacht, siehe dazu *Holz-Rau Christian*, Verkehrs und Siedlungsstruktur, eine dynamische Aufgabe, in: Raumforschung und Raumordnung, 4/2001, S. 264 ff. Eindrückliche Zahlen zum Flächenverbrauch pro Tag als Fakten und als Norm vermitteln *Jakubowski Peter / Zarth Michael*, Nur noch 30 Hektaren Flächenverbrauch pro Tag, in: Raumforschung und Raumordnung, 3/2003, S. 185 ff. Den Einfluss der Politik auf die Siedlungsentwicklung wurde sehr grundsätzlich untersucht von *Bökemann Dieter*, Siedlungsentwicklung: Evolution und Politik, in: ARL, Beiträge zur theoretischen Grundlegung der Raumentwicklung, Arbeitsmaterial, Hannover 2000, S. 69 ff. Den neusten Stand der schweizerischen Diskussion vermittelt *Wachter Daniel*, Raumentwicklung und Raumordnungspolitik in der Schweiz, in: Schweiz 2003, Geographische Rundschau 9/2003, S. 34 ff. Er behandelt u.a. die Metropolisierung, den Flächenverbrauch und auch Probleme der Siedlungsbegrenzung. Die Konsequenz des Verdichtens im Siedlungsraum wird beleuchtet von *Schweizer Heimatschutz (Hrsg.)*, Heimatschutz, Verdichten im Siedlungsraum, Forum Thema 2-15, Heft 4/03; daselbst *Rellstab Ursula*, Dichte Bauweise verlangt viel Landschaft, a. a. O., S. 6 ff.

3 Die Literatur zur Raumplanung zu erfassen, ist hier nicht angezeigt. Als allgemeiner Hinweis und Brücke diene: *Akademie für Raumforschung und Landesplanung (ARL)*, Handwörterbuch der Raumordnung, Hannover 1995; *eadem*, Methoden und Instru-

Wie unsicher die Doktrinen sind, erhellt aus zwei sich widersprechenden Ansätzen, einerseits der Verdichtung der bestehenden Siedlungsgebiete, verbunden mit einer absoluten Grenzziehung nach aussen, und anderseits der

---

mente räumlicher Planung, Hannover 1998; *eadem*, Grundriss der Landes- und Regionalplanung, Hannover 1999; *Jenkins Helmut W. (Hrsg.),* Raumordnung und Raumordnungspolitik, München 1996. Zu den schweizerischen Publikationen und Materialien siehe *Lendi Martin*, Politisch, sachlich und ethisch indizierte Raumplanung – am Beispiel der Schweiz, Wien 1998; *Lendi Martin / Elsasser Hans*, Raumplanung in der Schweiz, eine Einführung, 3. A., Zürich 1991; *Lendi Martin*, Grundriss einer Theorie der Raumplanung, 3. A., Zürich 1996. Eine eindrückliche Literaturzusammenstellung zum geltenden schweizerischen Recht der Raumplanung, aber auch umfassende Angaben zu den Rechtsquellen und Materialen, finden sich bei *Bertschi Martin*, Die Umsetzung von Art. 15 lit.b RPG über die Dimensionierung der Bauzonen: Bundesrecht, föderalistische Realität und ihre Wechselwirkung, Zürich 2001, S. XIX ff. Auch auf die greifbare Literatur zu marktwirtschaftlichen Instrumenten wird daselbst aufmerksam gemacht.

Ich verzichte unter diesen Umständen auf detaillierte Verweise, hebe allerdings noch zum hier behandelten Thema den grundsätzlichen Aufsatz von *Kuttler Alfred*, Wann ist für die Bewilligung von Bauten und Anlagen in Nichtbauzonen Art. 24 RPG anzuwenden?, in: Ruch / Hertig / Nef (Hrsg.), Das Recht in Raum und Zeit, Festschrift für Martin Lendi, Zürich 1998, S. 337 ff., hervor, weil dort die verfassungsrechtliche Grundlegung des Prinzips der Trennung von Siedlungs- und Nicht-Siedlungsgebiet besonders deutlich angesprochen wird. Sie leitet sich im übrigen aus der Zielsetzung der geordneten Besiedlung und der zweckmässigen sowie haushälterischen Nutzung des Bodens her (Art. 22quater aBV und Art. 75 BV) her. Siehe auch *idem*, Die Trennung der Bauzonen von den Nichtbauzonen – zur Tragweite eines verfassungsrechtlichen Prinzips, BlAR 1995, S. 10 ff. Die wichtigsten Aufsätze des Autors finden sich neu zusammengestellt in *Kuttler Alfred*, Beiträge zur Raumordnung als Weg und Ziel, Festgabe zum 80. Geburtstag des Verfassers, Zürich 2003

Nicht vergessen werden dürfen die Landesplanerischen Leitbilder des ORL-Instituts der ETH Zürich. Sie versuchten, eine materielle Antwort zur Frage nach der Siedlungsentwicklung zu vermitteln: *ORL-Institut*, Landesplanerische Leitbilder, 3 Bde. und Plankassette, Zürich 1971. Daselbst wird dem siedlungsseitigen Aspekt der räumlichen Entwicklung ein grosser Stellenwert beigemessen, offensichtlich deshalb, weil er der beeinflussbare ist.

Das Bundesgericht hat bereits vor dem Erlass des BG über die Raumplanung (RPG) vom 22. Juni 1979 klargestellt, dass der Inhalt des Grundeigentums ausserhalb der Bauzonen dem Grundsatz nach einheitlich festgelegt sei und dass bei einer Zuweisung grundsätzlich keine Entschädigungspflicht bestünde, wie umgekehrt kein Rechtsanspruch auf Zuweisung in eine Bauzone bestehe. Nur ausnahmsweise könnten einzelne Grundeigentümer enteignungsähnlich getroffen werden. (BGE 105 Ia 330). Diese Rechtsprechung, entwickelt im Zusammenhang mit dem Gewässerschutzrecht, wurde unter dem RPG in Anwendung von Art. 15 ff./24 fortgeführt. Sie reduziert die Zahl der Fälle mit Entschädigungspflicht, beachtet aber unter strengen Voraussetzungen der bundesgerichtlich definierten materiellen Enteignung die verfassungsrechtlich geforderte volle Entschädigung.

fliessenden Begegnung von Siedlungsräumen mit der offenen Landschaft in gegenseitiger, vor allem auch ökologischer Vernetzung. Die planerischen Intentionen neigen zu einer gewissen Begrenzung des Siedlungsgebietes samt Verdichtungen, allerdings bei gleichzeitiger Fähigkeit, das Siedlungsgebiet mit Freiräumen zu durchsetzen und eine ökologische Diffundierung zu begünstigen. Das geltende Recht – so nach schweizerischem Raumplanungsrecht – ist nicht weniger offen. Es drückt sich nur vermeintlich unmissverständlich aus. Bei genauerem Zusehen lässt es der räumlichen Planung erhebliche Spielräume: „Bund, Kantone und Gemeinden sorgen dafür, dass der Boden haushälterisch genutzt wird. Sie stimmen ihre raumwirksamen Tätigkeiten aufeinander ab und verwirklichen eine auf die erwünschte Entwicklung des Landes ausgerichtete Ordnung der Besiedlung. Sie achten dabei auf die natürlichen Gegebenheiten sowie auf die Bedürfnisse von Bevölkerung und Wirtschaft.“[4]

1. Zwei elementare Fragestellungen

Die konsequente Trennung von Siedlungs- und Nicht-Siedlungsgebiet kann als gesetzliche, wenn nicht sogar als verfassungsrechtliche Vorgabe, haushälterisch mit dem anvertrauten Boden und Lebensraum umzugehen, betrachtet werden. Die Verfassung betont denn auch die zweckmässige und haushälterische Nutzung des Bodens sowie die geordnete Besiedlung des Landes[5], das Bundesgesetz über die Raumplanung handelt sogar explizit von der Begrenzung der Ausdehnung der Siedlung.[6] Es operiert mit limitierten Bauzonen und einer ausgedehnten Landwirtschaftszone, die mit einem Bauverbot für nicht zonenkonforme Bauten belegt ist und in der nur über eine Ausnahme unter restriktiven Bedingungen eine Baubewilligung erlangt werden kann.[7] Dies Grundordnung des Raumplanungsrechts wird von der übrigen Rechtsordnung aufgenommen, so vom Bundesgesetz über das bäuerliche Boden-

---

4 Art. 1 Abs. 1 RPG. Selbstredend darf im Rahmen einer sorgfältigen Erfassung des positiven Rechts nicht nur diese eine Bestimmung konsultiert werden.

5 Art. 75 Abs. 1 BV (BV steht für die Bundesverfassung vom 18. April 1999, welche die Verfassung vom 29. Mai 1874 mit ihren rund 130 seitherigen Änderungen (aBV) sanft novelliert hat.

6 Art. 3 Abs. 3 RPG – im Geiste des Art. 1 Abs. 1 RPG.

7 Art. 15 ff. RPG, Art. 24 ff. RPG; die grundlegende Bestimmung des Art. 15 RPG hat folgenden Wortlaut:
„Art. 15 Bauzonen
Bauzonen umfassen Land, das sich für die Überbauung eignet und
a. weitgehend überbaut ist oder
b. voraussichtlich innert 15 Jahren benötigt und erschlossen wird.“

recht, das seinen Geltungsbereich nach ausserhalb und innerhalb der Bauzonen definiert.[8]

Ungeachtet dieser verbindlichen Regelung ist die Frage nach Notwendigkeit, Zweckmässigkeit und Art der Steuerung der Siedlungsentwicklung immer wieder neu zu stellen, bis und mit der Elementarkritik: Gewährleisten die geltenden rechtlichen Vorgaben die Zukunftsfähigkeit des Lebensraumes und damit des politischen, wirtschaftlichen, sozialen und natürlichen Lebens? Vermag die gesetzlich vorgesehene Steuerung der über die Lebensweisen und teilweise durch die Art von Bebauung und Verkehr städtisch geprägten Kulturlandschaft den sachlich angezeigten Stempel einer differenzierten, spannungsreichen Landschaft den Stempel aufzudrücken? Was beinhaltet Siedlungsentwicklung und was ist für das Nicht-Siedlungsgebiet vorzusehen – was ist deshalb ursächlich anzugehen und zielbezogen zu steuern?

Die auffallende Ungeduld, welche die Lehre von der Raumplanung seit kurzem erfasst hat, hält diese an, nach neuen Begründungen und Instrumenten der Raumordnung zu suchen.[9] Analysiert werden – nicht ohne Grund –

---

8 Art. 2 Bundesgesetz über das bäuerliche Bodenrecht vom 4. Oktober 1991. Im übrigen nehmen auch die Gewässerschutzgesetzgebung, das Erschliessungsrecht usw. auf die Grundordnung nach Raumplanungsrecht Bezug.

9 Diese Feststellung trifft vor allem auf Deutschland zu. Die Wissenschaft daselbst scheint sich in Teilen von der tradierten, gesetzlich erfassten Raumplanung zu lösen. Sie wendet sich einem neuem Regionalismus zu, der von der amerikanischen und englischen Literatur inspiriert sein dürfte. Beredtes Zeugnis davon legt beispielsweise ab das Lesebuch *von Schaffer / Spannowsky / Troeger-Weiss (Hrsg.)*, Implementation der Raumordnung, Wissenschaftliches Lesebuch für Konrad Goppel, Augsburg-Kaiserslautern 2003. Es handelt praktisch nur noch vom Regionalmanagement als Instrument der Klärungen wider die Unsicherheiten. Aufschlussreich ist daselbst der Aufsatz von *Kistenmacher Hans*, Evaluationen – ein Ansatz zur Weiterentwicklung der neuen Instrumente der Raumordnung, a. a. O., S. 105 ff. Zu den Veränderungen in der Raumplanung resp. ihres Verständnisses siehe auch, weitgehend im gleichen Sinn, *Fürst Dietrich*, Kontinuität und Wandel der Planungsphilosophie, in ARL / BBR (Hrsg.), Zukunftsforum RaumPlanung, Hannover 2002, S. 43 ff.; daselbst findet sich auch eine souveräne Zusammenfassung von *Ritter Ernst-Hasso*, Perspektiven der Raumentwicklung, a. a. O., S. 96 ff. Siehe dazu auch *idem*, Stellenwert der Planung in Staat und Gesellschaft, in: ARL, Methoden und Instrumente räumlicher Planung, Hannover 1998, S. 6 ff. Der Wechsel zur Region, verbunden mit einer Flucht aus den Institutionen samt Zuwendung zu neuen Methoden, scheint noch nicht in allen Teilen ausreichend bedacht zu sein. Nuanciert anders, eher auf die materiellen Probleme konzentriert und nach wie vor die zuständigen Behörden in Pflicht nehmend, *Lendi Martin*, Raumplanung im Umbruch. Auf dem Weg zu einer politisch, sachlich und ethisch indizierten Raumplanung, Schweizerisches für Staats- und Verwaltungsrecht (Zbl) 1999 (Jg. 100) Nr. 4, S. 193 ff.; *idem*, Gesellschaftlich vernetztes Recht, Zürich 1999, mit mehreren Abhandlungen zum Verständnis und zur Rechtfertigung der räumlichen Planung. Die neuste Standortbestimmung stammt von *Muggli Rudolf*, Wo steht die Schweizer Raumplanung, in: VLP / ASPAN, September 2003.

veränderte Rahmenbedingungen, die Grenzen staatlicher Steuerung in einer internationalisierten Welt und der fehlende Umsetzungsbezug. Deshalb werden gefügte Strukturen, tradierte Instrumente, genormte Zielsysteme usw. hinterfragt, bisweilen sogar mit Vorwürfen überhäuft. Der Ruf nach neuen Institutionen und Wegen – weg vom Hoheitlichen, vom Gesetzlichen, vom Verwaltungsmässigen, hin zu Kooperation, Mediation, Management – scheint zu dominieren. Davon erfasst werden, wen wundert's, auch so einfache, jahrelang unwidersprochene Massnahmen wie die Trennung von Siedlungsgebiet und Nicht-Siedlungsgebiet. Die Antworten weisen in Richtung von Managementmethoden vor dem Hintergrund der Region, was wohl nichts anderes bedeuten würde, die Siedlungsentwicklung einem regionalen Management anzuvertrauen. Nur bleibt darüber die Frage nach den Zielen unbeantwortet. Management ohne Ziele, ohne Konzepte, ohne Strategien, ohne Programme, ohne Umsetzungs- und Durchsetzungsmassnahmen, ohne Verbindlichkeiten tappt längerfristig ins Leere.[10] Wir versuchen deshalb, auf der Evaluierung sowie Festschreibung von Zielen zu insistieren und auf abgestimmten Instrumenten sowie Massnahmen zu beharren, aber doch bewehrt mit der kritischen Sonde, Varianten zu den bisherigen Methoden auf ihre Angemessenheit zu befragen.[11]

In die nachstehende Abhandlung vermengen sich, wie bereits angedeutet, mindestens zwei, wenn nicht sogar drei Fragen: Jene nach der zu erwartenden und lenkend zu begleitenden, also angestrebten Siedlungsentwicklung, jene nach der Art (Methode) der Steuerung und drittens jene nach dem „ob", d.h. ob es überhaupt einer Steuerung der Siedlungsentwicklung bedürfe. Die dritte Frage ist allerdings mit einem Hinweis, basierend auf kurzen Gedankengängen, relativ leicht zu beantworten: Es findet selbst dann eine Steuerung statt, wenn von einer förmlichen abgesehen wird. Der Verzicht auf eine bewusst vorgesehene Lenkung gibt nämlich dem Markt Raum und dieser bewirkt eine Steuerung nach Angebot und Nachfrage, somit über den Preis, und dies angesichts wirksamer Fakten, nämlich der anhaltend eher hohen Nachfrage für knappe Güter, nämlich des Bodens und des Raumes. Ob das Marktgeschehen eine geordnete Entwicklung bewirkt, dies ist ein anderes Problem. Immerhin kann vermutet werden, hohe Preise würden zu einer restriktiven,

10 Daraus kann geschlossen werden: Gute Raumplanung ist von Haus aus qualifiziertes, allerdings rechtsnahes Management. Statt Wortspielen „Von der Raumplanung zum Regionalmanagement" anzuhangen, wäre es besser zu fragen, wie wird gekonnt geplant resp. gemanaget.

11 Bei dieser Gelegenheit erinnern wir daran, dass der kantonale resp. regionale „Richtplan" nach RPG (Art. 6 ff. RPG) das äusserlich sichtbare Instrument eines räumlichen Managements ist, das sich nicht davor scheut, den ganzen Bezugsraum – und nicht nur einzelne Objekte / Projekte – zielbezogen in Betracht zu ziehen und unter dessen Berücksichtigung zu handeln.

gleichzeitig intensiven Nutzung beitragen, während umgekehrt nicht ganz von der Hand zu weisen wäre, dass für landschaftliche Unikate Liebhaber-Preise erzielt werden könnten, was zu deren Privatisierung beitragen würde – immerhin Anzeichen, dass nicht nur die hoheitliche Planung Steuerung impliziert, auch die Überantwortung an den Markt bewirkt Siedlungsentwicklung.[12] Ob dies dann eine erwünschte oder weniger vorteilhafte sei, dies ist offensichtlich eine Bewertungs-, eine Zielfrage.

Insofern ist davon auszugehen, dass eine Steuerung der Siedlungsentwicklung stattfindet.[13] Von besonderer Dringlichkeit sind somit die erwähnten Fragen nach der tatsächlichen resp. der geforderten Siedlungsentwicklung und jene nach der Methode der Steuerung. Die letztere schliesst die dritte nach der informellen des Marktes und der förmlich angelegten Steuerung ein. Insofern beschränken wir uns hier auf zwei Kernthemen, nämlich der Ziele und der Art der Steuerung der Siedlungsentwicklung, vergessen aber darüber den Hintergrund der massiven Veränderungen des Lebensraumes nicht. Statt vieler Begründungen mag der exemplarische Blick auf den schweizerischen Alpenraum – vermeintlich noch weitgehend unberührt – genügen: Urbane Lebensweisen sind bis ins Engadin vorgedrungen; Davos ist faktisch eine Stadt; Plateaus im Wallis sind überbaut; nicht nur der Transitverkehr am Gotthard belastet, der Ziel-, Quell- und Binnenverkehr des Berggebietes ist ebenfalls beachtlich. Vor allem lastet die Freizeitwelt mit ihrem Verkehrsaufkommen auf ihm. Parallel dazu: Das Vordringen des Städtischen in die periurbanen Räume, faktisch in den sog. ländlichen Raum, folgt ähnlichen Kräften.

---

12 Wird das Marktgeschehen nicht als „Steuerung“ der Boden- und Bodennutzungsverteilung verstanden, so müsste von einer dreifach vernetzen Fragestellung ausgegangen werden. Zur Problematik der Bodenfragen im europäischen Rahmen vgl. die Forschungen und Publikationen der Europäischen Fakultät für Bodenfragen, u. a.: *Fitsch D. B. S./Pikalo A. (Hrsg.)*, Soziale und ökonomische Aspekte der Bodennutzung, Bd. 12, Frankfurt a. Main/Bern/New York/Paris 1990; *Balekjian W. H./Kühne Josef (Hrsg.)*, Bodenordnung in ländlichen Gebieten, Bd. 15, a. a. O.; *Lendi M./Pikalo A. (Hrsg.)*, Die Entwicklung von Bodenpreisen in Europa, Bd. 17, a. a. O.

13 Die Theorie der Raumplanung, soweit sie sich nicht mit dem Verständnis der Raumplanung befasste, hat in ihrem Kern immer Aspekte der Siedlungsentwicklung mitverfolgt. Siehe dazu *Wolf Klaus*, Theoretische Aspekte der räumlichen Planung, in: ARL, Methoden und Instrumente räumlicher Planung, Hannover 1998, S. 39ff., und die dort zit. Literatur.

2. Die sachliche Problematik und deren Ursachen

Die Siedlungsentwicklung ist ein besonders dynamischer und einwirkender Faktor der Veränderungen des Landschaftsbildes und zwar nach Ausdehnung, Intensität und gestalterischem Vermögen resp. Unvermögen.[14] Sichtbar wird sie an den Rändern der Dörfer und Städte, dann aber vor allem am Rand der Agglomerationen, nicht minder eindrücklich mitten im Veränderungsprozess vermeintlich weitgehend überbauter Bauzonen. Einen Spezialfall bilden die einzelnen Neubauten und infrastrukturellen Einrichtungen sowie die intensiveren Nutzungen bestehender Bauten resp. Einrichtungen in der offenen Landschaft.

Die Ursachen für einen tendenziell nicht haushälterischen Umgang mit dem Boden sind vielfältig. Sie liegen vor allem in den steigenden menschlichen, gesellschaftlichen, wirtschaftlichen und öffentlichen Bedürfnissen, aber auch in den offenkundigen egozentrischen Neigungen, sich laufend mehr Wohn-, Verkehrs-, Einkaufs- und Freizeitraum zu verschaffen. Veränderte gesellschaftliche Strukturen (Einzelhaushalte, Freizeit-, Erlebniswelten) tragen das Ihre bei. Durch die Kräfte des ökonomischen Umfeldes, der Immobilienwirtschaft, der relativ billigen Kosten des privaten und öffentlichen Verkehrs, usw. wird die aufgezeigte Entwicklung hin zu einer Ausdehnung der Siedlungsflächen (Wohn-, Arbeits- Verkehrs- Versorgungs-/Entsorgungs- und Freizeitflächen) zusätzlich begünstigt. Vor allem die Wirkungen der anhaltend zunehmenden Mobilität, inkl. Freizeit- und Einkaufsverkehr, dürfen nicht unterschätzt werden. Die vorauseilenden Effekte des Leistungsangebotes des

---

14 Auf zahlenmässige Belege wird hier verzichtet. Sie werden regelmässig publiziert. Siehe dazu vor allem *Bundesamt für Statistik*, Die Raumgliederungen der Schweiz, Bern 1994, *idem*, Arealstatistik Schweiz –Bodennutzung im Wandel, Neuenburg 2001. Vgl. sodann die aufschlussreichen Angaben im Bericht des Bundesrates über die Grundzüge der Raumordnung Schweiz vom 22. Mai 1996, Separatdruck, Bern 1996, S. 13 ff. Siehe auch den Bericht des Bundesrates über die Massnahmen des Bundes zur Raumordnungspolitik: Realisierungsprogramm 2000-2003, vom 2. Oktober 2000, Separatdruck, Bern 2000, wo allerdings die Bodenfragen als roter Faden und eher am Rand diskutiert werden. Bekannt sind die Zahlen des quadratmetermässigen Verbrauchs von Land durch Überbauung und Einbezug in Infrastrukturanlagen pro Sekunde. Die Thematik der Versiegelung der Böden steht dabei im Vordergrund. Es wird im übrigen auf die entsprechenden statistischen Jahrbücher verwiesen, insbesondere auf die sog. Arealstatistik. Ein interessanter internationaler Vergleich ist einer demnächst erscheinenden Dissertation der Technischen Universität München zu entnehmen: *Markstein Melanie*, Instrumente und Strategien zur Baulandentwicklung und Baulandmobilisierung in Ländern Deutschlands, Osterreichs und der Schweiz – ein methodischer Vergleich mit Entwicklungsvorschlägen für das deutsche System, München 2003. Zur schweizerischen Bodemarktpolitik siehe beispielsweise *Wachter Daniel*, Bodenmarktpolitik, Bern 1993. Ein Aufriss aus volkswirtschaftlicher Sicht findet sich bei *Hotz-Hart Beat / Mäder Stefan / Vock Patrik*, Volkswirtschaft der Schweiz, Zürich 1995, S. 355 ff.

privaten und öffentlichen Verkehrs dürfen ihrerseits nicht marginalisiert werden. Ob, wann und wie intensiv die potentiell negative demographische Entwicklung Korrekturen setzen wird, ist mindestens für die schweizerischen Verhältnisse offen[15], da die erhöhten Bedürfnisse prognostizierte Einsparungen konsumieren, wenn nicht gar überbieten, wenn auch nicht immer an den gleichen Orten. Zudem ist die internationale Migration mit ihren positiven und negativen Begleiterscheinungen zu beachten. Dass die früheren Boom-Kräfte der Wirtschaft für eine gewisse Zeit abflachen und die Nachfrage insgesamt reduzieren, kann nicht ausgeschlossen werden. Gewisse Grenzen der zivilisatorischen Akzeleration sind jedenfalls erkennbar, doch bleibt ungewiss, welche Trends sich durchsetzen. Prognosen bleiben heikel.

Die Auswirkungen der Siedlungsentwicklung sind nicht allenthalben gleich ausgeprägt.[16] Sie fallen regional unterschiedlich aus. Aktuell sind, wie bereits erwähnt, vor allem die Agglomerationen betroffen – mit den bekannten Tendenzen in Richtung sich überlappender urbaner Siedlungsräume, vor allem im schweizerischen Mittelland vom Bodensee bis zum Genfersee. Mögliche vorantreibende Ursachen liegen im privaten und öffentlichen (!) Agglomerationsverkehr (S-Bahnen über die Zentren hinaus) sowie in den geringeren Bodenpreisen, wie sie sich in äusseren resp. sich sukzessive angliedernden Agglomerationsgürteln anbieten.[17] Gleichzeitig stellt die Freizeitgesellschaft mit der Nachfrage nach strukturierten und nicht-strukturierten Erlebniswelten wachsende Ansprüche sowohl an die Siedlungsgebiete als auch an die offene Landschaft, ausgeprägt gegenüber den agglomerationsnahen Räumen, soweit sie als Naherholungs- / Freizeit- / Einkaufsgebiete prädestiniert sind. Eigenartig ist die Entwicklung in den Tourismusregionen. Hier bahnen sich zulasten der Landschaft örtlich und teilweise regional Verstädterungen an – ausgelöst durch den Zweitwohnungsbau, vor allem aber durch urbane Lebensweisen der Gäste und teilweise auch der Ortsansässigen.

Bedenklich ist die ungenügende, wenn nicht sogar fehlende Datendichte für das Abschätzen des Siedlungsbedarfs. Wir wissen wenig (zu wenig) über

---

15 Die in der Schweiz lebende, dauernd anwesende Bevölkerung ist innert Jahresfrist um rund 50'000 auf aktuell 7,3 Mio gestiegen, wobei die Geburtenrate sank. Die entscheidende Rolle spielt die Zuwanderung. Zum Vergleich: 1798 1,68 Mio, 1900 3,31 Mio, 1941 4,26 Mio, 1960 5,42 Mio, 1980 6,36 Mio, 2000 7,1 Mio

16 Hier stehen die räumlichen Aspekte im Vordergrund. Die ökonomischen Folgen einer ausufernden Siedlungsentwicklung sind erheblich, dazu *Ecoplan*, Siedlungsentwicklung und Infrastrukturkosten, Bern 2000.

17 Der allgemeine Hinweis auf den privaten Verkehr, im Arbeits- und Freizeitbereich, trifft die Problemlage, doch darf die Bedeutung des öffentlichen Verkehrs für die Agglomerationsentwicklung vor allem dort nicht unterschätzt werden, wo S-Bahnen die Reisezeiten erheblich verkürzen und zwar bis in Siedlungsbereiche hinaus, die mit günstigeren Bodenpreisen aufwarten als sie in Kernstädten auszumachen sind. Selbstverständlich sind dies nicht die einzigen Gründe der ausufernden Agglomerationsentwicklung, doch relevante sind es allemal.

die demographische Entwicklung – mit oder ohne gelenkte Migration; die Auswirkungen der wirtschaftlich-technischen, der wirtschaftlich-dienstleistungsorientierten Arbeitswelt einer Informationsgesellschaft – mit oder ohne Technologieschübe – kennen wir nicht ausreichend; die Folgen der Verhältnisordnung zur EU/EG – mit oder ohne EU-Nähe der Schweiz – verharren mehr oder weniger im Dunkeln; auch über verlässliche Angaben des künftigen Flächenbedarfs bezüglich Arbeiten, Wohnen, Freizeit, Verkehr, Versorgung/Entsorgung – mit oder ohne sozialer Flucht aus der Familie hin zu den bald schon dominierenden Single-Haushalten, mit steigendem oder stagnierendem Wohlstand usw. – verfügen wir nicht: Die Siedlungsentwicklung ist mit andern Worten zur schwer diagnostizierbaren Grösse geworden, auch wenn wir aktuell durch äussere Faktoren der Bevölkerungszunahme, der Verstädterung des Mittellandes und der Urbanisierung der Lebensstile beeindruckt sind. Was belastet am meisten? – die ökologischen, die sozialen, die wirtschaftlichen, die politischen oder gar die lebensräumlichen Ungewissheiten? Diese Frage nach einzelnen Aspekten zu beantworten wäre wohl falsch, da sie sich gegenseitig bedingen. Erkennbar bleiben die Gefährdungen der natürlichen Lebensgrundlagen, die Zunahme des überbauten und des dereinst überbaubaren sowie des mit infrastrukturellen Einrichtungen belasteten Landes innerhalb und ausserhalb des Siedlungsgebietes, relative Knappheiten an Bauland an begehrten Lagen, sozial-menschlich strapazierte, verblühende und parallel aufblühende Quartiere in ein und derselben Stadt oder Agglomeration.

Dazu kommen ungenügende Zielansprachen der Steuerung der Siedlungsentwicklung. Je nach dem, in welche Richtung Ziele artikuliert werden, kann der Akzent gegen die Fortschreibung der Baugebiete resp. auf deren forcierte Entwicklung und/oder auf die Erhaltung der minimal notwendigen Räume natürlicher Lebensgrundlagen oder auf die Nutzung der Freiräume für die Freizeitaktivitäten gesetzt werden. Weitere Variationen sind denkbar. Über sie entscheidet letztlich der Gesetzgeber, so nicht der Markt allein agiert. Instrumentell können die bewusst postulierten Ziele über Gesetz oder Plan umgesetzt werden. Die Unterschiede sind markant. Das Gesetz fixiert, der Plan sucht Festschreibung und Dynamik zugleich. Ob es hilfreich sein würde, die Raumplanung zu überwachen, beispielsweise durch eine Raum-, eine Umweltverträglichkeits- oder gar durch eine Nachhaltigkeitsprüfung?[18] Zwei-

18 Mit solchen Fragen, vorweg von materiellen Anliegen angetrieben, hat sich Peter Saladin auseinandergesetzt und zwar an Tagungen der Schweizerischen Vereinigung für Landesplanung. Siehe dazu die letzte Fassung seiner Vorträge: *Saladin Peter*, Landesplanung – Lebensraumgestaltung für morgen, in: Knapp/Hertig/Saladin/Tschannen/Zimmerli (Hrsg.), Perspektiven des Raumplanungs- und des Bodenrechts, Beihefte zur Zeitschrift für Schweizerisches Recht, Heft 11, Basel 1990, S. 83 ff. Saladin befasste sich vor allem mit dem Raumplanungsbericht des Bundesrates von 1987, geht von einer begrenzten Wirkungsmöglichkeit der Raumplanung aus, bezweifelt die Notwendigkeit einer Verfassungsrevision und sucht die Lösungen in gesetzlichen und verfahrensmässigen Instrumenten.

fel sind angesagt. Die nachträgliche Kontrolle kann nicht mehr leisten als die Planung erbringen müsste, anders herum, sie müsste den gleichen Aufwand betreiben, um das zu tun, was gute Planung von sich aus tun würde. Kann es hilfreich sein, eine ideelle Verbandsbeschwerde einzusetzen? Sie weist erhebliche Nachteile auf, vor allem weil sie auf Einzelvorhaben gerichtet ist und nicht den Raum als solchen betrachtet, es sei denn, sie werde gegenüber Plänen zugelassen, was jedoch neue Probleme mit sich bringt. Enger betrachtet könnte eine Lösung darauf zielen, die Trennung von Siedlungsgebiet und Nicht-Siedlungsgebiet, ob planerisch oder gesetzlich festgeschrieben, justiziabel zu machen und den Richter für Grenzfragen zuständig zu erklären.[19] Auch dazu drängen sich grundsätzliche Vorbehalte auf, zumal der Richter über keine planerische Fachkenntnisse verfügt und da die Planung aus sich heraus reich an Ermessen ist, das der richterlichen Kognition grundsätzlich entzogen bleibt.[20]

Ob die anhaltend kritisch verlaufende Siedlungsentwicklung zu den grössten Problemen der räumlichen Entwicklung zählt, darf gefragt werden.[21] Ob sie Ursache, Wirkung oder nur Symptom ungünstiger Trends ist, muss bedacht werden. Möglicherweise ist sie Symptom, Ursache und Wirkung zugleich. Sie hat auf alle Fälle sehr viel zu tun mit der Bevölkerungsentwick-

---

19 Im Sinne einer justiziablen Trennung von Siedlungs- und Nicht-Siedlungsgebiet äussert sich *Zimmerli Ulrich*, Richterstaat in der Raumplanung?, in: Knapp / Hertig / Saladin / Tschannen / Zimmerli, Perspektiven des Raumplanungs- und Bodenrechts, a. a. O., S. 105 ff., insbesondere S. 122. Er vertritt die Auffassung, die Trennlinie zwischen Bauzone und Nicht-Bauzone sei justiziabel zu machen. Er setzt auf selbstvollziehendes Recht, wendet sich gegen Wachstumsplanung und redet einer Bestandesplanung, die richterlich überprüfbar wäre, das Wort. Bei dieser Gelegenheit hält er auch sinngemäss fest, es lasse sich vertreten, das RPG auf die Regelung der Nutzungsplanung zu beschränken, wobei es genügen würde, a) die Abgrenzung der Bauzonen zu definieren und festzuschreiben sowie b) das Regime im Gebiet ausserhalb der Bauzonen festzulegen. Es handelt sich um ein juristisch denkbares Konzept, aber eines, das die Dynamik des räumlichen Geschehens und die Vielgestaltigkeit der Kräfte völlig ausser Acht lässt. Bauzone wäre nach Ulrich Zimmerli nur noch das weitgehend überbaute Gebiet. Der Höchstumfang der Bauzonen müsste durch einen Bundessachplan festgelegt werden. Zimmerli übersieht dabei, dass ein Bundessachplan nicht davon dispensiert, die verfügbaren Baugebiete regional zu verteilen und dimensioniert zu lokalisieren –und dies alles vor dem Hintergrund von wirtschaftlichen und gesellschaftlichen Veränderungen, die in Metropolitanregionen anders ausfallen als in ländlichen Gebieten..

20 Vgl. dazu *Lendi Martin*, Die Bedeutung der Rechtsprechung für die Raumplanung, in: Cagianut / Geiger / Hangartner / Höhn (Hrsg.), Aktuelle Probleme des Staats- und Verwaltungsrechts, Festschrift für Otto K. Kaufmann, Bern / Stuttgart 1989, S. 295 ff.

21 Zur Tragweite der Zersiedelung äussert sich beispielsweise *Weber Gerlind*, Zersiedlungsabwehr durch Raumplanung – ein Kampf gegen Windmühlen?, in: Straube M. / Weimar R., Jurist und Technik zwischen Wissenschaft und Praxis, Festschrift für Josef Kühne zum 60. Geburtstag, Wien 1984, S. 231 ff.

lung, den wachsenden Ansprüchen und der hohen Mobilität, dann aber auch mit Planungsträgheiten, beispielsweise dort, wo Ortsplanungen nicht nachgeführt und regionale Vorgaben vernachlässigt werden. So besehen ist sie als Ursache aufzufangen, als Wirkung einzudämmen und als Symptom ernst zu nehmen. Gleichzeitig ist zu bedenken: Siedlungsplanung allein vermag die Raumordnung nicht zu gestalten. Es bedarf eben abgestimmter Landschafts-, Siedlungs- sowie Transport-/Versorgungsstrukturen. Bei alledem wird die Raumplanung nicht mehr als ein Ordnungsgefüge bewirken können, da sie – zu Recht und gut begründet – keine Gesellschafts- und keine Wirtschaftsplanung sein will und darf. Diese Zurückhaltung darf ihr nicht vorgeworfen werden. Sie hat die verfassungsrechtlichen Vorgaben des demokratischen, liberalen, föderalistischen und sozialen Rechtsstaates zu akzeptieren. Im übrigen würde sie sich übernehmen, wenn sie die Ordnungsfunktionen übersteigen würde.[22]

Feststeht: Szenarien oder Leitbilder der künftigen Siedlungsentwicklung fehlen, es sei denn, die Raumplanung verlasse sich auf die gesetzliche Vorgabe der Begrenzung des Baugebietes und erkenne darin ein Leitbild.[23] In

---

22 Für die Schweiz genügt ein Hinweis auf die Bundesverfassung. Massgebend ist die Verfassung vom 18. April 1999. Die schweizerische Raumplanung ist in deren Kontext zu verstehen, bis und mit ihren Vorgaben unter dem Titel der Grundrechte (Art. 7 ff. BV) und der Wirtschaftsordnung (Art. 94 ff. BV). Die gesellschaftliche Ordnung wird nicht näher ausgeführt, doch ist sie Ausdruck der freiheitlichen Ordnung, insbesondere der Privatautonomie, welche dem Privatrecht Richtung weist. Wissenschaftlich kann bei der Darstellung der Siedlungsentwicklung von der Rechtsordnung abgehoben werden, bis und mit der These, die Mobilität sei die wahre Ursache der negativen Siedlungsentwicklung, nur ist diese verfassungsrechtlich, wenn auch nicht ausdrücklich, geschützt, jedenfalls nicht verworfen. Siehe dazu *Lendi Martin*, Verkehrsfreiheit, in: Zen-Ruffinen Piermarco/Auer Andreas, De la Constitution, Études en l'honneur de Jean-François Aubert, Basel/Frankfurt am Main, 1996, S. 371 ff.; *idem,* Verkehr und Recht, Zürich 1998, S. 27 ff. Dementsprechend ist es utopisch, die Siedlungsentwicklung einseitig von der Bekämpfung der Mobilität her anzugehen. Zu einem Verständnis der Raumplanung, das sich vorweg der Mobilität widersetzt, siehe beispielsweise *Broggi Mario*, Raumplanung – Gedanken zur notwendigen Anpassung an die „Stadt-Schweiz“, in: Maisch M./Vonder Mühll D./ Monbaron M. (Hrsg.), Entwicklungstendenzen und Zukunftsperspektiven in der Geomorphologie, Geographisches Institut Universität Zürich, Zürich 2003, S. 75 ff.

23 Art. 15 RPG begrenzt Bauzonen auf das Land, das sich für die Überbauung eignet und a) weitgehend überbaut ist oder b) voraussichtlich innert 15 Jahren benötigt und erschlossen wird. Diese Regelung enthält nur indirekt materielle Vorstellungen. Sie weisen in Richtung der vorsichtigen Zurückhaltung bei der Ausdehnung des Bau- und damit des Siedlungsgebietes, ohne aus dem Wortlaut heraus ein Leitbild zu formulieren. Die Bestimmung muss in Verbindung mit Art. 3 Abs. 3 (Ingress) RPG gelesen werden. Vgl. dazu *Tschannen Pierre*, Kommentar zu Art. 3 RPG, in: Aemisegger/Kuttler/Moor/Ruch (Hrsg.), Kommentar zum Bundesgesetz über die Raumplanung, Zürich 1999; daselbst *Flückiger Alexandre*, Kommentar zu Art. 15 RPG,

einem übergeordneten Sinne wird dieses mit „konzentrierter Dezentralisation" oder „dezentralisierter Konzentration" beschrieben, was für die Richtplanebene und landesweite Konzepte zutreffen mag.[24] Für die Stufe der Nutzungsplanung wäre es angemessener, von einem Leitbild der Begrenzung des Baugebietes und des Schutzes des Nicht-Baugebietes als offene Landschaft zu sprechen. Das heute gängige Leitbild der nachhaltigen Raumentwicklung, wie es dem deutschen Raumordnungsrecht und eng verknüpft dem Baugesetzbuch eigen ist, betrifft sowohl die überörtliche, landesweite Raumordnung als auch die örtliche Bebauungsplanung. Das entsprechende Leitbild ist auch dem schweizerischen Recht, über den Weg der Interpretation, nicht fremd.[25]

Das Ausufern der Siedlungen unter Kontrolle zu behalten scheint angezeigt, soll es nicht zu negativen Folgen für die offene Landschaft (Nicht-Siedlungsgebiet), für die bebauten resp. zu überbauenden Räume (Siedlungsgebiet), für die Kosten der Erschliessung resp. Entsorgung und zu einer Verschwendung der Ressourcen „Boden" und „Landschaft" bei gleichzeitig wachsender Belastung des gesamten Ökosystems kommen. Das quantitativ und qualitativ anforderungsexponierte Siedlungsgebiet mit seiner bebauten Umwelt ist dabei genauso optimal zu nutzen und gestalterisch zu stützen wie die offene Landschaft vor baulichen Eingriffen und andern Einwirkungen, insbesondere gegenüber einem Verlust an Ressourcen und vor übermässigen Belastungen durch die wachsenden Freizeitaktivitäten, zu schützen ist. Gestaltungsaufgaben im einen wie andern Bereich rufen nach herausragender Ästhetik und sorgfältiger Einbindung in die Orts- und Landschaftsbilder. In jedem Fall sind

---

a. a. O. Die bundesgerichtliche Rechtsprechung tat sich anfänglich mit den regional und örtlich unterschiedlichen Akzentuierungen schwer, bis sie akzeptierte, dass die Siedlungsentwicklung zusätzlich durch überörtliche Aspekte mit behördenverbindlichen Vorgaben für die örtlich zuständigen Behörden gesteuert wird, nämlich durch die Richtplanung. Also muss Art. 15 RPG in Verbindung mit Art. 3 und Art. 6 ff. RPG gelesen werden.

24 Das RPG spricht in Art. 1 Abs. 2 lit. c unter dem Marginale Ziele von einer angemessenen Dezentralisation der Besiedlung, in Art. 3 Abs. 3 im Rahmen der Planungsgrundsätze handelt es von der Begrenzung der Ausdehnung der Siedlung. Gemeint sein dürfte eine dezentralisierte Konzentration – bei begrenzter Ausdehnung der Siedlungen. Ob sich die Formel der dezentralisierten Konzentration resp. der konzentrierten Dezentralisation unter den veränderten tatsächlichen Verhältnissen noch halten lässt, muss zur gegebenen Zeit geprüft werden. Hervorgegangen ist dieses Leitbild aus den Landesplanerischen Leitbildern des ORL-Instituts (a. a. O.). Der Öffentlichkeit wurde es unterbreitet *als EJPD (Hrsg.)*, Raumplanung Schweiz, Raumplanerisches Leitbild der Schweiz, CK 73, Bern 1973.

25 § 1 Abs. 2 Raumordnungsgesetz (ROG) vom 18.8. 1997; § 1 Abs. 5 Baugesetzbuch vom 27.8.1997. Für das schweizerische Recht ist der systematische Zusammenhang auf Verfassungsstufe Basis der Vorgabe der nachhaltigen Raumplanung (Art. 73 und Art. 75 BV)

Qualitäten gefordert. Es ist nicht abwegig, von Siedlungsqualität und von Qualitäten der offenen Landschaft als zentralen Aufgaben moderner Raumplanung zu sprechen.[26]

3. Die Ziele der Trennung von Siedlungs- und Nicht-Siedlungsgebiet

Die schweizerische Raumplanung hat sich seit ihren Anfängen nicht einseitig der „positiven Entwicklungsplanung" für Städte/Dörfer und Verkehr (Siedlung und Verkehr) verschrieben – selbst in den Zeiten voranstürmender Planungseuphorie (60ger Jahre des 20. Jahrhunderts) hat sie dies nicht getan. Sie hat die Positiv- stets mit einer begrenzenden Negativplanung verbunden und also dem planlosen Siedlungswachstum Grenzen gesetzt. Vom Ansatz her war sie sodann bereits in ihren Erstzeiten bemüht, den Raum als Einheit zu betrachten, nämlich die städtischen, dörflichen Siedlungen, den ländlichen Raum mit Einschluss von Wald, Gewässern und die Berggebiete als gesamthaft zu

26 Siehe jenseits der schweizerischen Literatur beispielsweise *Weber Gerlind*, Zersiedlungsabwehr durch Raumplanung – ein Kampf gegen Windmühlen?, in: Straube M./Weimar R. (Hrsg.), Jurist und Technik zwischen Wissenschaft und Praxis, Wien 1984, S. 231 ff. Die Autorin argumentiert von der Landwirtschaft her –unter Gesichtspunkten der Gesetzgebung, des Vollzugs und der Grundeigentümer. Mit der Zukunft der Kulturlandschaft setzt sich eingehend auseinander: *ARL/ÖGR (Hrsg.)*, Die Zukunft der Kulturlandschaft zwischen Verlust, Bewahrung und Gestaltung, Hannover 2001, daselbst vor allem der Titel von *Haber Wolfgang*, Kulturlandschaft zwischen Bild und Wirklichkeit, a. a. O., S. 6 ff. Die Publikation der ARL im Verbund mit der ÖRG beeindruckt und kann als Hintergrunddarstellung des mit der Trennung von Siedlungs- und Nicht-Siedlungsgebiet nicht einfangbaren sachlichen Umfeldes verstanden werden. Folgende Stichworte erhellen die Problemsituation: Europäische Urbanität, Verstädterte Landschaft, Perspektive Stadtlandschaft, Kulturlandschaft Alpen, Zwischenstädte, Netzstadt, Problemüberwindung aus natur- und humanwissenschaftlicher Sicht usw. Die Alpen als Kulturlandschaft im Spannungsfeld verschiedener Interessen schildert im gleichen Band *Ruppert Karl*, Die Alpen – Kulturlandschaft im Spannungsfeld unterschiedlicher Interessen, a. a. O., S. 127 ff. In demselben Werk der ARL/ÖGR ist aus dem schweizerischen Beitrag von *Egli Hans-Rudolf*, Bewertungen, Kompetenzen und Instrumente der Kulturlandschaftspflege am Beispiel der Schweiz, a. a. O., S. 182 ff. als schlussfolgerndes Stichwort zu entnehmen, die Raumplanung müsse sich zur Raummanagementplanung entwickeln, ohne dass ausgeführt würde, was dies zu bedeuten hat. Immerhin wird erkennbar, es bedürfe klarer Vorstellungen über die Entwicklung eines Landschaftsraumes, bevor Massnahmen ergriffen werden könnten. Auch bedürfe es einer entsprechenden wissenschaftlichen Fundierung. Wie dem auch sei: Hat tradierte Raumplanung nach schweizerischem Recht je etwas anderes verlangt?

erhaltende und zu gestaltende Landschaft.[27] Daraus entstand die zukunftsorientierte Orts-, Regional- und Landesplanung, seit 1969 durch die Verfassung „Raumplanung" genannt, welche den ganzen Lebensraum in Betracht zieht, alle raumwirksamen Massnahmen erfasst, nach einen andauernden Prozess verlangt, die Landschafts-, die Siedlungs- und die Transport- sowie die Versorgungsstrukturen aufeinander abstimmt und den Schutz der Ressourcen (Boden, Wasser usw.) wie auch die Belastbarkeit der Räume im Auge behält.[28]

Äusserlich wandte sich die Raumplanung seit der Wegbereitung als öffentliche Aufgabe zwei Hauptaufgaben zu, nämlich a) der baulichen Nutzungsplanung (Planung des Baugebietes, Bebauungsplanung usw.) und b) der Abgrenzung des Baugebietes zum „übrigen Gebiet" resp. zum Nicht-Siedlungsgebiet, zum Nicht-Baugebiet. Die Bekämpfung der Zersiedlung und insbesondere der Streusiedlungen, soweit sie nicht als natürliche zum Kulturerbe zählen, war eine der dringenden Aufgaben. Im Baugebiet wurden allgemeine Entwicklungsfragen und vor allem Überbauungsplanungen samt Verkehrserschliessungen vorangetrieben, im Nicht-Baugebiet galt die Aufmerksamkeit vorweg landschaftlichen, landwirtschaftlichen und ökologischen Schutzzwekken. Als politisch und rechtlich taugliches Rezept für eine prägende Grundstruktur bot sich die für jedermann nachvollziehbar zweckbewehrte Trennung von Siedlungs- und Nicht-Siedlungsgebiet[29] an, in einer ersten, frühen Phase

---

27 Vgl. dazu vor allem *Winkler / Winkler / Lendi (Hrsg.)*, Dokumente zu Geschichte der schweizerischen Landesplanung, Schriftenreihe zur Orts-, Regional- und Landesplanung Nr. 1, Zürich 1979. Der Begriff der Landschaft wurde, vor allem seitens der Geographen, nicht auf die offene Landschaft zurückgefahren, sondern immer bewusst ausgedehnt auf die Landschaft als solche – mit Einschluss der urbanen Räume.

28 Begrifflich ist klarzustellen, dass der Begriff des Baugebietes enger ist als jener des Siedlungsgebietes. Das letztere umfasst Bau- und Bauverbotsgebiete, die funktional zusammengehören, mit ihm also eine Einheit bilden. Das Nichtsiedlungsgebiet ist seinerseits komplex angelegt, weil es land- und forstwirtschaftliche Gebiete genau so einschliesst wie Schutzgebiete, Gewässer, Freizeiträume, Infrastruktureinrichtungen usw.

29 Als verfassungsrechtliches Prinzip ableitbar seit dem Erlass der Verfassungsartikel 22ter und 22quater aBV. Bei deren Erlass ging es darum, den Bauzonen die Landwirtschaftszonen gegenüber zu stellen, um auf diesem Weg das Nicht-Siedlungsgebiet, von Ausnahmen abgesehen, einem Bauverbot zu unterstellen, in der Regel ohne Entschädigungsfolgen. Zudem hatte die Verfassung die Aufgabe gestellt, eine geordnete Besiedlung anzustreben, was verpflichtete, das Siedlungsgebiet als solches zu erfassen (und mithin vom Nicht-Siedlungsgebiet zu separieren). Vgl. dazu *Jagmetti Riccardo*, Kommentar BV zu Art. 22quater aBV, in: Kommentar zur Bundesverfassung der Schweizerischen Eidgenossenschaft, Basel / Zürich / Bern 1987 ff. Die neue Verfassung vom 18. April 1999 hat dies nicht grundsätzlich verändert (Art. 75 BV); dazu *Lendi Martin*, Kommentar zu Art. 75 BV, in: Ehrenzeller/

verfolgt durch die Synchronisierung von Erschliessungsplanung und Baubewilligungspraxis[30] im Baugebiet, begleitet von einem Erschliessungsverbot ausserhalb des generellen Kanalisationsprojektes gemäss der damaligen Gewässerschutzgesetzgebung, was faktisch zu einem Bauverbot ausserhalb des Baugebietes im sog. Übrigen Gebiet führte, später überlagert durch einen richt- und nutzungsplanerischen Akt[31] gestützt auf neue gesetzliche Grundlagen der Bemessung der Bauzonen und der Vorgabe der Landwirtschaftszone, wie sie im Bundesgesetz über die Raumplanung vom 22. Juni 1979 vorgezeichnet

---

Mastronardi / Schweizer / Vallender (Hrsg.), Die Schweizerische Bundesverfassung, Kommentar, Zürich / Lachen 2002. Die weiteren immanenten Ziele der Trennung von Siedlungs- und Nicht-Siedlungsgebiet, wie sie im gleichen Abschnitt erwähnt werden, folgten dicht auf. Das Bundesgesetz über die Raumplanung vom 22. Juni 1979 erwähnt das Prinzip der Trennung von Siedlungs- und Nicht-Siedlungsgebiet resp. von Baugebiet und Nicht-Baugebiet – es spricht von der haushälterischen Bodennutzung (Art. 1 Abs. 1 Satz 1 RPG) und von der Begrenzung der Ausdehnung der Siedlungen (Art. 3 Abs. 3 RPG), was dem Trennungsprinzip nahe kommt – nicht explizit, stellt aber die Instrumente dazu zur Verfügung und regelt die Bau- und die Landwirtschaftszonen in Verbindung mit der Ausnahmeregelung für Bauten ausserhalb der Bauzonen derart, dass daraus die erwähnte Trennung resultieren muss (Art. 15, 16 und 24 RPG).

30 Damals vor allem gestützt auf die Gewässerschutzgesetzgebung, BG über den Schutz der Gewässer gegen Verunreinigung vom 8. Oktober 1971 (Art. 19/20) und dessen Vorläufer, dann aber auch in Ausführung des BG über die Massnahmen zur Förderung des Wohnungsbaues vom 19. März 1965, das in allgemeiner Art die Erschliessung (Grob- und Feinerschliessung) zu erfassen bemüht war.

31 Dieser wurde erst in Verbindung mit dem Erlass des BG über die Raumplanung (RPG) vom 22. Juni 1979 (SR 700) möglich, das durch Bundesrecht die Richt- und Nutzungspläne für die Realisierung der Trennung von Siedlungs- und Nicht-Siedlungsgebiet zur Verfügung stellte und stellt. Vgl. zu den rechtlichen Fragen insbesondere *Haller Walter / Karlen Peter*, Raumplanungs-, Bau- und Umweltrecht, Bd. I, 3. A., Zürich 1999. Dieses Werk vermittelt breit abgestützte, am geltenden Recht orientierte Informationen zum System der Raumplanung, ihrer Ziele, Instrumente und Massnahmen sowie Verfahren, insbesondere im Konnex zum Umweltrecht, ohne allerdings den Grundsatz der Trennung von Siedlungs- und Nicht-Siedlungsgebiet rechtlich eingehend und als grundlegend abzuhandeln. Ähnliche Vorbehalte sind zu *Schürmann Leo / Hänni Peter*, Planungs-, Bau- und besonderes Umweltschutzrecht, 3. A., Bern 1994, anzubringen, allerdings wird hier an einer zentralen Stelle klar formuliert, die Hauptaufgabe der Nutzungsplanung – was in dieser Form, bezogen auf die Nutzungsplanung, wohl kaum gesagt werden kann, ist diese doch in der Lokalisierung und Dimensionierung von Nutzungsarten und -intensitäten zu sehen – bestehe in der Trennung von Siedlungs- und Nicht-Siedlungsgebiet (S. 121 ff., übrigens mit Verweis auf zahlreiche BGE), was regelmässig bedeute, von der tatsächlichen Entwicklung überholte Bauzonen zu verkleinern.

sind.[32] Mit dem Trennungsakt sachlich verknüpft ist die Regelung, wonach ausserhalb der Bauzonen Baubewilligungen nur ausnahmsweise unter Beachtung strenger gesetzlicher Voraussetzungen erteilt werden können.[33] An die Trennung wurden in der Folge weitere bodenrechtliche Massnahmen (Landwirtschaftliches Bodenrecht, Steuerrecht usw.) geknüpft.[34]

Hinter der stringenten, örtlich, aber auch regional (durch kantonalen Richtplan) vorbedachten Trennung von Siedlungs- und Nichtsiedlungsgebiet[35] stehen resp. standen – über die Bekämpfung des Landschafts- und Ressourcenverbrauches hinaus – gleich mehrere planerische Absichten und gesetzliche Ziele, die allerdings in Teilen aus heutiger Sicht relativiert resp. neu bedacht werden müssen.[36] Die hier – nachstehend – resümierten spiegeln die häufigsten, ohne dass sie logischer Abfolge genügen würden oder frei von Über-

---

32 Vgl. die Art. 15/16 RPG mit ihren Ausführungen über die Bau- und Landwirtschaftszonen, vorweg hinsichtlich der Dimensionierung der Bauzonen. Die Neufassung des Art. 16 RPG über die Landwirtschaftszonen durch das novellierende BG vom 20. März 1998 führte zu einem multifunktionalen Verständnis der Landwirtschaftszone, doch wäre es der Sache nach um eine multifunktionale Sicht des Nicht-Siedlungsgebietes gegangen, was bei der Revision des RPG verpasst worden ist (vgl. dazu die bundesrätliche Botschaft, BBl 1996 III 513). Die multifunktionale Landwirtschaftszone ist und bleibt eine Landwirtschaftszone, deren Land sich für die landwirtschaftliche Nutzung eignet, für die Aufgaben der Landwirtschaft benötigt wird oder im Gesamtinteresse landwirtschaftlich bewirtschaftet werden soll – so der klare Wortlaut des revidierten Art. 16 RPG.

33 In Frage steht Art. 24 RPG (Ausnahmen ausserhalb der Bauzonen) mit seinen begleitenden Modifikationen auf Gesetzes- und Verordnungsebene. Die Rechtserlasse zum Raumplanungs- und Umweltrecht des Bundes sind leicht greifbar bei *Karlen Peter (Hrsg.)*, Raumplanungsrecht und Umweltrecht des Bundes, Textausgabe, Basel/Genf/München 1999.

34 Dies geschah vor allem mit der Absicht, die überhitzte Bodenpreisentwicklung gleich von mehreren Seiten her zu bekämpfen: *Bundesamt für Justiz*, Bodenrecht im Siedlungsbereich, Massnahmen des Bundes, Bern 1991.

35 Seine rechtliche Begründung folgt aus der Bemessung der Bauzonen (Art. 15 RPG), aus der Legaldefinition der Landwirtschaftszonen (Art. 16 RPG), aus der gesetzlichen Festschreibung der Erschliessung (Art. 19 RPG) und schlussendlich aus den tatbestandsmässigen Voraussetzungen der Erteilung der Baubewilligung (Art. 22 RPG).

36 Eine Übersicht über die der Bundesgesetzgebung zur Raumplanung und vor allem dem Erlass der Verfassungsartikel 22ter und 22quater aBV vorangehenden Absichten vermittelt der Bericht der Eidg. Expertenkommission für Fragen der Landesplanung vom 6. Oktober 1966, Bern 1967. Unterstützt durch die bundesgerichtliche Rechtsprechung (BGE 91 I 329) wurde in deren Rahmen das Postulat einer entschädigungslosen Landwirtschaftszone erhoben, so deutlich im eben zit. Bericht, a. a. O., S. 116. Vgl. dazu ferner die anschliessend ergangene Botschaft des Bundesrates an die Bundesversammlung über die Ergänzung der Bundesverfassung durch die Art. 22ter und 22quater vom 15. August 1967, BBl 1967 II, S. 133ff.

schneidungen wären. Abstrahiert die Analyse von den zu nennenden materiellen Zielvorstellungen, so verbleibt neben der Massnahme der Trennung als solcher – materiell bewertet und zusammengefasst – *die Vorsicht, der Siedlungsentwicklung nicht freien Raum zu gewähren, sie bestmöglich unter Kontrolle zu behalten, ohne ihr starre Grenzen zu setzen.* Dies ist ein durchaus beachtenswertes Anliegen, das dem Geist der Nachhaltigkeit nahe kommt.

Als besondere Ziele können erwähnt werden:

- *Bekämpfung der Zersiedlung des Landes*, die drohte, weil das ältere Recht und die Verwaltungspraxis in die Nutzungsfreiheit des Eigentums ganz allgemein die bauliche einschlossen und die private Erschliessung aus eigenen Mitteln zuliessen. Die frühe Gewässerschutzgesetzgebung und in der Folge das RPG haben, wie bereits erwähnt, Korrekturen gesetzt und die Zielsetzung aktiviert, indem sie das erschliessbare Baugebiet definierten.
- *Sicherung der Eigenernährungsbasis* durch Gewährleistung von landwirtschaftlich nutzbaren Flächen, verbunden mit der Einführung der Landwirtschaftszone. Die Erfahrungen des ersten und zweiten Weltkrieges standen Pate. Der Grad an Versorgungsautarkie hat vor dem Hintergrund sinkender militärischer Bedrohungen in Europa, der Produktivitätssteigerungen in der Landwirtschaft und des ausgedehnten internationalen Handels mit Versorgungsgütern graduell an Bedeutung eingebüsst, kann aber früher oder später aufgrund von weltweiten, regionalen und/oder sektoralen Krisen unterschiedlichster Ursachen wieder relevant werden. Das Problem der Landesversorgung besteht jedenfalls weiter, wenn auch unter anderen Bedingungen.
- *Dämpfung der Bodenpreisentwicklung*. Diese Zielsetzung wurde nicht in allen, aber doch zu grossen Teilen erreicht, nämlich zugunsten der mit einem Bauverbot belasteten landwirtschaftlich genutzten Flächen. Hingegen wurde der Baulandmarkt innerhalb des Siedlungsgebietes nicht in dem Masse verflüssigt, wie dies intendiert wurde, weil sich die Erschliessung der ausgewiesenen Bauzonen verzögerte und weil der Baulandmarkt durch zu zahlreiche, zu nuanciert eingesetzte Bauzonenarten unverhältnismässig segmentiert worden ist. Ausserdem wird Bauland stellenweise nach wie vor gehortet. Dementsprechend besteht da und dort ein preistreibender Mangel an baureifem Land.
- *Kostenbewusste Begrenzung der Versorgungs- und Entsorgungsnetze* (Kanalisations-, Strassennetze usw.) – gleichsam die ökonomisch-technisch initiierte Trennung von Siedlungs- und Nicht-Siedlungsgebiet. Diese Zielsetzung wurde dank präziser gesetzlicher Regelungen der Bauvoraussetzungen (Baureife aufgrund von Zonierung und Erschliessung) und der Finanzierung der Bau- und Betriebskosten (Kausalabgaben: Beiträge, Gebühren) weitgehend erreicht. Geblieben ist das Problem der Hortung von baureifem Land, wenn auch nicht mehr im früheren Ausmass.

- *Nicht-Siedlungsgebiet als Landwirtschaftsgebiet* (exkl. Wald, der separat geschützt wird, verstanden als Bauverbotsgebiet mit Ausnahmebewilligungen, überlagert allenfalls durch Schutzzonen). Diese Zielsetzung wurde bei Erlass des Verfassungsartikels über die Raumplanung im Jahre 1969 beinahe apodiktisch betont – sie führte zur Einführung der Landwirtschaftszone – und hielt sich auch im RPG von 1979.[37] Die Begründung lag in einer Kombination von a) Ernährungsplan, b) der Abkoppelung vom Baulandmarkt und c) der Gewährleistung einer allgemeinen Pflege der Landschaft. Die erste der beiden Begründungen vermag aktuell nicht mehr in allen Teilen zu überzeugen, weil die Landwirtschaftsflächen optimaler Nutzung flächenmässig kleiner sind als das Nicht-Siedlungsgebiet und weil die ökonomischen Bedingungen gemessen an den grösseren Märkten nicht allenthalben zur Alimentierung der hiesigen Landwirtschaft hinreichen. Diese Tatsachen bedingen eine subtile, neu zu konzipierende Planung des Nicht-Siedlungsgebietes samt neuen Zielvorstellungen und allenfalls neuen Planungsinstrumenten. Damit einher gehen Probleme der Pflege der Landschaft, soweit sie nicht oder monoton landwirtschaftlich genutzt wird und/oder einer besonderen Gestaltung bedarf. Auch sind der Ressourcenschutz und das ökologische Gleichgewicht zu bedenken.

---

37 Ursprünglich war seitens des Bundesrates sogar beabsichtigt, die Landwirtschaftszone in einem neu zu erlassenden Verfassungsartikel ausdrücklich zu erwähnen. In den parlamentarischen Beratungen wurde davon abgesehen, nicht aber in der Meinung, das Institut der Landwirtschaftszone aufzugeben. Es wurde vielmehr als notwendig erachtet, von einer den ganzen Raum erfassenden Orts-, Regional- und Landesplanung zu sprechen – gestützt auf diese Absicht wurde schlussendlich der Begriff der Raumplanung in den neuen Artikel 22quater aBV eingefügt, der auch Art. 75 BV eigen ist. Vgl. dazu Botschaft des Bundesrates an die Bundesversammlung über die Ergänzung der Bundesverfassung durch die Art. 22ter und 22quater, vom 15. August 1967, BBl 1967 II 133 ff., sowie die Botschaft zum Bundesgesetz über die Raumplanung vom 31. Mai 1972, BBl 1972 I 1435 ff. Nach wie vor lesenswert ist das grundlegende Gutachten vom 21. November 1964, verfasst von Hans Huber, zur Landwirtschaftszone und zur Zonenordnung im allgemeinen: *Huber Hans*, Gutachten vom 21. November 1964, erstattet an das Eidg. Justiz- und Polizeidepartement, Zur Verfassungsmässigkeit der Landwirtschaftszone, Rechtsgutachten über den Entwurf vom März 1963 über die Revision des Bundesgesetzes über die Erhaltung des ländlichen Grundbesitzes vom 12.6.1951, insbesondere die Verfassungsmässigkeit einer Zonenordnung, nachgedruckt in DISP Nr. 82, Zürich 1985, S. 5 ff. Der Bundesrat setzte sich nach diesem Gutachten für eine besondere Verfassungsvorlage ein, nämlich in der oben zit. Botschaft vom 15. August 1967, a. a. O., hatte doch Hans Huber begründet, dass es für die mit der Landwirtschaftszone einhergehende landesweite Beschränkung der Baufreiheit einer besonderen Verfassungsgrundlage bedürfe. Das Gutachten war nötig geworden, nachdem der Autor dieser Zeilen in einem Vortrag vor den schweizerischen Ortsbürgergemeinden das Fehlen einer Verfassungsgrundlage für die Landwirtschaftszone beanstandet hatte.

- *Haushälterische Bodennutzung*. Diese allgemeine Formel, die seit jeher zum RPG[38] gehört und nun auch der neuen Verfassung[39] eigen ist, steht im Dienst des Schutzes des Bodens als Ressource und der optimierten, konfliktbereinigten Nutzung des Bodens, unter anderem für die bauliche, die landwirtschaftliche und die regenerative. Sie fasst teilweise die vorstehenden gesetzgeberischen Intentionen zusammen.

- *Das Prinzip der nachhaltigen Raumentwicklung*[40] schliesst das Postulat der haushälterischen Nutzung des Bodens ein. Indem es wirtschaftliche, soziale und ökologische Aspekte – über den Ressourcenschutz des Bodens hinaus – forciert, fördert es den sorgfältigen Umgang auch mit den die wirtschaftliche und gesellschaftliche Entwicklung beeinflussenden Kräften, indem es für jeden Bereich und unter sich abgestimmte Nachhaltigkeit fordert. Die nachhaltige Entwicklung wird durch die Trennung von Siedlungs- und Nicht-Siedlungsgebiet zwar nicht gewährleistet, aber doch begünstigt, da das Prinzip der Trennung u. a. Zurückhaltung bei der Bemessung der Bauzonen auferlegt, die innere Nutzung des Siedlungsgebietes in den Vordergrund rückt, der Unsorgfalt im Umgang mit der Ressource „Boden" allenthalben wehrt, der Belastung von frei zu haltenden Räumen entgegenwirkt sowie vorsorglich ökologische Ausgleichsflächen planbar macht.[41] Vor allem veranlasst es, sich kritisch und alsdann planend mit der Zukunft zu befassen.

---

38 Art. 1 Abs. 1 Satz 1 leg. cit.

39 Art. 75 Abs. 2 BV

40 Art. 73 BV. Diese Bestimmung ist dem Abschnitt über Umwelt (Art. 74 BV) und Raumplanung (Art. 75 BV) vorangestellt. Das Prinzip der Nachhaltigkeit ist aber auch in vielen andern Verfassungsbestimmungen – von der Präambel bis zur Bestimmung über die Landwirtschaft (Art. 104 BV) angesprochen. Sein Sinn greift tiefer als Art. 73 BV vermuten lässt. Es geht nicht nur um die Erneuerungsfähigkeit der Natur, so gewichtig diese ist. Mindestens so massgebend ist die Langfristigkeitskomponente der intergenerationellen Gerechtigkeit – gleichzeitig ein ethisches Prinzip. Nachhaltigkeit führt damit zur Auseinandersetzung mit der Zukunft und also zur Planung. Siehe dazu *Lendi Martin*, Grundorientierungen für die Raumplanung/Raumordnung – eine Vorlesung, Hinweise auf ethische Anhaltspunkte für das Planen und Entscheiden wie auch das Handeln in der räumlichen Planung, Etappenbericht auf dem Weg einer offenen Diskussion, Gastvorlesung an der Universität für Bodenkultur in Wien vom 24. November 2003 (vervielfältigt).

41 Vgl. dazu die Botschaft des Bundesrates zum Bundesgesetz über die Raumplanung vom 27. Februar 1978, BBl 1978 I 1006 ff., aber auch die Botschaft zum abgelehnten BG über die Raumplanung: Botschaft des Bundesrates zum Bundesgesetz über die Raumplanung vom 31. Mai 1972, BBl 1972 I 1435 ff. Es ist vor allem die Umweltschutzgesetzgebung, die Beiträge an die ökologische Sicht des Bodens vermittelt. Die Raumplanung übernimmt vorsorgliche Aufgaben unter dem Titel zukunftsorientierter Planungen der Bodennutzung.

Diese Auflistung der Ziele ist unvollständig, weder logisch gegliedert noch widerspruchsfrei. Nicht beantwortet ist mit deren Erwähnung die Frage, ob die Trennung von Siedlungs- und Nicht-Siedlungsgebiet durch planerischen Akt mit Rechtsfolgen für das Bauen die bestmögliche Verfolgung der Ziele erlaubte und erlaubt. Noch nicht hinreichend geklärt ist bis hierher sodann, was das Nachhaltigkeitsprinzip aus sich heraus für die räumliche Planung, gerade auch im Hinblick auf die Trennung von Siedlungsgebiet und Nicht-Siedlungsgebiet, erfordert, insbesondere bei der Abstimmung von wirtschaftlichen, sozialen und ökologischen Anforderungen. Vor allem beunruhigt die Ungewissheit, ob die (einstigen) Ziele der Trennung von Siedlungs- und Nicht-Siedlungsgebiet morgen und übermorgen taugen.

4. Die grundlegenden instrumentellen Konzepte

Für die Steuerung der Siedlungsentwicklung innerhalb der Bauzonen und in Abgrenzung zur offenen Landschaft sind neben den bekannten Methoden der erschliessungstechnischen und der durch planerischen Akt vollzogenen Trennung von Siedlungsgebiet und Nicht-Siedlungsgebiet mehrere andere denkbar. Sie lassen sich, wie wir weiter unten sehen werden, aufreihen, doch ist gleich zu bedenken, dass sie letztlich nicht rein vorkommen, sondern in Kombinationen münden, und deshalb, so sie nachstehend erwähnt werden, nicht singulär gelesen werden dürfen. Die vorangehende Grundsatzfrage, ob eine förmliche Steuerung – jenseits des Marktes – angezeigt oder ob auf den Markt zurückgegriffen werden soll und welche Arten resp. Methoden zu wählen wären, ist letztlich durch den Gesetzgeber zu beantworten, wenn auch auf der Basis planerischer, rechtlicher, ökonomischer und ökologischer Erkenntnisse, erfasst im Bemühen, zusammenführende Gesichtspunkte herauszuarbeiten.[42]

Gesetz oder Plan, Gebot / Verbot oder Vertrag, Verhaltensnorm oder Verhaltensbeeinflussung durch Abgaben? – das ist das (vereinfachte) Grundmuster denkbarer Steuerungsinstrumente, das sich anbietet.[43] Dahinter steht die Ge-

---

42 Nicht näher erläutert werden in diesem Text die faktischen Kräfte räumlicher Entwicklung, welche in besonderem Masse die Siedlungsentwicklung beeinflussen und deren Steuerung resp. Steuerungsnotwendigkeit und Angemessenheit nach Art und Intensität herausfordern. Sie sind nicht minder relevant als die planerischen und rechtlichen wie auch die ökonomischen, doch dürfen sie als bekannt vorausgesetzt werden. In Umrissen wurden sie einleitend angesprochen. Ihre Bedeutung für die Wahl der optimalen Instrumente darf aber keinesfalls unterschätzt werden.

43 Zur grundsätzlichen Diskussion marktwirtschaftlicher Instrumente sind die Aussagen im Kontext des Umweltschutzes zahlreicher als im Umfeld der Raumplanung, grundlegend beispielsweise *Frey René L.,* Umweltschutz und Marktwirtschaft, in: Lendi Martin (Hrsg.), Umweltpolitik, Zürich 1991, S. 43 ff.; sodann *Frey René L. /*

genüberstellung von Gesetz und Markt. Betont ist hier das rechtlich verfügbare Instrumentarium. Wie der Wald durch Gesetz geschützt wird[44], so könnten auch das Bau- und/oder das Landwirtschaftsgebiet ausschliesslich gesetzlich – per Legaldefinition – erfasst werden. Über Pläne, zusätzlich oder originär eingesetzt, lässt sich eine Flexibilisierung über die Zeiten hinweg integrieren. Die Abgaben mit Lenkungswirkungen, von den Steuern bis zu den förmlichen Lenkungsgaben, heben ihrerseits von der starren Verhaltensnorm ab. Die Massfragen werden dabei eine Rolle spielen. Vor allem aber muss entschieden werden, in welcher Art und in welchem Mass die Türen dem Markt geöffnet werden sollen.

Wenn nun eine Auflistung möglicher Beiträge an die Siedlungssteuerung gewagt wird, so um zu zeigen, wie breit der Fächer ist. Gleichzeitig soll sichtbar werden, dass sich kein Modell aus sich heraus erklärt. Einzubeziehen sind auf jeden Fall die Wirkungen verfassungs-, bau- und planungsrechtlicher, planerischer und ökologischer, aber auch ökonomischer und gesellschaftlicher Ziele resp. von Randbedingungen, die unterschiedlich gewichtet werden können.

Im Prinzip können – im Rahmen der verfügbaren Instrumente – zwei Arten von Grundzutritten unterschieden werden, nämlich hoheitliche und nicht-hoheitliche, wobei die Marktöffnung bei den nicht-hoheitlichen vorausgesetzt wird, während sie hoheitliche begleiten kann. Selbstverständlich ist die Marktbeseitigung durch Gesetz mit der faktischen Wirkung der Verstaatlichung denkbar. Dies ist aber angesichts der enormen Bedeutung des Immobilienmarktes für die Volkswirtschaft – u. a. Mobilisierung des Bodenwertes, u. a. zur Sicherung von Forderungen – nicht sinnvoll und mit der vorausgesetzten Marktwirtschaft im Sinne der freiheitlichen Grundsätze der Wirtschaftsordnung verfassungsrechtlich nicht vertretbar.[45] Inhaltlich befassen sich die Instrumente mit Verfügungs- und der Nutzungsbeschränkungen. Die Verfügungsfreiheit umfasst vorweg Erwerb- und Veräusserung von Grundstücken und betrifft mithin das Marktgeschehen bezüglich Kauf und Verkauf von Liegenschaften innerhalb und ausserhalb der Bauzonen, innerhalb und ausserhalb des Siedlungsgebietes. Die Nutzungsfreiheit schliesst die bauliche oder anderweitige, beispielsweise die land-, die forstwirtschaftliche Nutzungen (ect.)

Staehelin Elke W. / Blöchlinger H. J. (Hrsg.), Mit Ökonomie zur Ökologie, 2. A. Basel / Frankfurt a. Main 1993

44 Art. 2/3 Waldgesetz vom 4. Oktober 1991. Es ist also nicht so, dass Raumplanung zwingend nach dem geographisch (kartenmässig) unterlegten Plan ruft. Nötig ist er dort, wo Lokalisierungen und örtliche Dimensionierung unumgänglich sind (Nutzungspläne). Umgekehrt kann sich ein Plan auch durch Sätze, Zahlen, Tabellen, Graphiken, Signaturen usw. ausdrücken, ohne eine Karte zu bedingen. Entscheidendes Merkwal eines Planes ist die zukunftsorientierte Richtungsweisung – unverbindlich, behördenverbindlich, grundeigentümerverbindlich oder allgemeinverbindlich.

45 Art. 94 ff. BV

ein. Beschränkt wird sie durch Nutzungsbeschränkungen (Zonen für Wohnen, Gewerbe, öffentliche Bauten etc., aber auch durch Begrenzungen der Intensität der Nutzungen mittels Bauvorschriften, ökologische Belastungsnormen etc.). Die marktwirtschaftlichen Instrumente können sich nun auf die Verfügungs- und/oder die Nutzungsfreiheit beziehen.[46] Heikel ist die Frage, ob die Grenze zwischen Bau- und Nichtbau- resp. zwischen Siedlungs- und Nichtsiedlungsgebiet vorgegeben sein muss. Auf alle Fälle betrifft sie nicht die Verfügungs-, sondern die Nutzungsfreiheit, allerdings in besonders intensiver Art, nämlich im Sinne des absoluten Überganges von der relativen Freiheit der baulichen Nutzung zu deren Verbot (unter Vorbehalt von Ausnahmen). Ob er relativiert und marktwirtschaftlichen Instrumenten überantwortet werden kann, muss geklärt werden. Stets im Auge zu behalten ist: Eine Diskussion einseitig auf die Massnahmen hoheitlicher oder marktwirtschaftlicher Art gerichtet genügt nicht. Es bedarf immer der Rückkoppelung zur Zieldiskussion und zur Zielerfüllung. Und diese ist primär politischer, konsolidiert rechtlicher und also verbindlicher Art.

4.1. Hoheitliche Zutritte, mit Marktöffnungen

Die hoheitlichen Massnahmen sind zahlreich. Sie folgen imperativen Geboten und Verboten, die allerdings nicht nur als Rechtssätze, sondern auch in der Form von Plänen (grundeigentümerverbindliche Erschliessungs-, Sondernutzungs- und Zonenpläne usw.) ergehen können. Zu erwähnen sind u. a.:

a) Steuerung durch hoheitliche Erschliessungsplanung (Basis-, Grob- und Feinerschliessung) unter Definition der Bauvoraussetzungen, mit oder ohne Überwälzung der Kosten des Gemeinwesens für die Grob- und Feinerschliessung auf den Baugesuchsteller;
b) Steuerung durch örtliche Zonierung (Flächenwidmungsplanung) des Baugebietes – aufgrund planerischer Prognosen und hoheitlicher Vorgaben – und deren etappenweise Erschliessung, allenfalls erleichtert durch Erschliessungsansprüche, durch Nutzungstransfers und Massnahmen der Parzellarordnung;
c) Steuerung durch regionale (kantonale) und/oder kommunale Vorgaben für die Dimensionierung der Baugebiete sowie durch gesetzliche und/oder planerische Anforderungen an deren Erschliessung;

---

46 Im schweizerischen geltenden Recht steht die Verfügungsfreiheit im Vordergrund. Ihr dienen der Baulandmarkt und der landwirtschaftliche Bodenmarkt, unter Vorbehalt des Expropriationsrechts zu, wobei der letztere durch das bäuerliche Bodenrecht begrenzt wird. Die Nutzungsfreiheit wird wesentlich stärker limitiert, doch ist auch sie auszumachen – innerhalb der Schranken des geltenden Bau-, Planungs- und Umweltrechts.

d) Förmliche Ausscheidung von Siedlungs- und Nichtsiedlungsgebieten aufgrund richtplanerischer Vorgaben der Eignung und des Bedarfs – nach Massgabe von Sachplänen der Eigenversorgung, der Fruchtfolgeflächen, des Landschaftsschutzes, des voraussichtlichen regionalen und örtlichen Baugebietsbedarfes, der Transport- und Versorgungsstrukturen, des Ressourcenhaushaltes und der Belastbarkeit von Räumen –, nachfolgend abgestimmte Zonen- und Sondernutzungsplanungen (samt Schutzzonenplanungen) für das Baugebiet und das Nichtbaugebiet, unter Berücksichtigung der Erschliessung und allenfalls weiterer Bauvoraussetzungen;
e) Zonenplanähnliche Etablierung von Marktbereichen für Bauland- und Nicht-Bauland, bei gleichzeitig erheblich reduzierten Anforderungen an das Bauen – insbesondere hinsichtlich formaler Begrenzungen der Nutzungsintensitäten – innerhalb des Baulandmarktes, verbunden mit dem Ziel, die optimale Nutzung des Bodens marktseitig zu beeinflussen;
f) Zonenplanähnliche räumliche Trennung von Siedlungs- und Nichtsiedlungsgebiet nach vorrangigen Kriterien von Baulandmarkt und Nicht-Baulandmarkt, zusätzlich hoheitliche Einflussnahme, sei es durch Erschliessungsplanung, durch Zonierungen und Definitionen der Bauvoraussetzungen, sei es durch steuerrechtliche und andere Arten von Abgaben (Verkehrswert-, Ertragsbesteuerung, Grundstückgewinnsteuer, Abschöpfung von Sondervorteilen, Gebühren usw.);
g) Beschränkungen der baulichen Nutzungsmöglichkeiten ausserhalb des Siedlungsgebietes (Allgemeines Bauverbot, Erschliessungsbeschränkungen, Baukonzessionen/Ausnahmebewilligungen, erhöhte Planungsanforderungen, stringente Baubewilligungsvoraussetzungen usw.);
h) Mobilisierung des Baulandmarktes innerhalb des Siedlungsgebietes durch rechtliche Baugebote, durch Verkehrswertbesteuerung für erschlossenes Bauland usw., erleichterte Bauvoraussetzungen für verdichtete Bauweisen und bessere Nutzungen bestehender Bausubstanzen, alles unter Abbau resp. Relativierung der Regelungsdichte, aber doch mit rechtlichen Vorgaben als Voraussetzungen der Erteilung einer Baubewilligung;
i) Stärkung der Baulandpolitik der Gemeinwesen auf dem an sich freien Baulandmarkt durch vertraglichen Handel, zusätzlich begleitet durch öffentlichrechtlich begründete Vorkaufsrechte zugunsten des Gemeinwesens resp. Expropriationsrechte wider die Baulandhortung resp. die nicht intensive Nutzung bestehender Bauten, alles zugunsten eines flüssigen Baulandmarktes innerhalb des Siedlungsgebietes, verbunden mit der Auflage, die erworbenen Grundstücke wieder dem Markt zuzuführen, innert nützlicher Frist, unterstützt durch Vorgaben für die Art und Intensität der Überbauung, wenn möglich mit präjudizieller Wirkung auf die Nutzung angrenzender, zusammenhängender Gebiete, begleitet von Gestaltungs- resp. Bebauungs-/Überbauungsplänen.

4.2. Marktnahe Instrumente, allenfalls begleitet von hoheitlichen

Die nicht hoheitlichen Wege zielen auf marktnahe Instrumente oder sie überlassen die Bildung von baureifem Land vollumfänglich oder für grössere Räume den Marktkräften.[47] Sie intendieren, der Markt trage von sich aus bis zu einem gewissen Grade zur geordneten Bedürfnisbefriedigung bei, wobei nötigenfalls Instrumente gesetzlicher Ausrichtung, vertragliche Absprachen mit dem Gemeinwesen und lenkungsakzentuierte Abgaben eingesetzt werden, um zielbezogene Anreize zu setzen. Die nicht hoheitlichen Wege, dies muss beigefügt sein, sind in aller Regel nicht ab ovo frei von gesetzlichen Regelungen. Die Marktordnung ist vom Ansatz her rechtlicher Natur. Sodann werden marktnahe Instrumente oft mit hoheitlichen, wenn auch knapp formulierten Vorgaben kombiniert, da ihre Wirkung mindestens teilweise von verbindlichen Zielsetzungen abhängt.

Als Modelle, nie in Reinkultur realisierbar, können – beispielsweise – hervorgehoben werden:

a) Freier Bodenmarkt mit baulicher Nutzungsfreiheit, verbunden mit Rechtsanspruch auf Erschliessung resp. Möglichkeit der Selbsterschliessung;
b) Freier Bodenmarkt unter voller Kostenanlastung für die durch das Gemeinwesen allein oder zusammen mit den Bauinteressenten geplante Erschliessung, Baureife usw.;
c) Freier Bodenmarkt mit polizeirechtlichen Begrenzungen der baulichen Nutzungsmöglichkeiten – bis und mit polizeilich-ästhetischen Gestaltungsvorschriften, ordnungsbetonenden Gestaltungsbaulinien, Strassenlinien usw.;
d) Freier Baulandmarkt innerhalb eines raumplanerisch/ökologisch definierten und gesetzlich resp. planerisch abgegrenzten Gebietes mit Rechts-

---

47 Vgl. dazu vor allem *Staehelin-Witt Elke / Plattner Rolf M.,* Wirtschaft und Raumplanung, Beziehungen – Konflikte – Lösungen, Bern 1998; *Staehelin-Witt Elke / Blöchliger Hansjörg*, Marktwirtschaftliche Reformvorschläge: Unterstützen sie die Raumplanung?, Bern 1995. Eine eingehende Darstellung ökonomischer Instrumente zum Schutz der Bodenflächen resp. zur Gestaltung der Flächenhaushaltspolitik findet sich bei *Losch / Lecke-Lopatta / Lehmberg / Wallbaum*, Ökonomische Instrumente zum Schutz der Bodenflächen, in: ARL (Hrsg.), Flächenhaushaltspolitik, a. a. O., S. 197 ff. Es werden vor allem Abgabearten und deren Lenkungskraft diskutiert, akzentuiert unter dem Gesichtspunkt des Bodenschutzes. Dietmar Scholich äussert sich in derselben Publikation der ARL zu informellen Planungsansätzen, wobei er darunter Kooperations- und Konsensstrategien sowie die Institutionalisierung von Managementaufgaben, Moderation, Beratung und Information versteht: *Scholich Dietmar*, Informelle Planungsansätze, in: ARL (Hrsg.), Flächenhaushaltspolitik, a. a. O., S. 142 ff. Das Beispiel der Flächennutzungszertifikate behandeln *Frey René L. / Gmünder Markus*, Marktwirtschaftliche Instrumente in der Raumplanung, Basel 2003 (vervielfältigt).

anspruch auf behördliche resp. private Erschliessung innerhalb dieses Gebietes, vertragliche Absprachen über Erschliessungsgrade sowie bauliche Nutzung und Gestaltungsanforderungen;

e) Vertragliches Aushandeln der Baumöglichkeiten mit der öffentlichen Hand bei gesetzlich offen, allenfalls final oder gar dispositiv definierten öffentlichen Aufgaben im Bereich der Erschliessung, des Baupolizeirechts, des Umweltschutzes, des Bauplanungsrechtes, insbesondere der flächenmässigen Bebauungsplanung – im Sinne einer Public/Private – Partnership (PPP);
f) Börsenähnlicher Handel von behördlich ausgegebenen, in globo dem Markt übergebenen, preistauglichen „Zertifikaten" baulicher Nutzungsmöglichkeiten (differenziert nach Art und vor allem nach Intensität wie auch nach Erschliessung resp. Erschliessbarkeit) für offen definierte angebotsvielfältige Räume oder gar für den Raum als solchen, wobei das aktuelle oder zu erwerbende Eigentum an einer adäquaten Parzelle und der Besitz von entsprechenden Nutzungszertifikaten Voraussetzung der konkreten Überbaubarkeit ist;
g) Differenzierte Abgabenerhebungen (Steuern, Beiträge, Gebühren) auf nicht überbauten oder unternutzten Grundstücken mit direkten oder indirekten Lenkungswirkungen in Richtung der Angebotsvermehrung zugunsten des Baulandmarktes, der Verdichtung der Bebauung, und umgekehrt der Nicht-Attraktivität der Überbauung im Interesse der Freihaltung usw., beispielsweise aufgebaut auf (gestuften) Perimetern mit örtlichen Bezügen;
h) Aufbau eines öffentlich/privaten Managementsystems für die Widmung der Bodennutung;
i) Entwickeln von Konsens- und/oder Kooperations- wie auch Informationsstrategien zur zweckmässigen und haushälterischen Bodennutzung mit dem Ziel der Allokation öffentlicher und privater Nutzungen resp. von Nutzungsbeschränkungen.

Die Probleme der eben angesprochenen Modelle liegen letztlich in der Frage, ob überhaupt und allenfalls welche Ziele mit der Wahl der Modelle verfolgt werden, und welche Zielerfüllungsgrade erwartet werden. Dahinter steht primär ein raumplanerisches Thema. Nicht minder heikel ist die Frage, ob die Grenze zwischen dem eigentlichen Baugebiet (Siedlungsgebiet) mit einem (mehr oder weniger) freien Baulandmarkt und dem Landwirtschaftsgebiet (Nichtsiedlungsgebiet) scharf gezogen sein muss oder fliessend gestaltet werden kann.

5. Das geltende Mischsystem

Das positivrechtliche System, also dasjenige nach dem RPG von 1979 und den Verfassungsartikeln von 1969 resp. gemäss der heute geltenden Verfassung von 1999 sowie abgestimmt darauf nach den kantonalen Planungs- und Baugesetzen, besteht in einem Verbund ordnungs- und planungs- wie auch bodenrechtlicher sowie ökologischer Massnahmen. Da es als differenzierte Regelung samt eingebundenen Zielsetzungen in seiner vielfältigen Substanz – trotz jahrelanger Geltung – letztlich wenig bekannt ist und im wesentlich mit Erwähnung des Prinzips der Trennung von Siedlungs- und Nicht-Siedlungsgebiet abgetan wird, muss es näher betrachtet werden und zwar sowohl hinsichtlich der Grundlinien als auch seiner Stärken und Schwächen.[48] Das offen, marktstützend angelegte RPG von 1979 wurde leider in der Folge von einer Flutwelle spezialgesetzlich normierter öffentlicher Interessen erfasst und durch die Verwaltungs- und Gerichtspraxis verrechtlicht. Der Blick richtet sich auf die (entschlackten) Grundintentionen des genannten Gesetzes.[49]

---

48 Das geltende schweizerische Recht lässt sich nicht ohne weiteres mit dem ausländischen vergleichen, weil die hiesigen Gegebenheiten mit einem ausgedehnten Berggebiet und einem seit dem 2. Weltkrieg während Jahrzehnten – aus Gründen der positiven Erfahrung mit der Sicherstellung der Eigenernährung – hoffierten ländlichen Raum von andern Staaten divergieren. Auch ist das Baurecht bundesstaatlich den Kantonen und nicht dem Bund anvertraut. Dennoch verweisen die gesetzlichen Regelungen, beispielsweise diejenigen Deutschlands, materiell in die nämliche Richtung. Siehe dazu das dt. Raumordnungsgesetz vom 18. 8. 1997, daselbst § 2 mit den Grundsätzen der Raumordnung. Darnach ist die Siedlungstätigkeit zu konzentrieren, ist die grossräumige Freiraumstruktur zu erhalten, sind Natur und Landschaft zu schützen, zu pflegen und zu entwickeln usw. Auch das Bundesbaugesetz, ergangen als Baugesetzbuch vom 27. August 1997, verweist auf verwandte Anliegen (§ 1). Weder das Raumordnungsgesetz noch das Baugesetzbuch schreiben den Grundsatz der Trennung von Baugebiet / Nicht-Baugebiet resp. von Siedlungs- und Nicht-Siedlungsgebiet expressis verbis fest, bewirken aber die Trennung und zwar vor allem über die zit. Grundsätze und deren Abstimmung (§ 1 und 2 Raumordnungsgesetz/ § 1 f. Baugesetzbuch), über die Festschreibung des Bebauungsplans (§ 5 ff. Baugesetzbuch) auf der Basis des Flächennutzungsplanes sowie über die Regelungen zum Bauen im sog. Aussenbereich (§ 35 Baugesetzbuch). Vgl. die oben erwähnte Dissertation von *Melanie Markstein*, die auch Österreich einbezieht, a. a. O. Eine knappe Übersicht zum dt. Raumordnungsrecht vermittelt *David Carl-Heinz*, Grundlagen des Raumordnungsrechts, in: Jenkins Helmut W., Raumordnung und Raumordnungspolitik, München 1996; ausführlicher und eher auf der Grundlage des Baugesetzbuches, aber mit vielen Hinweisen auf das Raumordnungsrecht *Brohm Winfried*, Öffentliches Baurecht, München 1997.

49 Es war – auch nachträglich gesehen – ein rechtswissenschaftlicher Fehler, die Erstkommentierung des RPG der Verwaltung zu überlassen. Siehe dazu: *EJPD / Bundesamt für Raumplanung*, Erläuterungen zum Bundesgesetz über die Raumplanung, Bern 1981.

Die geltende Ordnung bemüht sich, haushälterisch mit der Nutzung des Boden umzugehen, die Ressource Boden zu schonen, die Belastung des Ökosystems in Grenzen zu halten und innerhalb des Siedlungs-/Baugebietes dem Markt und somit wirtschaftlichen wie auch individuellen Bedürfnissen Beachtung zu schenken. Gleichzeitig widersetzt es sich einem Einfrieren des aktuellen Zustandes. Es lässt Möglichkeiten der Entwicklung offen, insbesondere zugunsten von Aktivitäten im planerisch zu umreissenden Baugebiet, dann aber auch, allerdings limitiert, im Nicht-Siedlungsgebiet, wobei für den Gesetzgeber gerade für diesen Raum Anlass bestünde, die planerischen Anforderungen, mit Begrenzungen und/oder positiven Entwicklungsmöglichkeiten, zu erhöhen.[50] Es geht also insgesamt um wesentlich mehr als das Problem der Landwirtschaftszone als Bauverbotszone samt Ausnahmeregelung, wie dies der Gesetzgeber mit der Akzentsetzung auf der Regelung der Landwirtschaftszone und des Bauens ausserhalb der Bauzonen suggerierte.[51]

Als charakteristische planerische Merkmale des geltenden Rechts können erwähnt werden:

a) das Siedlungsgebiet vom Nichtsiedlungsgebiet regional und örtlich durch positive Beschreibung der Planung des Siedlungsgebietes und des Nicht-Siedlungsgebietes zukunftsorientiert (Bedarfsprognose zulasten des Baugebietes usw.) zu trennen,
b) das Siedlungsgebiet knapp zu bemessen und nach übergeordneten Zielen landesweit, regional und örtlich gemäss raumplanerischen Kriterien – durch Richt- und nachfolgend durch Nutzungspläne – zu bemessen und zuzuweisen, und es später nur aufgrund neuer Bedarfsprognosen unter Berücksichtigung regional vorhandener Reserven innerhalb der Baugebiete wachsen oder eingrenzen zu lassen,
c) das Siedlungsgebiet von innen her und nach innen zu erschliessen sowie optimal zu nutzen, das Nicht-Siedlungsgebiet demgegenüber von Überbauungen möglichst frei zu behalten und Infrastrukturanlagen, soweit sie zeitlich, sachlich und nach der Lage im Nicht-Siedlungsgebiet unumgänglich sind, bestmöglich einzupassen,

---

50 Diskutiert werden müssten mögliche rechtsrelevante räumliche Gliederungen des Nicht-Siedlungsgebietes – unter Einbezug des Waldes – und sachgerechte Planungsinstrumente, zum Beispiel eine besondere Art des „Gestaltungsplans", gedacht u. a. für das sachgerechte Einbinden standortgebundener Infrastrukturanlagen (Strasse, Eisenbahnlinien, Flugplätze/Flugfelder, Starkstromleitungen, Stauseen usw.).

51 Art. 24 RPG samt seitherigen Ergänzungen. Leider liess sich die Rechtswissenschaft in Teilen verleiten, darin ein gewichtiges Thema, das es detailliert auszulegen gelte, zu erkennen. Sie reduzierte bisweilen die Problemfülle der Raumplanung auf diese Frage, wobei zuzugestehen ist, dass für Grundeigentümer in der Landwirtschaftszone (und ihre Banken) die Nutzungsoptimierung nicht von nebensächlicher Bedeutung ist. Die umfangreiche Rechtsprechung zu Art. 24 RPG trug ihrerseits zu einer Überbewertung bei.

d) zwei Markträume zu schaffen, einerseits für einen Baulandmarkt und anderseits für einen eher landwirtschaftlich orientierten Nicht-Baulandmarkt,
e) Belastung der Kosten und Abgeltung der Sondervorteile der Fein- und der Groberschliessung über Gebühren und Beiträge an resp. durch die Grundeigentümer des überbaubaren Baugebietes,
f) innerhalb des Baulandmarktes im Sinne der Verflüssigung des Baulandmarktes baureifes Land zu produzieren – Rechtsanspruch auf Erschliessung bei zulässigen Erschliessungs- und weitgehendem Verzicht auf Baugebietsetappierungen, Massnahmen der Parzellarordnung, Zuhilfenahme abgabenrechtlicher Vorkehrungen und weiterer gesetzlicher Massnahmen wider die Baulandhortung, baurechtliche Vorschriften für intensivere Nutzungen innerhalb des Baugebietes, ohne Ausgleichsräume zu verdrängen,
g) ausrichten des Siedlungs- und das Nichtsiedlungsgebietes auf die Anforderungen einer integrierten Siedlungs-, Landschafts- und Transport- / Versorgungsstruktur, insbesondere unter Abstimmung von Siedlungs- und Verkehrsplanung – landesweit, regional und örtlich,
h) sowohl im Siedlungsgebiet als auch im Nicht-Siedlungsgebiet hohen gestalterischen als auch ökologischen Anforderungen zu genügen, Gefährdungen der natürlichen Lebensgrundlagen entgegenzuwirken und die Ressourcen, insbesondere Luft, Ruhe, Boden und Wasser, zu schützen.

Diese Übersicht ist unvollständig – und erst noch (vermeintlich) kompliziert. Sie macht aber deutlich, dass die geltende Regelung positive Züge unseres überkommenen und verfassungsrechtlich festgelegten Rechtssystems aufweist – sie akzeptiert den Markt und demokratisch legitimierte, rechtsstaatlich bewertete gesetzliche und planmässige Vorgaben. Es ist also weder marktverachtend noch einseitig legalistisch. In diesem Sinne kann knapp formuliert werden:

1. Die gesetzlich vorgesehene Begrenzung des Siedlungsgebietes hebt die Freiheit des Bodenmarktes[52] nicht aus den Angeln, führt allerdings zu dessen Zweiteilung und innerhalb des Siedlungsgebietes zu einer Differenzierung der Nutzungsmöglichkeiten und damit der Voraussetzungen der Preisbildung.
2. Das Institut des Eigentums[53] bleibt gewahrt und gewährleistet die Verfügungs- und zusätzlich die Nutzungsfreiheit, auch wenn die letztere im öf-

52 Der Bodenmarkt ist privatwirtschaftlich erfasst und untersteht dem Grundsatz der Wirtschaftsfreiheit (Art. 94 BV) und setzt die Privatautonomie sowie die Vertragsfreiheit voraus. Es gilt insbesondere das Regime der Nutzungs- und Veräusserungsfreiheit, nicht hingegen der Formfreiheit. Schranken ergeben sich auf Bundesebene aufgrund der Kompetenzen zum Privatrecht (Art. 122 BV), zur Raumplanung (Art. 75 BV), zur Landwirtschaft (Art 104 BV) und vielen weiteren Bestimmungen, auf kantonaler beispielsweise aufgrund des Baurechts.

53 Art. 23 BV gewährleistet das Eigentum als Institut, als Bestand von Rechten und als Vermögenswert. Es ist Teil der marktwirtschaftlichen Ordnung und untersteht sowohl Regelungen des Privat- als auch des öffentlichen Rechts.

fentlichen Interesse (auf gesetzlicher Grundlage und unter Beachtung der Verhältnismässigkeit) sowohl im Siedlungs- wie auch im Nicht-Siedlungsgebiet beschränkt werden muss. Die Verfügungsfreiheit bleibt, abgesehen von Ausnahmen wie der Expropriation, die der (formellen) gesetzlichen Grundlage bedürfen, erhalten.

3. Die Siedlungsgebietsbegrenzung erfolgt nicht nach einem starren Massstab, sondern nach Massgabe eines zukunftsorientierten, räumlich und ökologisch bedachten, Lebensraum bezogenen, planerischen Aktes.[54] Räumliche Entwicklungen bleiben möglich, ohne die vielen Funktionen des Nicht-Siedlungsgebietes zu vernachlässigen und ohne die Ziele des Ressourcenschutzes, des ökologischem Ausgleichs, der Freiräume für das Erholen, der Sicherstellung standortgebundene Infrastrukturen wie Strassen, Stauseen usw. zu beeinträchtigen.

6. Unzulänglichkeiten der geltenden Ordnung

Das geltende System der Trennung von Siedlungs- und Nichtsiedlungsgebiet vermag, so einleuchtend es angelegt ist, nicht in allen Teilen zu befriedigen, aber nicht primär wegen des dahinterstehenden Konzeptes, sondern wegen der hohen Anforderungen, die es an eine raumplanerisch kohärente Politik stellt. Offen ist zudem die Frage, ob es sich gegenüber anderen als den bis dato bekannten Trends zu bewähren vermöchte. Immerhin scheint es gegenüber einer Umkehrung in Richtung stagnierenden oder gar sinkenden Bedarfs, sogar bei regionalen Unterschieden, wenigstens theoretisch relativ resistent zu sein, weil kein Zwang besteht, ausweitende Dimensionierungen der Bauzonen ohne planerischen Akt, gegründet auf raumrelevante, gesetzliche und planerischen Kriterien vorzunehmen.[55]

Zu den akuten Mängeln können gezählt werden: a) Der sachlich-politische Widerstand gegenüber einer restriktiven Auslegung der Bauzonen ist in der Regel seitens der Grundeigentümer beträchtlich, so sie nicht durch Restriktionen begünstigt werden.[56] b) Implementations- und Anwendungsprobleme,

---

54 Der Grundsatz der Trennung von Siedlungs- und Nicht-Siedlungsgebiet geht aus der Kompetenz des Bundes zur Grundsatzgesetzgebung über die Raumplanung und deren Anerkennung als öffentliche Aufgabe der haushälterischen Nutzung des Bodens der geordneten Besiedlung des Landes hervor (Art. 75 BV). Seine gesetzliche Konkretisierung hat er durch das RPG erfahren.

55 Art. 15 ff. RPG; zusätzlich sind die Planungsgrundsätze und die Aussagen der Richtpläne zu beachten (Art. 3 RPG und Art. 6 ff. RPG

56 Baubeschränkungen zulasten Dritter können sich für Grundeigentümer wertsteigernd auswirken, während sie für diese belastend sind. Schwierigkeiten der Prognostizierung der Entschädigungslosigkeit resp. der Entschädigungskonsequenz sind kaum

kurzum Vollzugsprobleme sind der sachlichen Widersprüche und unterschiedlicher Interessen wegen vorprogrammiert. c) Die Regelung des Art 24 RPG über das Bauen ausserhalb der Bauzonen greift dort ins Leere, wo es nicht um Relationen zur Landwirtschaft, sondern um andere berechtigte Anforderungen an das Nicht-Siedlungsgebiet geht.[57] d) Die gesetzlich und planerisch an sich sinnvolle Möglichkeit, das Baugebiet nötigenfalls nach Massgabe vorgegebener Kriterien in einem späteren Zeitpunkt geplant auszudehnen[58], verleitet Grundeigentümer und Gemeinden wie auch Kantone zu oft zum Ausscheiden resp. dem Beibehalten zu grosser Bauzonen. e) Von der angezeigten Redimensionierung der Bauzonen im Interesse der Versorgung des Baulandmarktes mit baureifem Land und der optimalen Lokalisierung der Bau- und Nicht-Siedlungsgebiete wird zu häufig abgesehen. Dies kann zu einer Vernachlässigung der Bereitstellung von baureifem Land beitragen. f) Fragwürdig sind sodann, möglicherweise, die in die Trennungsintention verwobenen gleich mehrfachen Zielsetzungen, die mindestens teilweise neu zu erwägen oder gar zu verwerfen sind. Dies trägt zu Plausibilitätsdefiziten bei.

Grundlegend problematisch sind vor allem die von Anfang an bestehenden Lücken im RPG hinsichtlich der Planung des Baugebietes und des Nicht-Siedlungsgebietes. Ohne abgestimmte, breitere und gleichzeitig zweckmässig

---

zu vermeiden, da die bundesgerichtliche Rechtsprechung zur sog. materiellen Enteignung in sich zwar klar und nachvollziehbar ist, aber gegenüber konkreten Einzelfällen Ungewissheiten einschliesst. Siehe dazu statt vieler Autoren *Haller Walter/ Karlen Peter*, Raumplanungs-, Bau- und Umweltrecht, a. a. O., S. 121 ff.

57 Art. 24 (und ff.) RPG (Ausnahmen für Bauten und Anlagen ausserhalb der Bauzonen) hat eine völlig falsche Überhöhung erfahren, mindestens in theoretischer Hinsicht, teilweise auch in der Rechtsprechung und vor allem bei der jüngsten Gesetzesrevision (1998). Sie kann als „Ausnahmebestimmung" konzipiert den zu regelnden Sachverhalten nicht in allen Teilen gerecht werden. Selbstverständlich: Die Bestimmung musste und muss durch Behörden, Gerichte und sogar durch das Bundesgericht seit 1980 (Inkrafttreten RPG) mit Sorgfalt und Umsicht angewandt werden, doch hätte mindestens im Verlauf der Praxis erkannt werden müssen, dass die Regelung des Art. 24 RPG mitsamt seinen seitherigen Weiterungen nicht hinreichend auf die Problematik des Nicht-Siedlungsgebietes antwortet: M. E. war nicht Art. 24 RPG falsch konzipiert, sondern das Bezugsgebiet des Nicht-Siedlungsgebietes ist breiter als die echte Landwirtschaftszone, auf welche die genannte Bestimmung an sich ausgerichtet war. Deshalb hat die Revision des RPG vom 20. März 1998 am falschen Punkt angesetzt (BBl 1996 III 513). Im Mittelpunkt hätte nicht Art. 24 RPG stehen müssen, sondern die Debatte über das planerische Erfassen des Nicht-Siedlungsgebietes, in dem sich verschiedenste Nutzungsarten, sich ausgrenzend oder überlagernd, begegnen: Landwirtschafts-, Infrastruktur- (Stauseen, Strassen, Eisenbahnen, Leitungen, Schiessplätze, Skipisten, Luftseilbahnen usw.), Schutz-, Erholungs-, Grundwassergebiete-, Gletscher, unwegsame Berggegenden, Wasserläufe (Seen, Flüsse, Bäche), Wald, Allmenden, Streusiedlungen, Aussenwachten, Freizeiteinrichtungen usw.

58 Art. 15 RPG

Regelung der Nutzungsplanung – für beide Bereiche! – bleibt die (wesentlich besser geglückte) Regelung der Richtplanung ein Torso. Es mangelt an differenzierenden Aussagen zum Baugebiet, es fehlt an elementaren planerischen Vorstellungen und Instrumenten für das Nicht-Siedlungsgebiet.[59] Zu wenig grosszügig angelegt sind die Markträume für das Gut „Boden" im Siedlungsgebiet; vor allem unter der Herrschaft zu differenzierter Zonierungen im Baugebiet ist die Segmentierung zu ausgeprägt – eine Folge der kantonal- und kommunalrechtlichen „Zonitis".[60] Sie schränkt die Markträume und damit den Markt ein. Zudem verleiten die eng gefassten Zonen zu restriktiven Normierungen und deren zurücknehmende Auslegung. Dies trifft beispielsweise auf den Gestaltungsbereich zu, wo aus Anliegen der Verhinderung der Verunstaltung Anforderungen der Gestaltung werden. Auf der andern Seite kann die Zurücknahme der Ausdehnung des Baugebietes nicht ohne ästhetische, siedlungsqualitative Aspekte verantwortet werden, zumal die miteinhergehende Verdichtung erhöhte gestalterische Anforderungen stellt, was nicht ein Plus an Vorschriften, wohl aber architektonische Leistung bedingt.

Sodann sind die nutzungsplanungsrechtlichen Regelungen in sachlicher Hinsicht ungenügend, ist doch die so wichtige Stadtplanung resp. die Ortsplanung nicht resp. nur unterschwellig geregelt, und sind die materiellen Anforderungen an den ländlichen Raum weder erfasst noch hinreichend strukturiert. Auch bleibt die elementar notwendige Baugesetzgebung als immanenter Teil der Nutzungsplanung bedauerlicherweise – auf Bundesebene – weitgehend ausgeklammert. Sicherheit, Hygiene, Nutzungsarten, Bemessung der Nutzungsintensitäten, Gestaltung usw. sind ihrer Relevanz für die räumliche Ordnung wegen vom Raumplanungsrecht nicht zu trennen – alles Bereiche, die nicht nur formell zu meistern, sondern so anzulegen sind, dass es materiell und korrespondierend mit den marktseitigen Bedürfnissen gelingt, eine positive Gesamtwirkung zu erzielen. Sinnvoll wäre ein Bundesbaugesetz mit Spielräumen für die regionalen Gestaltungsanforderungen, vor allem geleitet von der Zielsetzung, der (ordnungspolitisch eingebundenen) Baufreiheit stimulierende Kraft zuzuerkennen.[61] Ferner sind die Bundeskompetenzen der Raumplanung – begrenzt auf eine Grundsatzgesetzgebung – schon vom Ansatz her vergleichsweise zu knapp, zumal sich die koordiniert anzuwendende Umwelt-

---

59 Die hilflos anmutenden Regelungen, eingeschoben zu Art. 24 RPG, sind das Spiegelbild der planerischen Nicht-Erfassung des Nicht-Siedlungsgebietes.

60 Sie ist bundesrechtlich nicht gestoppt, weder im Rahmen der Planungsgrundsätze (Art. 3 RPG) noch im Rahmen der gesetzlichen Regelung der Nutzungsplanung (Art. 14 ff. RPG).

61 Die jüngst erhobene Kritik von Hansjörg Seiler am geltenden Planungsrecht und an den zu restriktiven Sichtweisen der Anwender des öffentlichen Rechts ist zu hören: *Seiler Hansjörg*, Das öffentliche Recht als Schikane der KMU: Erscheinungsbilder, Erklärungsversuche und Therapievorschlage, in: Girsberger Daniel/Schmid Jörg (Hrsg.), Rechtsfragen um die KMU, Luzerner Beiträge zur Rechtswissenschaft, Bd. 1, Zürich 2003, S. 85 ff.

schutzgesetzgebung auf umfassende Bundeskompetenzen berufen kann; dies gilt auch für den Gewässerschutz[62], bis zu einem gewissen Grade sogar für den Verkehr.[63] Auf jeden Fall müsste sichtbar werden, wie und nach welchen Kriterien sich Landwirtschaft, Landschaftsschutz, Infrastrukturanlagen (Verkehr, Energie usw.), Freizeitbelastungen und Ressourcenschutz, ökologischer Ausgleich usw. im Nicht-Siedlungsgebiet zu vertragen haben. Bedeutsam ist vor allem die Relationsfrage zur Ökologie (Ressourcen, Stoffhaushalt, Verbrauch nicht erneuerbarer Energie, Belastbarkeit von Räumen usw.). Selbst im Bereich der Agglomerationsentwicklung, die als so bedeutsam für die Raumentwicklung herausgestellt wird, sind die Bundeskompetenzen, auch wenn dies aktuell andersherum betont wird, gering, wenn nicht sogar marginal.[64] Alles in allem: Gesetzliche, sogar verfassungsrechtliche Novellierungen sind früher oder später im Bereich der Raumplanung angezeigt. Sie betreffen auch Qualitätsfragen, insbesondere Gestaltungsbelange.[65]

Als ein weiterer Mangel erweist sich die nachhinkende Planung von Siedlungsgebiet und Nicht-Siedlungsgebiet. Weder auf der kantonalen Ebene der Richtplanung – im Rahmen der zweiten Generation der Richtpläne besser als in demjenigen der ersten – noch auf der Ebene der Nutzungsplanungen wird die Anpassung der Planung an veränderte Verhältnisse intensiv genug vorangetrieben. Zum Wesen der Planung würde es gehören, Konstanz zu pflegen und nach geraumer Zeit gefestigter Berechenbarkeit Dynamik nicht zu scheuen. Vor allem auf der Ebene kommunaler Planungen können sich die Ansprüche nach Wohnbauland, nach Dienstleistungszentren, Industrieland, nach Freizeiteinrichtungen usw. erheblich verändern. Dies bedingt neue Planungen und gegebenenfalls sogar neue Grenzziehungen zwischen Siedlungs- und Nicht-Siedlungsgebiet. Bisweilen kann sich sogar bereits innerhalb der Periode der Beständigkeit die Notwendigkeit räumlicher Nutzungstransfers (zum Beispiel

---

62 Art. 76 Abs. 3 BV

63 Art. 74 und Art. 87 BV. Es ist so, das Raumplanungsrecht und das Baurecht, vor allem aber auch das Umweltrecht werden zu pointiert als Inkarnation öffentlicher Interessen gelesen, dabei gilt es, ex constitutione die Eigentumsbeschränkungen auf die Wahrung der öffentlichen Interessen in Abwägung gegenüber den privaten wie auch auf die Verhältnismässigkeit hin bedenken.

64 Art. 50 BV – der Bund verfügt über keine allgemeinen und speziellen Agglomerationskompetenzen, doch kann er im Rahmen des gesetzlich eingeräumten Ermessens, wie es seinen anderweitigen Aufgaben da und dorrt inhärent sein mag, agglomerative Aspekte einfliessen lassen, so unter dem Titel der Forschungsförderung (Art. 64 BV). Vorausgesetzt werden kann auch eine immanente Kompetenz zugunsten der Bundesbehörden, finanzielle Beiträge an Bestrebungen zu leisten, die helfen, Probleme der Agglomerationen zu erkennen und das Verständnis für sie zu fördern.

65 Auf sie kann hier nicht näher eingetreten werden, doch versteht es sich, dass verdichtete Siedlungen und intensiviere Nutzungen bestehender Bauten im Baugebiet sowie von Infrastruktureinrichtungen und anderen Anlagen im Nicht-Siedlungsgebiet nach erhöhten ästhetischen Anforderungen rufen.

die Verlegung nicht genutzter Bauzonen an nachgefragte Standorte nach planerischen Kriterien) einstellen. Darauf ist die geltende Ordnung noch weniger vorbereitet als auf die periodischen Plananpassungen, die ihrerseits dynamischer an die Hand genommen werden könnten: Eine gewisse Flexibilität gehört zum Wesen der Planung.

Im Lichte einer Gesamtwürdigung: Die geltende gesetzliche Trennung von Siedlungs- und Nicht-Siedlungsgebiet ist in den Grundzügen vorteilhaft konzipiert. Sie belässt auf alle Fälle dem Bodenmarkt erhebliche Spielräume, wirkt der Zersiedlung grundsätzlich entgegen und dient vorsorglich dem Landschafts-, Natur- und Umweltschutz zu. Ausserdem fordert sie auf, das bestehende Siedlungsgebiet und bestehende Bauten ausserhalb des Siedlungsgebietes zonenkonform bestmöglich zu nutzen. Die planerische Flexibilität, bei regional und örtlich unterschiedlichen Entwicklungen, ist gewährleistet.

Zu unterstreichen bleibt, dass die geltende Regelung nicht so marktfern ist, wie bisweilen behauptet wird. Allerdings sieht die Wirklichkeit etwas anders aus als die gesetzlichen Aussagen nahe legen, weil es an der konsequenten Umsetzung teilweise gebricht und weil die Ausnahmeregelung für das Bauen ausserhalb des Baugebietes mangels strukturiertem Nicht-Siedlungsgebiet letztlich nicht problemgerecht greifen kann. Auch die weiteren Defizite der bundesgesetzlichen Regelung der Raumplanung dürfen nicht unterschätzt werden. Die Frage ist, ob politische Bereitschaft besteht, sie zu beheben. Wenn sie nicht in Angriff genommen werden, dann droht die Zielsetzung der Trennung unterlaufen zu werden. Es muss deshalb gelingen, im Sinne der vorstehenden Ausführungen

a) die richtplanmässigen Vorgaben zu festigen, die nutzungsplanungsrechtlichen Vorschriften im RPG für das Siedlungs- und das Nicht-Siedlungsgebiet näher auszuführen und umzusetzen,
b) den bodenrechtlichen Gehalt sowie den Ressourcenschutz des Raumplanungs- mit dem Umweltrecht abzustimmen,
c) auf das Steuer- und ganz allgemein auf das Abgaberecht hinsichtlich seiner bodenrechtlichen Relevanz, vor allem zugunsten der Verflüssigung des Baulandmarktes wie auch des Bodenschutzes tiefer einzuwirken sowie
d) die siedlungsqualitativen und landschaftserhaltenden/-gestaltenden Anforderungen erheblich zu steigern.

Ein allfälliger Verzicht auf die geltende Art der rechtlich/planerischen Trennung von Siedlungs- und Nicht-Siedlungsgebiet müsste sorgfältig hinterfragt werden. Falls als Ersatz marktwirtschaftliche Instrumente, wie „Lenkungsabgaben", realisiert werden sollen, so müssten diese nicht nur theoretisch und durchführungstechnisch erörtert, sondern vor allem hinsichtlich der Randbedingungen, der involvierten Zielsetzungen, der erwarteten Wirkungen und der konkreten Auswirkungen im Graubereich, der sich bei einem Verzicht auf eine scharfe Grenzziehung einstellen könnte, bedacht werden. Die Modelle dürfen nicht nur als Modelle diskutiert werden – sie müssen, ich unterstreiche dies, auf ihre Auswirkungen unter wechselnden Bedürfnissen befragt werden. Vor allem müssen sie den verfassungsrechtlichen Anforderungen der nachhaltigen Raumordnung, insbesondere der haushälterischen Bodennutzung und der geordneten Besiedlung des Landes, was eine Gesamtordnung bedingt, gerecht werden.[66]

7. Raumordnungspolitik

Die Probleme der gesetzestechnischen Seite, die offenen Flanken hinsichtlich der zu wählenden Steuerungsinstrumente, sind heikel, doch letztlich – ob befriedigend oder unbefriedigend, das ist eine andere Thematik – in der Art der aufgelisteten Optionen lösbar. Offen ist hingegen die andere Unklarheit: Vor welche Siedlungsentwicklung sieht sich die Raumplanung morgen und übermorgen gestellt? Reicht es je nach den denkbaren Stossrichtungen hin, mit dem einen und einzigen Ziel der Trennung von Siedlungsgebiet und Nicht-Siedlungsgebiet raumplanerisch zu agieren? Ist die Begrenzung der Bauzonen wirklich das sachgerechte Anliegen der Raumordnungspolitik, der Raumplanung? Was ist zusätzlich vorzukehren, damit eine räumliche Entwicklung anhebt, die den Bedürfnissen der Bevölkerung, der Wirtschaft und dem natürlichen Lebensraum zudient? Anders gefragt: In welche Raumordnungspolitik ist die haushälterische Nutzung des Bodens einzubinden?

Auf positive und negative Ursachen sowie mögliche Entwicklungstendenzen wurde bereits einleitend hingewiesen.[67] Eins scheint festzustehen, die Impulse der *Städte*, der *Agglomerationen*, resp. der in ihnen wirkenden wirtschaftlichen und gesellschaftlichen Kräfte, messen den urbanen Räumen einen zentralen, einen höheren Stellenwert zu als dies früher der Fall war. Dies führt tendenzmässig zu einer Neubewertung des städtischen Siedlungsgebietes, möglicherweise sogar der Agglomerationen und des Wachstums ihrer Gürtel. Die Lehre von der Planung neigt jedenfalls dazu, den Städten neue Chancen zu eröffnen.[68] Der sich anbietende Schritt, die Begrenzungen städtischer Siedlungsgebiete zugunsten von Vernetzungen mit ländlichen und angrenzenden

66 Art. 75 BV, in Verbindung mit Art. 73 BV.

67 Siehe Ziff. 2

68 Sie dazu *Flückiger Hans/Frey René L. (Hrsg.)*, Eine neue Raumordnungspolitik für neue Räume, Zürich 2001; *ORL-Institut (Koch Michael/Schmid Willy A., Hrsg.)*, Die Stadt in der Schweizer Raumplanung, ein Lesebuch, Martin Lendi gewidmet, Zürich 1999. Siehe dazu auch *Wachter Daniel*, Raumentwicklung und Raumordnungspolitik in der Schweiz, in: Schweiz 2003, Geographische Rundschau, Heft 9, S. 34 ff. Die eben zitierte Sondernummer befasst sich in hohem Masse mit der Stadt in der Schweiz (Zürich, Bern usw.). Siehe dazu auch *Lendi Martin/Elsasser Hans*, Raumplanung in der Schweiz, eine Einführung, 3. A., Zürich 1991, S. 46 ff.

urbanen Räumen zu relativieren oder gar zurückzunehmen, ist schnell getan. Vorsicht ist geboten. Dennoch wird die tradierte kommunale Zonenplanung in einem Bereich, der von den territorialen Grenzen der Gemeinden abhebt, problematisch. Die räumliche Planung kommt nicht darum herum, die Städte und die Agglomeration als solche nach aussen und nach innen zu betrachten und die Siedlungsentwicklung grossräumig, mindestens regional, zu würdigen. Die Dimensionierung der örtlichen Baugebiete wird darüber an Bedeutung einbüssen, es sei denn, der kantonale Richtplan greife ein und suche nach einer weiträumigeren Bilanzierung unter Vorgaben an die einzelnen Gemeinden innerhalb der Agglomeration.

Werfen wir über die ökonomischen und raumplanerischen Stadtplanungen hinaus einen Blick in die Welt der *Ökologie*, die einen Umbau der urbanen Systeme und begleitend ein anderes Verhältnis zu den ländlichen Räumen erwarten lässt, dann ist die Frage nach der angemessenen gesetzlichen Regelung der Begrenzung der Siedlungsgebiete auch von dieser Seite aufzuwerfen. Was ergibt sich aus einer sich über die Zeiten ändernde Öko-Bilanz für die Bemessung der Siedlungs- und Nicht-Siedlungsgebiete?[69] Die Antworten darauf sind (noch) nicht schlüssig. Sie können in Richtung einer Durchgrünung der Siedlungen und damit in eine flächenmässige Ausdehnung führen, sie können aber auch die Konzentration des Baugebietes und die grüne Lunge der Ausgleichsgebiete fordern, vor allem aber könnten sie nach einer Vernetzung mit naturbedingten Übergangsgebieten verlangen. So oder so ist die Frage der Trennung von Siedlungs- und Nicht-Siedlungsgebiet jedenfalls kritisch zu überprüfen.

Gemessen an der Stadt- und Agglomerationsfokussierung wie auch an den ökologischen Verflechtungen verliert die Planung des Nicht-Siedlungsgebietes die früher für sich reklamierte Bevorzugung nach Situierung und Ausmass gesicherten, integralen Schutz. Das Nicht-Siedlungsgebiet wird in Zukunft am Rande und teilweise sogar bis in den innersten Kern andere und gleichzeitig mehrfache Funktionen zu erfüllen haben, die von den Städten und Agglomerationen resp. von den dort lebenden Menschen her mitgeprägt werden. Dies heisst nicht, das Nicht-Siedlungsgebiet sei nicht zu bewahren und nicht als ökologischer Ausgleichraum zu stärken; dies verlangt aber nach einem Nachdenken darüber, wie das Nicht-Siedlungsgebiet für sich und in Relation zum Siedlungsgebiet raumplanerisch von Grund auf – nach seinen Funktionen her – zu verstehen und differenziert zu planen ist. Ob die kommende Raumordnungspolitik resp. Raumplanung unter diesen Umständen auch eine andere Art der Steuerung der Siedlungsentwicklung bedingt?[70] Die Fra-

69 Grundlegend dazu – im Sinne eines ersten Ansatz – *Baccini Peter/Oswald Franz (Hrsg.)*, Netzstadt, Transdisziplinäre Methoden zum Umbau urbaner Systeme, Zürich 1997.

70 Voraussetzung ist mindestens eine Änderung von Art. 15 und 16 RPG; vorteilhafter wäre es, das ganze Kapitel über die Nutzungspläne neu zu fassen.

ge ist zu klären, vermutlich bleibt es bei einer Trennung, doch müssen die Planungen auf die Differenzierungen der Funktionen und der sich aufdrängenden örtlichen Vernetzungen unter regionalen Gesichtspunkten vermehrt Rücksicht nehmen. Vor allem die Agglomerationsränder, die nach Belastung und Empfindlichkeit besonderer Aufmerksamkeit rufen, müssen planerisch vielseitig und problembewusster angegangen werden.

Wie dem auch sei, im Siedlungsgebiet, das aus neuen Funktionen heraus möglicherweise nach neuen Kriterien zu dimensionieren und zu verlegen ist, werden schon morgen die Qualitätsanforderungen gross geschrieben werden – sie reichen von der baulich-ästhetischen über die verkehrlichen und infrastrukturellen bis zu den ökologischen.[71] *Die Stadtplanung ruft der Stadtgestaltung.* Aber auch im Nicht-Siedlungsgebiet geht es nicht nur um Flächen und Flächenanteile sowie den Schutz der Ressourcen nach Quantitäten, sondern ebenfalls um Qualitäten der Gestaltung und des Erhaltens wie auch des Bewahrens – von Landschaften, schutzwürdigen Landschaftselementen, Ressourcen aller Arten. Sodann: *Der Landschaftsschutz ruft der Landschaftsgestaltung.* Ihm beigeordnet ist der breit und sorgfältig zu bedenkende *Ressourcenschutz*, im Hinblick auf erneuerbare und vor allem nicht erneuerbare.[72] Diese Aspekte belegen von einer andern Seite her, allerdings mit beinahe identischen Konsequenzen, die Fragwürdigkeiten der geltenden, allzu summarischen gesetzlichen Ordnung nach Raumplanungsgesetz, die gemessen an den vielfältigen Problemen zu rudimentär ausgefallen ist. Gleichzeitig ist aber zu betonen: *Zielsetzungen und Massnahme der Trennung von Siedlungsgebiet und Nicht-Siedlungsgebiet taugen insofern, als sie zur Vorsicht mahnen und Zurückhaltung nahe legen, bei gleichzeitiger planerischer Flexibilität – in Grenzen.*[73]

---

71 Problematisch wird es auf alle Fälle, die Verdichtung nach innen ohne Brückenschlag zum Nicht-Siedlungsgebiet prinzipientreu zu forcieren. Die Ökologie kann nach ökologischen Brücken verlangen. Zur Siedungsökologie neuerdings *Eigenmann Thomas/Hugentobler Ignaz*, Handbuch Siedlungsökologie, St. Gallen 2003; zur rechtlichen Seite der Siedlungsqualität *Lendi Martin*, Städtebauliche Gestaltungsvorgaben aus rechtlicher Sicht, Zürich, DISP Nr. 103, S. 14ff., Zürich 1990, nachgedruckt in *idem*, Bewährung des Rechts, Zürich 1992, S. 383 ff. Stadtplanung und Stadtgestaltung, Landschaftsplanung und Landschaftsgestaltung werden an der ETH erst seit wenigen Jahren auf wissenschaftlicher Ebene vor dem Hintergrund systematischer Analysen gepflegt. Im Rahmen des einstigen ORL-Instituts waren es immerhin zwei Sektionen, die sich dieser Fragen annahmen, eine für Landschaftsbelange (Ernst Winkler als erster Leiter, später Willy A. Schmid), eine für Stadt und Architektur (konzipiert von Benedikt Huber, fortgeführt von Franz Oswald).

72 Vgl. dazu *Magel Holger (Hrsg.),* Vom Biotop- zum ganzheitlichen Ressourcenschutz – Neue Herausforderungen an die Landentwicklung in Deutschland, Materialiensammlung Heft 27/2002 (Lehrstuhl für Bodenordnung und Landentwicklung TU München), München 2002.

73 Die Vorsicht ist dem Vorsorge- (Art. 74 Abs. 2 BV) und dieses dem Nachhaltigkeitsprinzip (Art. 73 BV) nahe.

Dies ist ein Vorteil, den es unter allen Umständen zu respektieren gilt. Er darf nicht blindlings fallen gelassen werden.

Die schweizerische Raumplanung handelt im Bundesgesetz über die Raumplanung unter dem Titel der Nutzungsplanung (Art. 14 ff. RPG) – dem Wortlaut nach – schwergewichtig nicht von den eben anklingenden Aufgaben, sondern von einer Flächenhaushaltspolitik, gebündelt durch das Prinzip der Trennung von Siedlungs- und Nicht-Siedlungsgebiet. Die gesetzlich schematisch wirkende Formulierung zu den Bau- und den Landwirtschaftszonen hat, so scheint es, von einer kreativen Auseinandersetzung mit einer Politik des zweckmässigen und haushälterischen Umganges mit dem Boden und gleichzeitig mit dem örtlichen, regionalen und landesweiten Raum sowie dem räumlichen Geschehen dispensiert. Bei genauerem Hinsehen ist dies nicht – mindestens nicht in allen Teilen – der Fall, auch wenn die phantasielose Regelung der Nutzungsplanung blutleer wirkt. Sie wird durch andere Planungsbereiche wenigstens teilweise korrigiert –nur muss dies die Praxis wahrnehmen. Die Richtplanung (Art. 6 ff. RPG), die der Nutzungsplanung vorangeht, weitet nämlich die Optik aus und vermittelt der Nutzungsplanung ausholende Dimensionen. Sie verknüpft die verfassungsrechtlich verlangte haushälterische Nutzung des Boden und die geordnete Besiedlung des Landes (Art. 75 Abs. 1 BV).[74] Sie verbindet Bodennutzungs- mit der Raumordnungspolitik, deutlicher ausgedrückt, *die Bodennutzungspolitik ist ein Teil der Raumordnungspolitik*. Die Ziele (Art. 1 RPG) und Planungsgrundsätze (Art. 3 RPG) machen dies deutlich. Dennoch: Unter dem Titel der Nutzungsplanung muss einiges nachgeholt und aufgearbeitet werden. Es geht im wesentlichen um Antworten auf neuere Entwicklungen im Siedlungs- und Nicht-Siedlungsgebiet – immer bedacht unter raumordnungspolitischen Gesichtspunkten.

Die Problematik der Notwendigkeit und allenfalls der Art sowie der Intensität und der Örtlichkeiten der Steuerung der Siedlungsentwicklung muss vor diesem Hintergrund auf die Grundfragen des Sinns und der Ziele zurückgeführt werden. Erst wenn diese geklärt sind, kann nach den Instrumenten und deren Anwendung gefragt werden. Das, was der Ratio nach an Grundsätzlichem zu bedenken ist, wird gewichtiger als die Modalitäten der Handhabung. Voraussetzung ist dabei das Erfassen der Grundlagen – vom Ist- bis zum Soll-Zustand. Zu denken ist an Prognosen zum Siedlungsbedarf (Flächen für Wohnen, Arbeiten, Freizeit, Versorgung, Verkehr), an Aussagen zu den sozial verträglichen Konzentrationen, zur räumlich-regionalen Verteilung – und dies alles auf der Basis nationaler sowie internationaler, wirtschaftlicher, sozialer, ökologischer und politischer Entwicklungen – in ständiger

---

74 Der Begriff der geordneten Besiedlung steht als pars pro toto für die räumliche Ordnung, da die Art der Besiedlung und deren Dimensionierung und Lokalisierung ein Erfassen des ganzen Raumes bedingt. Dazu nach wie vor lesenswert die Botschaft des Bundesrates zum BG über die Raumplanung vom 31. Mai 1972 (BBl 1972, S. 1453 ff.).

Veränderung.[75] Die Fakten gilt es zu bewerten, planerisch zu erfassen und auszurichten. Keine einfache Aufgabe, auch deshalb, weil nicht nur örtliche, sondern überörtliche Gewichtungen zu beachten sind. Aus diesen Vorgaben sind die Ziele der anzustrebenden Siedlungsentwicklung sowie die abgestimmten Instrumente und Massnahmen zu entwickeln. Auf alle Fälle steht fest: Die Steuerung der Siedlungsentwicklung ist mehr als ein Aspekt der Flächenhaushaltspolitik, sie bildet ein Kernelement der Raumordnungspolitik.

Sind die Ziele bekannt, darf es keine Prestigefrage sein, ob die Siedlungssteuerung hoheitlich nach Geboten und Verboten resp. nach Rechtssätzen und Plänen oder marktwirtschaftlich resp. in Kombination zu konzipieren sei. Es kommt auf die Relation zwischen Zielen und Mitteln sowie die bestmöglichen Wirkungen unter Vermeidung von ungünstigen Nebenwirkungen an. Die Bedeutung des Rechts darf dabei nicht unterschätzt werden. Selbst marktwirtschaftliche Prinzipien und Instrumente setzen rechtliche Grundlegungen, die gleichzeitig demokratische Legitimierung vermitteln, voraus.[76] Werden die Ziele und zusätzlich die Bauvoraussetzungen – tatbestandsmässige Voraussetzungen der Erteilung einer Baubewilligung – besonders deutlich festgehalten, dann braucht die Siedlungssteuerung nicht über Gebühr direkt angesprochen und instrumentell strapaziert zu werden. Es genügt dann, die Instrumente bereitzustellen und auf die Konkretisierungen der Ziele durch übergeordnete Pläne zu verweisen.

Abzusagen ist einer Doktrin, welche die Frage der Siedlungsentwicklung auf eine Flächenhaushaltpolitik reduziert. Es geht nicht nur um „Mengen" an verfügbaren Flächen, es geht immer auch um „Qualitäten", hinein sogar in die Gestaltung. Die Siedlungsentwicklung und deren Steuerung sind deshalb auf die Landschafts-, Siedlungs- sowie Transport- und Versorgungsstrukturen auszurichten, vor allem aber auf die Prinzipien der Nachhaltigkeit mit ihren Akzentsetzungen zugunsten der wirtschaftlichen Leistungsfähigkeit, der gesellschaftlichen Solidarität und des ökologischen Gleichgewichts sowie der

---

75 Die Frage lautet: Ist der Zeitpunkt für die Neubearbeitung der elementaren Grundlagen der räumlichen Entwicklung gekommen? – jener der Landesplanerischer Leitbilder des ORL-Instituts (ETH Zürich), auch jener der zukunftsorientierten Prognostik aus der Feder von Francesco Kneschaurek (zu seiner Zeit Universität St. Gallen, Zentrum für Zukunftsforschung)? Sie liegen nun rund 35 Jahre zurück. Selbstverständlich geht es nicht um eine Wiederholung oder gar Imitation. Mit dem Leugnen von Zukunftsfragen ist es aber auch nicht getan. Illusionen aufgrund von irgendwelchen Zukunftsbildern zu wecken, reicht ebenfalls nicht hin. Es bedürfte mindestens sorgfältig unterlegter und bedachter Szenarien.

76 Vgl. dazu vor allem *Ruch Alexander*, Das Recht in der Raumordnung, Basel 1997. Er behandelt ausdrücklich das Problem des Bodenmarktes, ohne aber dessen rechtlichen Voraussetzungen und dessen Strukturierungen durch das Recht zu befragen (S. 42 ff.). Auf der andern Seite setzt er sich eingehend mit der Lenkungsschwäche der Nutzungsplanung auseinander, die für die Trennung von Siedlungs- und Nicht-Siedlungsgebiet nicht irrelevant ist (S. 93 ff.)

Freiheit zugunsten der künftigen Generationen. Das „Schöne", das „Gestaltete", das Natürliche in seiner Aussage – sie sind mitzunehmen. Das Ästhetische verschafft der offenen Landschaft und der verdichteten Siedlung Ausdruck und Identität.

Mit diesen Überlegungen kündigt sich eine Revision des Bundesgesetzes über die Raumplanung an, vor allem der Regelungen der Nutzungsplanung innerhalb des zitierten Gesetzes – selbstredend in Verbindung mit der Richtplanung.[77] Gesetzgebung ist dabei mehr als das Formulieren eines Erlasses.

---

77 Art. 14 ff. RPG.
Vorgängig müsste eine Ausweitung der Bundeskompetenzen mit der Ermächtigung zu einer umfassenden der Raumplanungs- / Baugesetzgebung diskutiert werden. Diese Forderung aus meiner Feder, verwundert aufgrund früherer Interventionen, insbesondere zu Fragen eines Bundesbaugesetzes, nicht. Vgl. dazu *Lendi Martin*, Bundesbaugesetz – eine Notwendigkeit, SJZ 97 (2001) Nr. 9, S. 189 ff. Zu bedenken sind aber auch die Raumplanungsaufgaben, die von internationaler Seite her eine bundesweite Planung erfordern, was allerdings nicht eine Planung von oben nach unten „zur Folge haben darf. Das „Gegenstromprinzip" ist nach wie vor für einen Bundesstaat gegeben (Planung von unten nach oben und Planung von oben nach unten unter Abstimmung der Intentionen und Folgerungen). Die geltenden, limitierenden verfassungsrechtlichen Vorgaben dürfen nicht blindlings als gegeben akzeptiert, sondern müssen sachlich auf ihre Angemessenheit, vis à vis der anstehenden Probleme, diskutiert und gewichtet werden. Zum Beispiel: Das Raumplanungsrecht wird gegenüber dem Umweltrecht in Rückstand geraten, weil der Bund immer dann, wenn er eigenen Handlungsbedarf sieht, auf das Umweltrecht, basierend auf einer umfassenden Bundeskompetenz (Art. 74 BV) greifen muss, zumal die Durchgriffsmöglichkeiten im Bereich der Raumplanung der limitierenden Grundsatzgesetzgebungskompetenz wegen beschränkt sind. Der Druck auf den Umweg über das Umweltrecht kommt auch vom EG-Recht her, dass sich seinerseits mangels Raumplanungs- auf Umweltkompetenzen berufen muss und deshalb unter diesem Titel handelt.

Müsste im Rahmen des geltenden Rechts, also des aktuellen RPG novelliert werden, so müssten die Planungsgrundsätze neu bedacht werden, müssten die Richtplanvorschriften in dem Sinne diskutiert werden, dass ihre Funktionen in Richtung der planerischen Erfassung der Agglomerationen und der planerischen Durchdringung des Nicht-Siedlungsgebietes zu betonen wären. Alsdann wäre unter dem Titel der Nutzungsplanung diese für das Baugebiet grosszügig nach innen und unter Verwendung von gestaltungsplanerischen Anforderungen zu strukturieren, wäre für das Nicht-Baugebiet die Landwirtschaftszone in der Art der Fruchtfolgeflächen enger zu definieren und müssten die andern relevant werdenden Zonenarten – unter Beachtung des Ressourcen- und Landschaftsschutzes – präziser umschrieben und bezüglich der Überlagerungen resp. der Grundfunktion neu erfasst werden. Selbstverständlich müssten Bauverbote, die planerischen Anforderungen an die Nutzung bestehender Bauten und vor allem planerisch gestaltende Einbindung von Infrastruktureinrichtungen usw. erfasst werden.

Die nähere, detaillierte Beschreibung einer denkbaren Revision muss hier abgebrochen werden, weil sonst von der verfassungsrechtlichen Diskussion der sachlich und zeitlich dringlich erforderlichen umfassenden, allenfalls fragmentarischen,

Es bedarf, ich wiederhole mich bewusst, wenn auch nun juristisch formuliert, der vorgängigen Rechtstatsachenforschung. Planer und Juristen sollten de lege ferenda in Umrissen „erahnen" resp. zur Kenntnis nehmen können, mit welchen denkbaren Entwicklungen, die auf dem gegebenen Lebensraum lasten könnten, sie zu rechnen haben. Daraus werden sie abzuleiten haben, mit welchen Normen und Plänen seitens des Gesetzgebers den laufenden Prozessen entgegenzutreten ist. Neben den quantitativen Aspekten sind, dies ist vorhersehbar, die qualitativen zu betonen. Auch sollten die Chancen und Grenzen der Steuerung und Lenkung mit Umsicht bedacht sein. Zu bedenken wird auch sein, dass das Recht allein schon als abstrakte Ordnung zukunftsorientierte Anwendungswirkungen und damit Steuerung und Lenkung zeitigen wird. Pläne vermögen diese Effekte (erheblich) zu verstärken. Unter Beachtung der Ziele und der erwünschten Wirkungen wird alsdann zu entscheiden sein, ob mittels des Gesetzes, des Plans oder beider Instrumente, allenfalls unter Beizug des Marktes, die Trennung von Siedlungs- und Nicht-Siedlungsgebiet, die nach wie vor aktuell ist, möglicherweise aber sachlich und örtlich erheblich modifiziert werden muss, herbeigeführt werden soll.

8. Übersicht zu einigen Kernaussagen

Im Sinne einer Übersicht vermittelnden Zusammenfassung können folgende Aussagen gewagt werden:

1. Das Problem der ausufernden Siedlungsentwicklung und des Drucks auf die offene Landschaft, sichtbar vor allem am Rand von Agglomerationen, dann aber auch ablesbar an der Verstädterung und an den Eingriffen in die offene Landschaft, offenbart erhöhte, *wachsende Ansprüche an den Lebensraum* unter den Titeln des Wohnens, des Arbeitens, des Verkehrs, der Freizeit, der Veränderungen der Lebensstile usw. Voraussichtlich kompensieren sie – in der Schweiz – sogar sich abzeichnende demographische Veränderungen, die als solche eine gewisse Entlastung bringen könnten.

2. Ob und wie die Siedlungsentwicklung gesteuert werden kann und soll, ist Gegenstand anhaltender Diskussionen. Feststeht, dass mit dem Boden und dessen baulicher Nutzung haushälterisch umgegangen werden soll und

---

Kompetenzen des Bundes abgelenkt würde. Sie muss in erster Linie geführt werden. Gelingt deren Begründung nicht, dann wird die Revision notgedrungenerweise im Rahmen der Kompetenz zur Grundsatzgesetzgebung an die Hand zu nehmen sein. Allerdings darf souveräner, als dies bis anhin der Fall war, von der Summe der dem Bund unter andern Titeln übertragenen öffentlichen Aufgaben her legiferiert werden.

dass der Boden als Ressource zu schützen ist. Bei den Instrumenten der Steuerung der Siedlungsentwicklung wird üblicherweise zwischen *hoheitlichen und marktorientierten Instrumenten* unterschieden, doch sind sie kaum in Reinkultur auszumachen.

3. Das geltende Recht wehrt der ausufernden Siedlungsentwicklung und stützt die Raumordnung mittels der *Trennung von Siedlungs- und Nichtsiedlungsgebiet.* Es spricht sich für eine Begrenzung des Siedlungsgebietes unter Vorgabe gesetzlicher und planerischer Bemessungsgrundsätze für die Bauzonen aus. Diese werden relativ knapp für überschaubare Zeiträume dimensioniert. Dabei sind zusätzliche Massnahmen der Siedlungsqualität, der besseren Nutzung des Bestehenden und der Verdichtung angesagt. Das Nichtsiedlungsgebiet wird dominiert von der Landwirtschaftszone, ausserdem von Wald- und Schutzgebieten. Es erfüllt neben Ernährungs- landschaftserhaltende und ökologische Funktionen, dann aber ist es auch Raum für Freizeit, Tourismus, Infrastrukturen usw. Es ist mit einem (relativ) rigorosen Bauverbot belegt.

4. Innerhalb des Baugebietes gilt im Rahmen gesetzlicher und planerischer Vorgaben *Nutzungsfreiheit*, begrenzt durch bau-, planungs- und umweltrechtliche Vorgaben und Vorschriften. Zudem gilt *Verfügungsfreiheit* (Erwerbs- und Veräusserungsfreiheit), also Zuordnung des Bodens durch den Markt. Im Nichtsiedlungsgebiet herrscht bei beschränkter Nutzungsfreiheit ebenfalls das Marktprinzip vor, allerdings limitiert durch das bäuerliche Bodenrecht.

5. Die Mischung von rechtlichen und planerischen Massnahmen einerseits und marktwirtschaftlichen Prinzipien anderseits wurde im Verlauf der Zeit zusätzlich durch intensivierte und überdifferenzierte Regelungs- und Planungsanordnungen (Bauvorschriften, Bauzonen- Bebauungspläne usw.) bedrängt. Es ist aber nicht so, dass die geltende Ordnung einseitig und ausschliesslich hoheitlich angelegt wäre. Es gibt ex lege einen *Bodenmarkt* (besser gesagt, einen Bauland- und einen landwirt- / forstwirtschaftlichen Bodenmarkt, sodann Übergangs- und Sondermärkte, beispielsweise für das Bauerwartungsland, für Land in der Industriezone).

6. Der marktwirtschaftlichen Ordnung entzogen sind nach geltendem Recht in der Regel die *Nutzungsbeschränkungen* und, soweit vorhanden, die *Verfügungsrestriktionen*. Eine marktmässige Flexibilisierung der Steuerung der Siedlungsentwicklung hätte neben der Minimierung der Verfügungs- vor allem die Relativierung der hoheitlichen Nutzungsbeschränkungen einzuschliessen, dies bedeutet beispielsweise, dass die Nutzungsintensitäten samt Situierung vom Preis her beeinflusst würden.

7. Ob marktwirtschaftliche Instrumente der Steuerung der Siedlungsentwicklung im Baugebiet zur *Zielerfüllung* der Verdichtung, der besseren Nut-

zung des Bestehenden und zur Mehrung der Siedlungsqualität beitrügen, ist nicht hinreichend geklärt. Noch schwieriger abzuschätzen ist die Unterstützung der disparaten Zielerfüllungen im Nichtsiedlungsgebiet durch einen liberalisierten Bodenmarkt. Auf alle Fälle müssten die (gesetzlichen) Randbedingungen, wie sie marktmässig (über Angebot/Nachfrage samt Preisentfaltung) umgesetzt werden sollen, vorbedacht sein.

8. Die Expansion des Anwendungsbereiches marktwirtschaftlicher Instrumenten, u.a. bis hin zu handelbaren Nutzungszertifikaten, dispensiert jedenfalls nicht von der *Festlegung von Zielen*. Diese muss auf jeden Fall hoheitlich erfolgen, da der Markt in Kenntnis von Zielen zwar Beiträge an die Zielerfüllung leisten kann, sich aber schwer tut, die Ziele, welche die Interessenwahrung der am Marktgeschehen Beteiligten übersteigen, zu bestimmen.

9. Zu klären ist insbesondere, ob die *Grenzziehung zwischen Baugebiet und Nicht-Baugebiet resp. zwischen Siedlungs- und Nichtsiedlungsgebiet* weiterhin hoheitlich durch Gesetz oder Plan festgelegt werden muss oder ob darauf aus sachlichen Gründen oder aufgrund der Ausgestaltung und Leistungsfähigkeit der marktwirtschaftlichen Instrumente verzichtet werden kann. Ohne nähere Untersuchungen kann dazu keine Aussage gemacht werden. Zu vermuten ist hingegen, dass die Ziele, die ohnehin gesetzlich vorzugeben sind, räumlich abgestützt sein müssten.

10. So besehen ist – erstens – eine *Zieldiskussion* zu führen und – zweitens – zu prüfen, ob und in welchem Masse welche Instrumente zur optimalen *Zielerfüllung* beitragen. Dazu sollten Modelle und Testversuche entwickelt werden. Voraussetzung dafür ist eine Vordiskussion der im Nichtsiedlungsgebiet zu verfolgenden Ziele, da diese zurückgeblieben ist. Auch für das Baugebiet ist eine entsprechende Debatte angezeigt, wobei aber davon ausgegangen werden kann, dass das Wissen darüber wenigstens in Teilen, wenn auch vielfältig und bisweilen unausgereift, verfügbar ist.

11. Inhaltlich darf die Steuerung der Siedlungsentwicklung *nicht allein mit der Notwendigkeit einer rationalen, marktorientierten Flächenhaushaltpolitik* begründet werden. Es stehen politische, wirtschaftliche, soziale und ökologische Anforderungen mit im Spiel. Sodann ist sie im Kontext der räumlichen Entwicklung der Landschafts-, der Siedlungs- sowie der Transport- und Versorgungsstrukturen, der Siedlungs- und Landschaftsqualitäten und ganz allgemein der Anforderungen des *Prinzips der Nachhaltigkeit* mit der Akzentuierung der wirtschaftlichen Leistungsfähigkeit, der gesellschaftlichen Solidarität und des ökologischen Gleichgewichts zu sehen. Folglich ist eine raumordnungspolitische Ausrichtung und Einbettung oder Rückkoppelung angezeigt.

12. Im Hinblick auf die gesetzliche Ordnung ist es – mit oder ohne Vertiefung resp. Ausweitung der marktwirtschaftlichen Komponenten – unabdingbar, das Bau-, das Raumplanungs-, das Umweltrecht usw. auf der gleichen Staatsebene materiell abgestimmt zu regeln Auf alle Fälle muss der Bodenmarkt, wie immer er ausgestaltet und instrumentell begünstigt wird, mit dem Wirtschaftsraum räumlich synchron und sachlich abgestimmt einhergehen. Somit ist eine *eidgenössische Ordnung der marktbeeinflussenden Randbedingungen (mindestens den Grundzügen nach) unabdingbar.* Sie ist aber auch aus sachlichen Gründen nötig: *Die Einheit des Lebensraumes bedingt die Einheit der Rechtsordnung.*[78]

---

78 Siehe dazu *Lendi Martin,* Das Recht des Lebensraumes, in: Schluep Walter R. (Hrsg.), Recht, Staat und Politik am Ende des zweiten Jahrtausends, Festschrift für Bundesrat Arnold Koller zum 60. Geburtstag, Bern / Wien / Stuttgart 1993, S. 107 ff. *Idem*, Rechtsordnung, 3. A., Zürich 2001, S. 61 ff. Sodann *Hepperle Erwin / Lendi Martin*, Leben Raum Umwelt, Recht und Rechtspraxis, Zürich 1993.

Teil II
Beiträge zu den Entwicklungen in Litauen

E. K. Zavadskas, A. Kaklauskas, S. Raslanas

Land Taxation and Sustainable Urban Development

Introduction

For many years, taxation is the main source of revenue in every state keeping it alive and allowing its functions to be realised. In particular, the oldest history belongs to the property taxation, without which many states including Lithuania in today's world cannot manage.

The objective of this paper is to analyse the importance of real property tax within the taxation system in Lithuania and the plans how to fulfil the reform in the future. The alternative version of the reform is presented, including the rejection of the taxing of buildings and suggestion to levy land tax only on land. Land tax and the principles of its operation are analysed in theory. Additionally, the review is presented concerning the experience of foreign countries, which are already applying this tax (Australia, Denmark) and which are going to apply (Germany). Land tax is presented as one of the reform alternatives of the real property taxation system in Lithuania based on the experience of some foreign countries. The attention is also paid to the possible attitudes and impacts of this tax.

1. Impacts of land taxation

Land has a unique and substantial impact for the business system as a whole and for personal life as well. If we have in mind the entire mankind, land is of significant value, because it is the only place for all the nations and generations to live. Moreover, land is the fundamental and natural factor in any area of business, which takes part directly or indirectly in the production of all goods and services [1].

When taxes are levied on land or buildings, investment meets difficulties. Besides, the assessment of buildings that differ in their usage purpose, condition, life, etc. needs a lot of expenditures. Lithuania is a developing country, so it should strive to attract as much investors as it is possible, i. e. to promote investment. However, neither current real property taxation system nor currently proposed real property tax is possible to allow such effect [2, 3, 4]. The main idea of land taxation lies in the necessity to reject the link between land and building as well as levy a charge only on land value.

Land taxation will work in the most effective way only if this tax is paid by landowners. If it is possible to transfer this tax to users, tenants or price of the commodity, then owners are not encouraged to use the land more effectively [5]. The transfer of land tax proceeds when a landowner (a tax payer), for example, increases the rent and transfers the burden of taxation to a tenant, which becomes a "bearer" of a tax. In such a way the impacts of land tax decrease. If there are no possibilities to transfer this tax, the burden of taxation goes to the owner. Customer will calculate (capitalize) the tax into the price, which he proposes for the land plot. Consequently, the price will be lower. In a perfect case, reduction of land plot value corresponds to the amount of land tax.

Land taxation impact on territorial land planning

1. Municipalities have not enough recourse to cover the infrastructure costs of inhabited localities, so they cannot give land on gage. Land taxation can improve the financing of the measures of urban development, because this tax is taken starting from the beginning of the construction's „maturity" moment, irrespective of the constructional progress of private investors (Fig. 1). Thus the tax is usually received earlier, income becomes more reliable and the preparation costs of land intended for construction are directly attached to those who has benefit from it.
2. Often the decisions regarding territorial planning are distorted due to some personal interests. Land taxation can increase the neutrality of planning. If land tax will be introduced, it will be capitalised and prices for land will be on the decrease. Consequently, due to planning the increment of value will decrease as well as income of those owners who keep land plots only for future increment of value. Thus, planning should become impartial because the pretext to influence venial decisions of territorial planning will be diminished.

Land taxation impact on usage of land for building-up

1. Land, which is intended to be built up very often is not used at all or it is used not at its best, since costs of uneconomical usage are invisible. At this case, land taxation operates positively, because if it is paid constantly, costs of uneconomical land usage become visible. In particular, it is applied to those owners who do not use their land plots optimally due to ignorance or some other reasons. Land usage will improve, because land tax is being paid considering land value as well as possible income in the same locality.
2. Unused land often is considered only as capital investment. Land taxation can stimulate the owners to behave otherwise, because this fee decreases income from unexploited or uneconomically used land plot. Those land

plots, which are not on built up or used not optimally, bear major burden of taxation. Consequently, possession of a land plot just hoping for the future increase of its value should be less profitable. Besides, if evaluation of land is carried out regularly, increments of values are taxed [5].

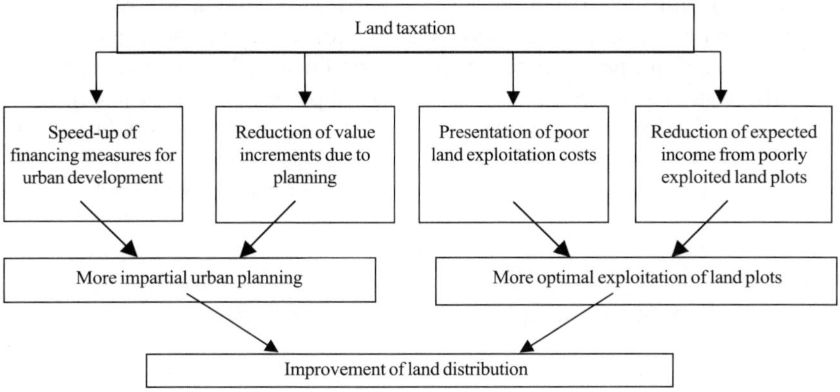

Fig. 1. How does land taxation works

Taxation techniques and fiscal impacts

1. Improvement of the correctness of evaluation. The current assessment has so many defects that land taxation does not experience any criticism in this area. Land appraisal presents results that are more precise what we cannot say about wholesale evaluation of buildings.
2. Reduction of evaluation costs. As it has been proven, evaluation costs of land tax are definitely lesser to compare with appraisal of real property tax values.

Influence of the tax over land prices

1. The attention should be paid to the impacts of land taxation, which depends on its tariff. The bigger rate the more reaction is given to the fee. Land taxation increases supply of land for building-up, because the biggest part of owners tries to sell or better use their unexploited land plots. True, the effect of mobilisation will be valid until unexploited or poorly exploited land plots will exist as well as newly established land for building. Pricing will be influenced only by impact of this effect on land demand. Due to increased exploitation costs land tax should diminish the demand for land intended to be built up. Thus, this fee will improve land

supply and decrease land demand, so lesser prices could be expected. Land prices will not increase until the municipalities provide land for development according to demand. However, if local authorities will take into consideration land tax and reduce the amount of land for urban development, then land supply decreases and prices are higher.

2. Land taxation helps fighting against corruption with land. While these problems are not finally solved by this fee, it cushions them reasonably. Partially „skimming" growth of values, practically there is no motive for corruption to receive benefit from change of prices. Land taxation ensures that infrastructural costs will be attached to their users and supports proper urban planning as well. Enduring capital investment regarding land will lose its attractiveness and because of that, speculation with land will be reduced.

Urban development aspects

1. Encouragement of the most effective economical exploitation. Land taxation will encourage landowners to take advantage of possibilities, which are legitimated by planning and realise the best and most effective economical exploitation.
2. Mobilisation of land for development, renewal of old industrial localities. Unbuilt land plots should be taxed with land tax much more than with real property tax. Consequently, it would be strongly recommended to build up newly established land plots as well as unbuilt ones as soon as possible. This effect of land taxation will be invalid or poorly valid in the old industrial localities, which are often built up with out-of-date buildings and constructions.
3. Consolidation of urban development. After establishment of land tax, the ultimate consolidation for building-up will be defined the same way as it is appointed now. This fee will encourage the realisation of possible building-up.
4. Preservation and restoration. Land taxation will also stimulate preservation and restoration of current buildings and constructions in order to ensure enduring potential profit. Of course, it could happen that some creditable but economically profitless buildings would be demolished and replaced by new ones.
5. Relationship between the city and the land surrounding it. Land taxation causes more intensive land exploitation, which should diminish supply of land designed for new constructions. Meanwhile the growth of real property tax standard indeed increases the urban space, which leads to urban development. Replacement of real property tax with land tax will influence reduction of urban space [6]. In particular, the demand for extensively used land for building-up will be reduced. Those owners of land plots

in valuable urban locations receiving low profit from land plot item will be encouraged to move into less prized places. Therefore, the demand of land for urban development in suburbs can be increased.

Economical aspects

Unlike other taxes, land tax will not be „placed“ on household income and profit. On the contrary, both buildings and investment will remain untaxed. Land taxation will also improve distribution of production factor and land as well. Moreover, investment will be supported and intensity of capital for land exploitation will be increased and it will have a great impact on the entire economics.

Social aspects

1. As it was mentioned before, land taxation will increase the supply of land for building-up as well as reduce its demand. Thus, it will be easier to obtain some property. Investors, who will appear in the land market after introduction of land tax and obtain land plot, will not taste the new tax, because the prices for land will not rise.
2. Land taxation will decrease burden of taxation of those land plots, which are already effectively in use. Land plots for rent are just as was mentioned before, so tenants will not experience the burden of taxation. Besides, even this load of taxation is increased, prices for rent practically remain the same. Burden of taxation will fall onto the person who is an owner, because the tenant will not be ready to pay a higher price caused by the tax.

Ecological aspects

Land taxation supports an intensive land use as well as the objectives of environmental politics such as reduction of land area for settlement. Extensive use of land for building-up with settlement and business purposes will be much more expensive. Due to land taxation it makes sense economically to reduce the area of usable land, for example, by erecting higher buildings. Consequently, the nature and landscape will be more appreciated. After consolidation of construction, the consumption of energy and roads will be reduced.

2. Experience of land taxation in foreign countries

Australia

Today in Australia land tax is levied by distinct states and local authorities. In 1995-1996 about 1.843 mln. dollars were received and it makes 2,4% of the common income of the distinct states. Real property tax in local authorities are weighty resources of taxes. For example, in 1995-1996 m. about 5.010 mln. dollars were received from these taxes, i.e. it makes 31% of common income of municipalities [5, 7].

Net land value is the basis for taxing in all the states. Besides, all the states foresee big untaxed values. Since the value of the average residentiary land plot in Australia is lesser than untaxed value, so those land plots remain almost untaxed. In many states land tax is progressive, i.e. it grows together with the land value. There are some marginal and average rates of land tax for different land values presented in Table 1.

Table 1. Comparison of land taxation rates in different states of Australia

State	Victoria	New South Wells	South Australia	Queensland	West Australia	Tasmania
The lowest limit of tax rate	0,20 %	1,50 %	0,35 %	0,20 %	0,35 %	0,75 %
The highest limit of tax rate	3,60 %	1,50 %	1,90 %	2,10 %	2,00 %	2,50 %
Average tax rate for land values (Aus. Dol.)						
200 000	0,08 %	0,35 %	0,21 %	1,10 %	1,47 %	1,53 %
500 000	0,48 %	1,04 %	0,75 %	1,34 %	1,79 %	1,79 %
1 000 000	1,14 %	1,27 %	1,13 %	1,69 %	1,89 %	2,16 %
5 000 000	2,69 %	1,45 %	1,75 %	1,47 %	1,98 %	2,43 %
10 000 000	3,14 %	1,48 %	1,82 %	1,79 %	1,99 %	2,46 %

Local land taxes differ much more than taxes of distinct states. The states establish laws and the particular taxation systems are formed out of them. Not all land plots are taxed with local taxes as well as with state taxes. Land for general use and for social purposes is exempted from taxation under the same conditions as in the states.

There were many different investigations carried out regarding the impacts of land tax in Australia, which stated the positive impacts of land taxation on land market as follows:

The amount of new constructions in local authorities with land taxation is much bigger than in the rest of local authorities. It is obvious, that this tax stimulates faster and consolidated building-up. In local authorities with relatively high rates of land taxation, land prices grow the most slowly than in other municipalities, because these rates reduce land prices and stimulate investment accordingly.

Denmark

Net land value is the basis for evaluation of land taxation. Local authorities can define the taxes by themselves within the limits from 0,6 to 2,4%, which are indicated by the State. The larger administrative units have the limit of 1 %, which is also established by the State. Danish land tax has no discounts or special rates of taxation for different uses. However, there are some land plots that are definitely exempted from land tax. Population in Denmark is very well informed about the value of land plots. The State even produces special brochures and explains how the evaluation is carried out. Besides, everyone has the possibility to learn about the ownership of any land plot and how much does it costs. Perhaps due to such informative way designed to society, there are very seldom complaints regarding taxing in Denmark [1, 8, 9].

High enough land tax in Denmark (2,4 %) has a significant impact on land use. New land plots designed for building-up are built up very quickly. There are no much unbuilt land plots in urban territories; speculation with land is not popular. Besides, it is estimated that land taxation in Denmark reasonably reduces land prices.

3. Real property taxation in Lithuania

Real property taxation in Lithuania involves land tax and real property tax of enterprises and organisations. Legal persons who own some real property in the Republic of Lithuania under the right of ownership are taxpayers of the real property tax. The object of the tax is the real property in the territory of the Republic of Lithuania to which the owner (legal entity) of the property has ownership rights excluding land, aircraft and ships. The rate of the tax makes 1% of the taxation value of real property.

The owners of private land plots are taxpayers of land tax. The object of the tax is private land. Annual rate of the tax makes 1,5 % of the land price (forestland price excluding the value of trees).

Land tax is based on land (water body) value, which is estimated according to the Resolution of the Republic of Lithuania Government 24 February 1999 NR. 205 "On approval of procedure of land appraisal" [10]. In 2003, computation of land value has changed. Today land value is estimated according to the maps of land values issued by State enterprise „The Centre of Registers". According to these maps, the estimated land value is close to the market value of private land plots for selling.

Annually the comparative part of revenue from real property taxes within national budget slightly increases: in 1996 it made 1,8 %, and in 2002 – 2,3 % of national revenue (Fig. 2). It should be stressed, that the national budget part, which was made by the real property tax of enterprises and organisations in 2002 was much as third to compare with 1996. The real property tax of enterprises and organisations was double in the period of 1996-2002. The comparative part of land tax within the national budget remained the same in 1996-2002: it was 0,2 % in 1996 as well as in 2002. Meanwhile, revenue from this tax had increased from 11613 thousand Litas in 1996 to 25400 thousand Litas in 2002, i. e. even 219 % [11].

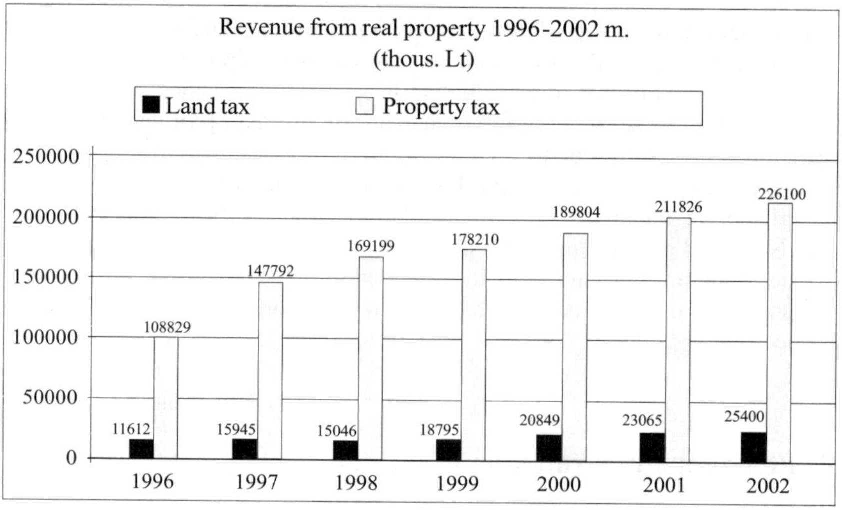

Fig. 2. Revenue from real property taxation

The comparative part of real property taxation in the budgets of municipalities is much bigger than in the national budget. Real property taxation is the second resource in size of taxing income for the municipalities. For the period of 1996-2000, the comparative part of these taxes in the budgets of municipalities had increased half as much again, i.e. from 4,2% in 1996 m. to 6,4% in 2000.

While Lithuania is in process of integration into the European Union, the taxation system should experience the reform accordingly. Therefore, real property taxation will be the subject of the reform as well. The project concerning the Law on real property taxation is being prepared, which should cancel both the Law on real property tax of enterprises and organisations as well as the Law on land tax by putting them together and adding some new material [4]. The rate of land tax should be from 0,5 percent to 1,5 percent of the taxing value of real property. The particular rate of the tax shall be established by municipality board. This project is being rated as controversial. It is suggested to reject the implementation of this project or even retract the current tax application for land as well as real property of enterprises [12].

4. Peculiarities of land tax application

During the transitive process of land tax application when real property tax is replaced by land tax, land plots with lower part of land value (intensively used, densely built-up land plots) should carry a lesser burden of taxation than now because buildings should be tax-free and receive some profit. On the contrary, a burden of taxation for those land plots, which has a high part of land value, will increase (Fig. 3). The bigger part of real property value, the bigger reduction of the burden. Today low real property taxes are applied to little built-up land plots or inefficiently exploited ones, so land taxation will tax them heavily. After the reform, burden of real property tax, which has equal parts of land and building values will remain the same [5].

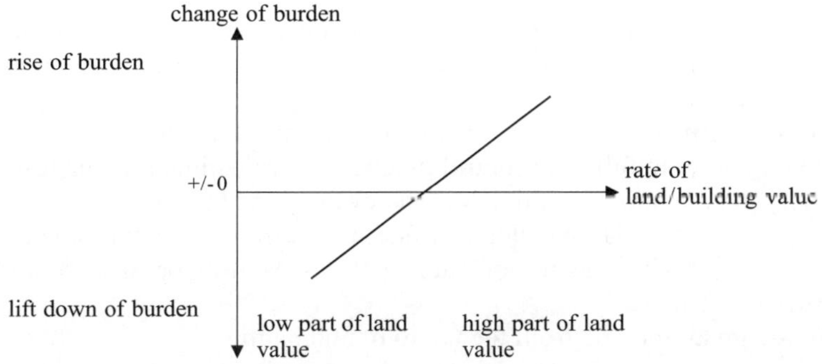

Fig. 3. Transfer of burden of taxation during the process of land tax establishment

With reference to the experience of foreign countries applying land tax, it is suggested for Lithuania:

1. To establish rates of the taxes keeping municipalities current income, which they receive from real property taxation. In order to reach this, it is suggested to establish the rate of the tax according to the current rate of income from real property tax and land values:

$$\frac{\text{Real property tax + land value tax}}{\text{Land value}} = \text{Rate of the land tax}$$

For example, after the calculation for the whole city (total urban income received from real property to divide by overall value of urban land), the rate of land tax is derivable. If to tax all land values with this tax, taxing income should remain the same as today (neutral rate of land tax with regard to income).

2. In order to avoid a significant increase of burden of taxation for land plots designed for living and a significant reduction of the burden for land plots owned by enterprises and organisations, it is suggested to apply different rates of the tax for land used by enterprises and organisations (higher rates) and for land designed for living (lower rates).
3. To establish the highest limits of the taxation rate to a point of government in order to eliminate possible misuse of municipalities.
4. To allow municipalities themselves to choose the rates of taxation within the limits foreseen.
5. Do not apply non-taxable minimum because it should load and make more expensive the structure of taxation system and equality of people rights might suffer.
6. Do not tax a land, which is indicated in the project of the Law on real property tax, excluding land, which belongs to failed enterprises.
7. Exempt from land tax shall be landowners that are group I and group II invalids, old-age pensioners (starting from 65) until land purchase or the case of land inheritance.
8. Increase of the tax should be divided for several years (3-5 years) in case the burden of taxation will increase more than twice during the period of tax establishment.
9. To use the maps of land values for land appraisal, i.e. wholesale evaluation. In order to fulfil the control function of evaluation, it is suggested to apply the method of multiple criteria evaluation [11].
10. In order to avoid large gallop of values between evaluation moments and increasing taxing income of municipalities, land reappraisal should be carried out annually.
11. To assign all income from the tax to municipalities.

5. A method of multiple criteria complex proportional evaluation of the land

This method assumes direct and proportional dependence of significance and priority of investigated versions on a system of criteria adequately describing the alternatives and on values and significances of the criteria [11]. The system of criteria is determined and the values and initial significances of criteria are calculated by experts. All this information can be corrected by interested parties (customer, users, etc.) taking into consideration their pursued goals and existing capabilities. Hence, the assessment results of alternatives fully reflect the initial data jointly submitted by experts and interested parties.

The determination of significance and priority of alternatives is carried out in four stages.

Stage 1. The weighted normalized decision making matrix D is formed. The purpose of this stage is to receive dimensionless weighted values from the comparative indexes. When the dimensionless values of the indexes are known, all criteria, originally having different dimensions, can be compared. The following formula is used for this purpose:

$$d_{ij} = \frac{x_{ij} \cdot q_i}{\sum_{j=1}^{n} x_{ij}}, \quad i = \overline{1,m}; \quad j = \overline{1,n}. \tag{1}$$

where x_{ij} - the value of the i-th criterion in the j-th alternative of a solution; m – the number of criteria; n – the number of the alternatives compared; q_i – significance of i-th criterion.

The sum of dimensionless weighted index values d_{ij} of each criterion x_i is always equal to the significance q_i of this criterion:

$$q_i = \sum_{j=1}^{n} d_{ij}, \quad i = \overline{1,m}; j = \overline{1, n}. \tag{2}$$

In other words, the value of significance q_i of the investigated criterion is proportionally distributed among all alternative versions a_j according to their values x_{ij}.

Stage 2. The sums of weighted normalized indexes describing the j-th version are calculated. The versions are described by minimizing indexes S_{-j} and maximizing indexes S_{+j}. The lower value of minimizing indexes is better (price of the plot). The greater value of maximizing indexes is better. The sums are calculated according to the formula:

$$S_{+j} = \sum_{i=1}^{m} d_{+ij}; \quad S_{-j} = \sum_{i=1}^{m} d_{-ij}, \quad i = \overline{1,m}; j = \overline{1, n}. \tag{3}$$

In this case, the values S_{+j} (the greater is this value, project „pluses“, the more satisfied are the interested parties) and S_{-j} (the lower is this value ,project “minuses”, the better is the goal attainment by the interested parties) express the degree of goals attained by the interested parties in each alternative project. In any case the sums of “pluses” S_{+j} and “minuses” S_{-j} of all alternative projects are always respectively equal to all sums of significances of maximizing and minimizing criteria:

$$S_{+}=\sum_{j=1}^{n} S_{+j}=\sum_{i=1}^{m}\sum_{j=1}^{n} d_{+ij},$$
$$S_{-}=\sum_{j=1}^{n} S_{-j}=\sum_{i=1}^{m}\sum_{j=1}^{n} d_{-ij}, \quad i=\overline{1,m};\ j=\overline{1,n}. \tag{4}$$

In this way, the calculations made may be additionally checked.

Stage 3. The significance (efficiency) of comparative versions is determined on the basis of describing positive projects ("pluses") and negative projects ("minuses") characteristics. Relative significance Q_j of each project a_j is found according to the formula:

$$Q_j=S_{+j}+\frac{S_{-\min}\cdot\sum_{j=1}^{n} S_{-j}}{S_{-j}\cdot\sum_{j=1}^{n}\frac{S_{-\min}}{S_{-j}}}, \quad j=\overline{1,n}. \tag{5}$$

Stage 4. Priority determination of projects. The greater is the Q_j the higher is the efficiency (priority) of the project.

The analysis of the method presented makes it possible to the State that it may be easily applied to evaluating the projects and selecting most efficient of them, being fully aware of a physical meaning of the process. Moreover, it allowed to formulate a reduced criterion Q_j which is directly proportional to the relative effect of the compared criteria values x_{ij} and significances q_i on the end result.

Significance Q_j of the land a_j indicates satisfaction degree of demands and goals pursued by the interested parties – the greater is the Q_j the higher is the efficiency of the land. In this case, the significance Q_{max} of the most rational land will always be the highest. The significances of all remaining land are lower as compared with the most rational one. This means that total demands and goals of interested parties will be satisfied to a smaller extent than it would be in case of the best land.

The degree of land utility is directly associated with quantitative and conceptual information related to it. If one land is characterized by the best comfortability, aesthetics, price indices, while the other shows better maintenance and facilities management characteristics, both having obtained the same significance values as a result of multiple criteria evaluation, this means that their utility degree is also the same. With the increase (decrease) of the significance of the land analyzed, its degree of utility also increases (decreases). The degree of land utility is determined by comparing the land analysed with the most efficient land. In this case, all the utility degree values related to the land analyzed will be ranged from 0% to 100%. This will facilitate visual assessment of land efficiency.

The degrees of utility of the land considered as well as the market value of the land being valuated are determined in seven stages.

Stage 1. The formula used for the calculation of the land a_j utility degree N_j is given below:

$$N_j = \left(Q_j : Q_{max}\right) \cdot 100\%. \qquad (6)$$

here Q_j and Q_{max} are the significances of the land.

The degree of utility N_j of the land a_j indicates the level of satisfying the needs of the parties interested in the land. The more goals are achieved and the more important they are, the higher is the degree of the land utility. Since clients are mostly interested in how much more efficient particular land are than the others (which ones can better satisfy their needs), then it is more advisable to use the concept of land utility rather than significance when choosing the most efficient solution.

A degree of land utility reflects the extent to which the goals pursued by the interested parties are attained. Therefore, it may be used as a basis for determining land market value. The more objectives are attained and the more significant they are the higher will be land degree of utility and its market value.

Thus, having determined in such a way the ratio of degree of utility and market value of land, one can see what complex effect can be obtained by investing money into anyone of the land. There is a complete clarity where it pays better to invest the money and what is the efficiency degree of the investment.

Stage 2. The efficiency degree E_{ji} of money invested into the land a_j is calculated. It shows by how many percent it is better (worse) to invest money into the land a_j compared with the land a_i. E_{ji} is obtained by comparing the degrees of utility of the land considered:

$$E_{ji} = N_j - N_i . \qquad (7)$$

The received results are presented as a matrix clearly showing utility differences of the land (see Table 2).

Stage 3. The average deviation k_j of the utility degree N_j of the land a_j from the same index of other land (n-1) is being calculated.

$$k_j = \sum_{i=1}^{n} E_{ji} : (n-1). \qquad (8)$$

Table 2. Calculation of average deviations of the land utility degrees

Land considered	Utility degree deviation of land analyzed compared to other land, %					Average deviation k_j of utility degree N_j of land a_j compared to other (n-1) land, %
	a_1	a_2	a_3	a_i	a_n	
a_1	0	E_{12}	E_{13}	...	E_{1n}	k_1
a_2	E_{21}	0	E_{23}	...	E_{2n}	k_2
a_3	E_{31}	E_{32}	0	...	E_{3n}	k_3
...	...	...	...	...	...	...
a_j	E_{j1}	E_{j2}	E_{j3}	...	E_{jn}	k_j
...	...	...	...	...	...	...
a_n	E_{n1}	E_{n2}	E_{n3}	...	0	k_n

Stage 4. The development of a grouped decision making matrix for land multiple criteria analysis. The market value of land being valuated is calculated according to a block-diagram presented in Fig.4. At the beginning, a grouped decision making matrix for land multiple criteria analysis is developed (see Table 3), the first criterion of which is based on the actual purchasing/selling prices of the land compared and the value of land being valuated.

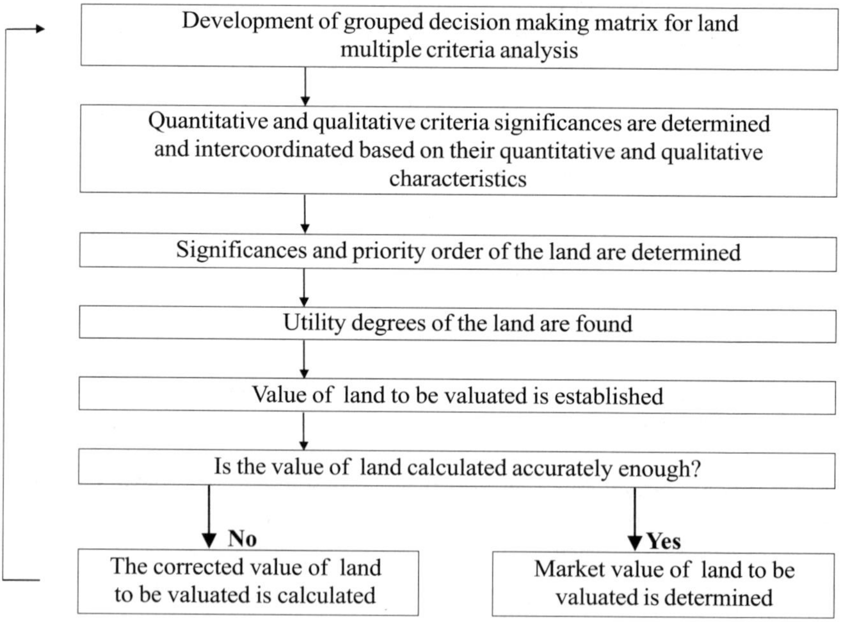

Fig. 4. Block-diagram of land market value estimation

The initial value of the land being valuated is obtained from the following equation:

$$x_{11} = \sum_{j=2}^{n} x_{1j} : (n-1). \qquad (9)$$

In this matrix, the land a_1 to be valuated should be assigned the market value ($x_{11\text{-}R}$). Other comparison standard land (a_2-a_n) were sold, their purchasing/ selling prices (x_{12}-x_{1n}) known. All the values and significances of the criteria relating to other land are also known (see Table 3).

Table 3. A grouped decision making matrix for land multiple criteria analysis

Criteria describing the compared land	*	Significance	Measuring units	Land to be valuated and comparison standard land						
				a_1	a_2	...	a_j	...	a_n
Price of the land a_1 being valuated and actual purchasing/selling prices of comparison standard land ($a_2 - a_n$)	z_1	q_1	m_1	x_{11}	x_{12}	...	x_{1j}	...	x_{1n}
	z_2	q_2	m_2	x_{21}	x_{22}	...	x_{2j}	...	x_{2n}
	...	...	...	...	...	...	...	...	...
Quantitative criteria	z_i	q_i	m_i	x_{i1}	x_{i2}	...	x_{ij}	...	x_{in}
	...	...	...	...	...	...	...	...	...
	z_t	q_t	m_t	x_{t1}	x_{t2}	...	x_{tj}	...	x_{tn}
	z_{t+1}	q_{t+1}	m_{t+1}	$x_{t+1\,1}$	$x_{t+1\,2}$	...	$x_{t+1\,j}$	...	$x_{t+1\,n}$
Qualitative criteria	z_{t+2}	q_{t+2}	m_{t+2}	$x_{t+2\,1}$	$x_{t+2\,2}$	...	$x_{t+2\,j}$	...	$x_{t+2\,n}$
	...	...	...	...	...	...	...	...	...
	z_i	q_i	m_i	x_{i1}	x_{i2}	...	x_{ij}	...	x_{in}
	...	...	...	...	...	...	...	...	...
	z_m	q_m	m_m	x_{m1}	x_{m2}	...	x_{mj}	...	x_{mn}

* – The sign z_i (+ (–)) indicates that a greater (less) criterion value corresponds to a greater significance for a client.

The problem may be stated as follows: what market value $x_{11\text{-}R}$ of a valuated the land a_1 will make it equally competitive on the market with comparison standard land (a_2-a_n)? This may be determined if a complex analysis of the benefits and drawbacks of the land is made. Using a grouped decision making matrix (see Table 3) and the equations 1-9 the calculations are made.

Stage 5. The corrected value $x_{11\text{-}p}$ of the land to be valuated a_1 is calculated:

$$x_{11\text{-}p} = x_{11} * (1 + k_1 : 100). \qquad (10)$$

Stage 6. It is determined whether the corrected value $x_{11\text{-}R}$ of the land being valuated a_1 had been calculated accurately enough:

$$|k_1| < s, \quad (11)$$

where s is the accuracy, %, to be achieved in calculating the market value $x_{11\text{-}p}$ of the land a_1. For example, given $s = 0{,}5\%$, the number of approximations in calculation will be lower than at $s = 0{,}1\%$.

Stage 7. The market value $x_{11\text{-}R}$ of the land a_1 to be valuated is determined. If inequality 11 is satisfied the market value of the land a_1 may be found as follows:

$$x_{11\text{-}R} = x_{11\text{-}p} \quad (12)$$

If inequality 12 is not satisfied this means that the value of the land being valuated had not been calculated accurately enough and the approximation cycle should be repeated. In this case, the corrected value $x_{11} = x_{11\text{-}p}$ of the land being valuated is substituted into a grouped decision making matrix of land multiple criteria analysis and the calculations according to the formulae 1-12 should be repeated until the inequation 11 is satisfied.

Solving the problem of determining the market value $x_{11\text{-}R}$ of the land a_1 being valuated, which would make it equally competitive on the market compared with the land (a_2 - a_n) already sold, a particular method of defining the utility degree and market value of the land was suggested. This was based on a complex analysis of all the benefits and drawbacks of the land considered. According to this method the land utility degree and the market value of the land being estimated are directly proportional to the system of the criteria adequately describing them and the values and significances of these criteria.

Conclusions

The current taxation system in Lithuania is being constantly improved in order to adjust it to the requirements of the European Union. Seeking to reform the real property taxation, the project of the law on real property tax is prepared, which is being rated as controversial. There are some doubts about its relevance in reformation of the taxation system of real property. Besides, the authority delays to confirm it.

The authors propose another variant of the reform. It is suggested to refuse of building taxation and levy land tax only on land. Such a system of real property taxation works successfully in some states of Australia, in Denmark, Canada, and Estonia. Moreover, Germany is going to establish this tax, too.

After careful analysis of current and future impacts of land tax in foreign countries, we can say, that these results should be received after introduction of this tax in Lithuania:

1. Evaluation costs will be reduced substantially.
2. Investment for buildings should be stimulated.
3. Land reform will go much faster and the speculation with land will be reduced
4. Land would be build-up more densely and faster.
5. It would be encouraged to renovate or demolish and rebuild shabby and unkept buildings.
6. Territorial planning and its impartiality would be better.
7. Land would be spared and used more effectively.
8. Increase of roads would be slower as well as energy consumption. Nature and landscape would be better preserved.
9. Growth of land prices should be lower.

References

1. Dieterich H., Dieterich B. Boden – Wem nutzt er? Wen stützt er? Wiesbaden: Lengericher Handelsdruckerei, 1997. 325 S.
2. Lietuvos Respublikos įmonių ir organizacijų nekilnojamojo turto mokesčio įstatymas. Valstybės žinios, 1994, Nr. 59-1156.
3. Lietuvos Respublikos žemės mokesčio įstatymas. Valstybės žinios, 1992, Nr. 21-612.
4. Lietuvos Respublikos nekilnojamojo turto mokesčio įstatymo projektas. Vilnius, 2002.
5. Josten R. Bodenwertsteuer – eine praxisorientierte Untersuchung zur Reform der Grundsteuer. Stuttgart: Verlag W. Kohlhammer Gmbh, 2000. 290 S.
6. Geoffrey A. Forster. Real estate (land) taxation – Australia. American Journal of Economics & Sociology, Nov. 2000, Vol. 59 Issue 5, p. 399, 18p, 2bw.
7. Ole Lefmann; Karsten K. Larsen. Land taxation – *Denmark.* American Journal of Economics & Sociology, Nov. 2000, Vol. 59 Issue 5, p. 185, 20 p.
8. Lietuvos Respublikos Vyriausybės 1999 02 24 nutarimas Nr. 205 "Dėl žemės įvertinimo tvarkos". Valstybės žinios, 1999, Nr. 21-597.
9. Lietuvos ekonominė ir socialinė raida 2003/3. Vilnius: Statistikos departamentas, 2003. 130 p.
10. Vainienė R. Dėl nekilnojamo turto mokesčio taikymo tikslingumo. 1999. http://www.lrinka.lt/Tyrimai/NTurtas/Nekur.phtml
11. E. K. Zavadskas, A. Kaklauskas, N. Banaitienė. Pastato gyvavimo proceso daugiakriterinė analizė. Vilnius: Technika, 2001. 380 p.

Summary

For many years, taxation is the main source of revenue in every state keeping it alive and allowing its functions to be realised. In particular, the oldest history belongs to the property taxation. In this paper, the importance of the real property tax within the taxation system is analysed as well as general reform of property taxation in Lithuania. Another version of this system is suggested here when the taxing of buildings is rejected and land is only taxed with the land taxation. Land tax and the principles of its operation are analysed in theory. Additionally, the review is presented concerning the experience of foreign countries, which are already applying this tax (Australia, Denmark) and which are going to apply (Germany).

VIDA MALIENĖ

Die Förderung des Umweltschutzes durch die Bodenordnung in Litauen

Einführung

Die geopolitischen und historischen Bedingungen haben es mit sich gebracht, dass bis heute die wirtschaftliche und soziale Bedeutung des Bodens und der Landwirtschaft in Litauen größer sind als in den Nachbarländern und in den Mitglieds-Staaten der Europäischen Union. Auf dem Lande leben 32% der Bevölkerung; in den Staaten der Europäischen Union sind es etwa 17,5%. Die mittlere Dichte der Landbevölkerung in Litauen beträgt nur 18,2 Menschen pro km^2 und ist damit kleiner als in den westeuropäischen Staaten. In der Landwirtschaft und in der Forstwirtschaft sind etwa 22% aller Erwerbstätigen des Landes beschäftigt; in den Staaten der Europäischen Union sind es etwa 5%.

Das litauische Dorf ist ein wichtiger Produzent des Bruttoinlandprodukts, der nationalen Wertschöpfung. Die Landwirtschaft trägt mehr als 10% des Bruttoinlandsproduktes bei [1]. Die Erzeugung der gesamten landwirtschaftlichen Produktion steigt beständig. Das Dorf erfüllt außer einer ökonomischen Funktion auch wichtige soziale, ethnokulturelle, Landschafts- und Naturschutzfunktionen.

Die neuesten Hauptrichtungen der Wirtschaftsentwicklung des Landes sind im Gesetze der Entwicklung der Landwirtschaft und des Landes der Litauischen Republik [2], in den Richtlinien des Gesamtplans des Landes [3] sowie in anderen Dokumenten vorgesehen. Bei der Durchführung der Staatspolitik in der Landwirtschaft und bei der Sicherung der rationalen Nutzung des Bodens werden folgende Ziele vorgesehen:

- Die Schaffung des fruchtbaren, dynamischen, auf den Markt gerichteten, konkurrenzfähigen Sektors der Landwirtschaft;
- Die Schaffung eines effektiven Informationssystems zur Bedienung des landwirtschaftlichen Sektors;
- Die Schaffung der rechtlichen und organisatorischen Bedingungen zur Strukturierung der Bodennutzungen der rationellen Größe und Konfiguration und zur Konsolidierung der separaten Bodenbesitze;
- Umweltverbesserung durch Anwendung agro-umweltschützender Maßnahmen und durch Förderung der Entwicklung der ökologischen Landwirtschaft u. s. w.

Bei der Verwirklichung der vorgesehenen Ziele kann die Regierung der Republik Litauen oder eine von ihr beauftragte Institution folgende Maßnahmen der Marktregulierung anwenden: den Ziel- bzw. den Interventionspreis festsetzen, die Erzeugung begrenzen und Maßnahmen zur Intervention, zur Hilfe beim Verbrauch und zur Regelung der Einfuhr bzw. Ausfuhr zum Schutz der agrarischen Umwelt treffen [2].

Im Gesetz zur Entwicklung der Landwirtschaft und des Landes oder Bodens sind für das Ministerium für Landwirtschaft, das die rationelle Bodennutzung, den Schutz der Böden und der Umwelt sichern muss, folgende Aufgaben vorgesehen:

- Die Erarbeitung und Bestätigung der Regeln zur Vorbereitung der Projekte der Flurbereinigung;
- Die Bekanntgabe der Maßnahmen zur Erfassung der landwirtschaftlichen Nutzflächen und zur Einschätzung ihrer Bodeneigenschaften und Ertragsfähigkeit;
- Die Festlegung der Verordnung zur Ermittlung der Angaben über die Benutzung industriell oder agrarisch genutzter Böden;
- Bestimmung der Organisation zur Ordnung der staatlichen Kontrolle über die Bodennutzung;
- Die Erarbeitung der Programme des agrarischen Umweltschutzes, der Regeln und Ratschläge progressiver Bewirtschaftung, die Beabsichtigung der Ordnung ihrer Durchsetzung u. s. w. [2].

Nach dem Gesetz muß die Landentwicklung anhand des Dorfentwicklungsplanes, der von dem Ministerium für Landwirtschaft genehmigt wird, durchgeführt werden. Dieser Plan wird mit den Unterlagen der Raumplanung – Gesamtplänen oder Bodenordnungsschemen – vereinbart.

Das Hauptziel des Landentwicklungsplanes ist die Schaffung einer günstigen Umgebung für die wirtschaftliche, soziale, naturschonende und ethnokulturelle Dorfentwicklung. Dies sind Voraussetzungen für eine konsequente Erhöhung der Lebensqualität bei der ländlichen Bevölkerung. Nach der Verwirklichung der strategischen Ziele und Aufgaben des Programmes für die Dorfentwicklung werden folgende Ergebnisse erwartet:

- Rationellere Struktur der Bodennutzung und des Agrarunternehmens;
- Die Entwicklung der neuen Unternehmenszweige und Wirtschaftssubjekte auf dem Lande;
- Eine kleinere Abhängigkeit der Bevölkerung von der Landwirtschaft, eine äquivalente Entwicklung der ländlichen Regionen;
- Eine größere Konkurrenzfähigkeit der Produktion der ländlichen Unternehmen und der Dienstleistungen;
- Die Verbesserung des ländlichen Umweltschutzes, die Wiederherstellung des wirtschaftlichen und ökologischen Gleichgewichtes u. a. m. [2].

Wie man sieht, wird in den von dem Staat vorgesehenen Richtlinien für die Entwicklung der Landwirtschaft festgesetzt, dass die Landwirtschaft als ein privilegiertes, durch den Staat subventioniertes Gebiet erhalten bleiben muss.

Es wird angestrebt, einen ertragsfähigen, dynamischen und konkurrenzfähigen landwirtschaftlichen Sektor und ein ihn bedienendes effektives Informationssystem zu schaffen. Dabei wird auch angestrebt, die Bedingungen für die Strukturierung der rationellen Bodennutzungen, für die Konsolidierung der Bodenbesitze, für Umweltschutz und für die rationelle Ausnutzung der Naturressourcen zu schaffen.

1. Bodenordnung als eine Massnahme zur Durchführung der Politik der Landentwicklung

Im Bodengesetz der Republik Litauen wird definiert, daß die Bodenordnung – die Festlegung und die Veränderung der Grenzen der Grundstücke, reglementiert durch rechtliche Akte, des Bestandes der Flächennutzungen, des Ortes der zugehörigen Objekte der Grundstücke, der Art und Weise der Bodennutzung – unter Berücksichtigung der wirtschaftlichen, umweltschützenden und anderer privaten und öffentlichen Interessen durchgeführt wird [4]. Die Grundsätze der Bodenordnung sind die Hauptregeln für die rationelle Nutzung und den Schutz der natürlichen Umwelt des natürlichen Bodens und des Eigentumsobjektes. Diese Regeln werden durch entsprechende methodische Anforderungen der Gesetze, der Beschlüsse der Regierung sowie der Projektierung der Bodenordnung zur Geltung gebracht. Nach den gegenwärtig in Litauen geltenden Gesetzen werden folgende Bodenordnungsgrundsätze unterschieden:

- Schutz des Bodeneigentums und der Nutzungsrechte;
- Die rationelle Nutzung der Naturressourcen (Naturressourcen werden dann rationell genutzt, wenn kein Schaden der Natur zugefügt wird. Nach diesem Prinzip ist eine solche Art der Bodennutzung zu planen, daß der Boden nicht abgetragen wird, und in Ackerländern die Humusmenge nicht kleiner wird);
- Die Gestaltung der Kulturlandschaft (Die Projekte der Bodenordnung geben das Recht für die Bodennutzung, die vorgesehenen Arbeiten der Gebietsordnung durchzuführen. Die Maßnahmen dürfen der Natur nicht schaden und müssen die Komponenten der Landschaft harmonisch abstimmen sowie das ökologische Gleichgewicht in der umgestalteten Landschaft erhalten);
- Schutz der landwirtschaftlichen Nutzflächen (Der Mutterboden formierte sich in tausenden Jahren und ist schwer wiederherzustellen, deshalb sind die landwirtschaftlichen Flächen zu schützen, damit diese Flächen nicht kleiner werden und sich die Eigenschaften des Bodens nicht verschlechtern);
- Schutz der natürlichen Biozönosen (Die natürlichen Biozönosen helfen zur Bewahrung der Landschaft, den natürlichen Beziehungen der Ökosysteme. Deshalb dürfen sie in andere Nutzungsflächen nur dann umge-

staltet werden, wenn nach gründlichen Untersuchungen festgelegt wird, daß nach dieser Umgestaltung der ökologische Zustand der Umwelt nicht schlechter wird);

– Die rationelle Organisation der wirtschaftlichen Tätigkeit des Bodennutzers (Die Bodenordnung hat die Bewirtschaftung dem Bodennutzer zu gewährleisten oder zu verbessern).

Die Grundsätze der Bodenordnung werden gemäß den Anforderungen der Gesetze und Beschlüsse in Bodenordnungsmassnahmen durchgeführt.

2. Die Veränderungen der Bodenordnung in Lita

Litauen, ein Agrarland, erlebte die Veränderungen des Bodenbesitzes, die den Wechsel der Bodenordnung und Nutzung beeinflußten, in der Vergangenheit immer sensibel.

Bei der gegenwärtigen Durchführung der Planung und Ordnung der Bodennutzung ist es sehr wichtig, wie sich die Bodennutzung in Litauen im Laufe der Geschichte änderte.

Erste Bodenordnungsarbeiten fanden in Litauen schon im 14.-15. Jahrhundert statt, als man die zugeteilten oder geschenkten Bodenbesitze am Ort zu vermessen begann, jedoch im ganzen Staat wurde die Bodenordnung im 16. Jahrhundert begonnen [5].

Man kann sieben Bodenreformen in Litauen unterscheiden:

1. Valakenreform. Im 16. Jahrhundert verkündigte der Grossfürst Litauens Žygimantas Augustas das Gesetz über Valakenreform; im Jahre 1557 erließ er die Regeln der Valaken [6]. Die zersplitterten kleinen Grundstücke wurden in drei große Fruchtfolgefelder, geteilt in Streifen, verbunden. So entstanden sogenannte Straßen – Streifendörfer. Bei der örtlichen Vermessung und Vermarkung der Grundbesitze wurde der benutzte Boden umgestaltet: Die Grenzen der Dörfer und der Grundstücke der Bodenbesitze wurden gerichtet, die für die Bauernfamilien bestimmten Bodenflächen gleich groß zugeteilt. Jede Bauernfamilie erhielt 20 ha, also 1 Valak des Bodens. Die Bauernfelder wurden möglichst rechteckig gemessen und diese Fläche in drei gleiche Felder geteilt. In der mittleren Flur wurde der Grund für die Bauten des Grundstükkes bestimmt. Der übrige Boden in den Fluren wurde nach Streifen vermessen. So geordneter Boden wurde nach Dreifeldfrucht wie folgt genutzt: Eine Flur wurde durch Winterfrucht (nach der Brache) besät, die zweite mit Sommergetreide und die dritte als Brachland belassen.

2. Die Bodenordnung in dem vom zaristischen Rußland besetzten Litauen (oder die Reform der Abschaffung der Leibeigenschaft). Im 19. Jahrhundert wurde die Flurbereinigung begonnen. Nach der Abschaffung der Leibeigen-

schaft im Jahre 1861, als die von der Leibeigenschaft befreiten Bauern das Recht zum Kauf des von ihnen benutzten, zugeteilten Bodens erhielten, wurden die Flurbereinigungsprozesse schneller. Während dieser Reform wurden die Grenzen der Dörfer geordnet und in den Karten vermerkt, die Nutzungsordnung des Bodens präzisiert [7]. Die Bauern, die während der Reform den Boden zum Privateigentum erwarben, wurden zu den vollberechtigen Besitzern. Jedoch gehörte der große Teil des Territoriums (etwa 40-45%) den Gutsbesitzern.

3. Die Reform von Stolypin. Im Jahre 1906 wurde der Erlaß des Zaren genehmigt, der die Verteilung der Dorfgrundstücke zu Einzelhöfen beschleunigte. Es wurden die Grunderwerbsteuern für den zugeteilten Boden abgeschafft. Dabei wurden in Gouvernements und Kreisen die Bodenordnungskommissionen gegründet, die die Dörfer in Einzelhöfe verteilten, die Zwischengrundstücke abschafften, die Gemarkungen der Bauern und Güter verteilten, die Grenzen der Bodenbesitze festlegten und erneuerten. In der Regel wurden im alten Dorf 5-6 Gehöfte gelassen, die übrigen richteten die Einzelhöfe auf der Fläche des eigenen Bodens ein [5].

Vor dem ersten Weltkrieg lebte auf dem Lande etwa 87% der Bevölkerung Litauens. Den Bauern gehörte mehr als die Häfte, den Gutsbesitzern 36%, den Behörden 10% des Bodens.

4. Die Bodenreform zwischen den Weltkriegen (Reform von M. Krupavicius). Diese Bodenreform fand in Litauen, nach der Erlangung der Unabhängigkeit, im Jahre 1922 statt. Am 3. April erließ Steigiamasis Seimas (Gründungsparlament) das Gesetz über die Bodenreform. Das Ziel der Reform war die Enteignung und Aufteilung des staatlichen Bodens und des Bodens der großen Güter an die Bauern sowie die Beendigung der Aufteilung der Dörfer in Einzelhöfe. Für jeden Gutsbesitzer verblieben 80 ha (später 150 ha) der Bodenfläche. Der Boden des staatlichen Bodenfonds wurde an litauische Freiwillige und an Bauern – Neusiedlern (je 8-20 ha), an Gutsarbeiter und Dorfhandwerker (je 1,5-2 ha), an Kleinbauern, Behörden, Forstwirtschaften und für andere Ziele verteilt. Die Aufteilung der Dörfer in Einzelhöfe stärkte die Familie des Landwirtes, vergrößerte ihre Bedeutung als eine ökonomische Einheit. In den Jahren 1919-1940 waren insgesamt 1443 Güter aufgeteilt, ihr Boden an 68.000 Bauern verteilt, 6.955 Dörfer in Einzelhöfe (etwa 159,1 Tausend Bodenbesitze) aufgeteilt und auf etwa 100.000 Grundstücken die Dienstbarkeiten abgeschafft [8].

5. Die sozialistische Bodenreform. Nach dem rechtswidrigen Anschluß Litauens zum Bestand der Sowjetunion wurde im Jahre 1940 wieder eine neue Reform begonnen. Mit dieser Reform wurde eine neue Agrarpolitik durchgeführt. Ihr Ziel war die Versorgung aller Landlosen und Kleinbauern mit dem Boden durch Enteignung des Bodens der bisherigen Eigentümer. Der Boden wurde zum Staatseigentum erklärt. Zum staatlichen Bodenfond wurde der Boden der Gutsbesitzer, der Großbauern (ihnen wurden je 30 ha gelassen),

der Kirchen, Klöster und staatlichen Organisationen genommen. Von der gesamten Fläche des enteigneten Bodens (608.000 ha) wurde an(etwa 72.500) Landlose und Kleinbauern 381.000 ha verteilt [5].

6. Die Bodenreform der Kollektivierung der Bauernwirtschaften. Diese Reform begann im Jahre 1949, als die Gründung der Kolchosen und Sowchosen (Staatsgüter) ihren Anfang nahm. Die Kollektivierung war 1952 beendet. In dieser Periode wurden Geldmittel zur Erhöhung der Bodenertrages eingesetzt. In großen Territorien des Landes wurden die Meliorationsarbeiten durchgeführt, die Felder der Fruchtfolgen vergrößert und geordnet, die Weideflächen zum Kulturland gebracht. Die Produktionskapazitäten auf dem Lande sind auch gestiegen. Jedoch bedingte in ökologisch sensiblen Territorien die Vergrößerung der bebauten Flächen die Erosionsprozesse der Böden. Während der Periode der Errichtung und Entwicklung der großen Wirtschaften haben die Landmesser die Grenzen der Bodennutzungen vermessen und vermarkt, das kartographische Material, Bodenordnungsschemen und Projekte erstellt und die Untersuchungen der Böden durchgeführt.

7. Die Bodenreform nach der Wiederherstellung der Unabhängigkeit. Nach der Wiederherstellung der Unabhängigkeit Litauens wurde die bis heute dauernde siebente Bodenreform begonnen, die als die Restitution des Bodeneigentums genannt werden kann. Ihr Ziel ist die Wiederherstellung des Eigentumrechtes auf den Boden für die ehemaligen Bodeneigentümer oder ihre Kinder bzw. Enkelkinder. Die Arbeiten der Bodenreform werden seit dem Jahre 1992 durchgeführt. Nach der Meinung verschiedener Gutachter wird durch diese Bodenreform keine rationelle Transformation der Bodennutzungen in ökonomisch effektivere Strukturen erlangt, sondern nur die mosaikartige Bodennutzung der privaten Grundstücke geschaffen: die Grundstücksgrösse beträgt etwa 0,5-2 ha in den Siedlungen und etwa 6-15 ha in den übrigen Gebieten. Nach dem Ende dieser Reform wird eine neue Etappe der Bodenreform – die Formierung der größeren und großen Wirtschaften beginnen. Die neue Landwirtschaft soll sich am Markt orientieren und eine effektiv arbeitende Wirtschaft der großen und spezialisierten Produktion sein.

3. Die künftigen Aufgaben der Bodenordnung

In dem letzten Jahrzehnt, während der Zeit der Rückgabe des Bodeneigentums und der Bodenprivatisierung, gestaltete sich in Litauen eine sehr unzweckmäßige Bodennutzung, die sich durch kleine und ungünstig gelegene Grundstücke auszeichnet. Dies führt zu einer uneffektiven Bewirtschaftung und erschwert den Bodenverkehr am Markt. Eine andere Hürde für die Entwicklung des Bodenmarktes sind die nach dem Recht des Gesamteigentums verwalteten Grundstücke. In diesem Falle ist der Abtritt des Bodens nur mit

Zustimmung aller Eigentümer möglich, was meistens nur schwer erlangt wird. Eine ähnliche Situation war in West- und Nordeuropa auf dem Bodenmarkt und im Prozess der Vererbung des Bodens während einer langen Zeitperiode entstanden. Deshalb haben diese Länder schon vor mehreren Jahrzehnten den Prozess der Boden(neu)ordnung begonnen. Jedoch war früher das Hauptziel der Bodenumgestaltung die Schaffung effektiverer Produktionsbedingungen für die Landwirtschaft. In der letzten Zeit, nach einem Wechsel der Prioritäten in der Landwirtschaft, und in der demographischen Situation ist das Hauptziel solcher Projekte neben der Schaffung einer bequemeren Bodennutzung die Durchsetzung der Maßnahmen des Umweltschutzes, die Schaffung der Bedingungen für andere Unternehmungsarten, die Entwicklung der Landschaft. Solche Projekte werden Konsolidierungsprojekte oder Flurbereinigungsmaßnahmen genannt. In vielen Staaten West- und Nordeuropas wird die Flurbereinigung durch ein besonderes Gesetz geregelt, weil die Flurbereinigung einen komplexen Prozess darstellt.

Die Flurbereinigung steht in jedem Lande vor spezifischen Aufgaben, abhängig von den Eigenschaften, Bedürfnissen, Traditionen. Zum Beispiel fußt in Dänemark die Flurbereinigung auf der Freiwilligkeit der an diesem Prozess beteiligten Bodeneigentümer und auf den demokratischen Standpunkt zur Ordnung der Grundstücke. In Deutschland wird dieser Prozess mehr von entsprechenden staatlichen Stellen geregelt, die zur Verwirklichung der staatlichen Politik der Bodenordnung die Vollmachten der Regierung haben. Jedoch umfasst sie in vielen Staaten Mittel- und Osteuropas, und besonders dort, wo die Konsolidierung des Bodens erst vor kurzem begonnen wurde, mehr Aufgaben als die einfache Neuordnung der Grundstücke. Bei der Lösung der Fragen der Bodenordnung werden gleichzeitig auch andere lokale Probleme gelöst, die mit dem Umweltschutz, mit der Verbesserung der Infrastruktur und allgemein mit der Verbesserung der Lebensverhältnisse auf dem Lande verbunden sind.

Die Projekte der Flurbereinigung sind ein zweiter Teil der Reform, mit dem Ziel der Strukturierung konkurrenzfähigen Wirtschaftens und der Befriedigung eines weiteren Bedarfs der Bodennutzer. Bei der Durchführung der Bodenreform ist schon heute abzusehen, daß der Boden der Landwirte in mehreren Grundstücken liegen wird.

Von August 2000 bis Dezember 2002 wurde in Dotnuva, Bezirk Kėdainiai, nach dem von der Abteilung für Flurbereinigung des dänischen Ministeriums für Nahrung, Landwirtschaft und Fischerei und vom Nationalen Bodenamt beim Ministerium für Landwirtschaft Litauens finanzierten Programm das erste Pilotprojekt der Flurbereinigung in Litauen durchgeführt. Das Hauptziel dieses Projektes war die Zusammenlegung von Grundstücken, die für die Bodennutzung geeignet sind. Im Territorium des Projektes waren 79 Bodeneigentümer. Das Territorium nimmt die Fläche von 392 ha ein, in der 46 ha des freien staatlichen Bodens sind.

Bei der Betrachtung der Situation der Bodennutzung war zu sehen, daß in der Tat nur einige Bodeneigentümer den Boden nutzten und andere Bodeneigentümer den Boden an andere Landwirte verpachteten. Die durchgeführte Umfrage zeigte, daß die Bodeneigentümer ein Interesse für das Projekt zeigten und etwa 20 Bodeneigentümer eine Beteiligung an dem Projekt beabsichtigten. Das Projekt endete (im Dezember 2002) mit der Unterzeichnung der notariellen Verträge durch die Bodeneigentümer und mit der Eintragung der Veränderungen im Register.

Nach der Erarbeitung des neuen Projektes war klar, dass die produzierenden Landwirte mehr Boden näher an ihrem Wohnort erhielten, einige Flächen nach ihrer Planung Zugang zu Wegen hatten und einige Streugrundstücke zusammengelegt waren. Daraus ist zu ersehen, dass die Flurbereinigung in Litauen erforderlich ist.

Die rechtliche Basis Litauens ist noch nicht für die Durchführung von Flurbereinigungen angepasst und musste während des Projektes der Flurbereinigung in Dotnuva mit der Bodenreform Schritt halten. Bei der Vorbereitung des Flurbereinigungsprojektes in Dotnuva gab es bei der Durchführung des Kaufes – Verkaufes unter Privatpersonen die wenigsten Probleme: der staatliche Boden, der im Territorium dieses Projektes lag, wurde nach dem neuen Zivilgesetzbuch geordnet. Da es keine Ordnung zur Regelung des Verkaufs des staatlichen Bodens für Bürger gibt, wurden im durchgeführten Projekt der Flurbereinigung die Grundstücke nach dem 10ten Artikel des Bodenreformgesetzes [9] neugeordnet; die Verträge über den Kauf der Grundstücke des staatlichen Bodens wurden unter Einhaltung der Vorschriften des Zivilgesetzbuches, das den Verkauf der Liegenschaft(en) reglementiert, abgeschlossen. Dieses erste Projekt der Flurbereinigung wurde an Hand unserer gegenwärtigen Gesetze durchgeführt, deshalb war es kompliziert, den Austausch der Grundstükke und den Bodenkauf aus dem freien staatlichen Bodenfonds vorzunehmen.

Schlussfolgerungen und Vorschläge

1. Obwohl der Prozess der Wiederherstellung der Bürgereigentumsrechte auf Boden, Wald und Gewässer noch nicht abgeschlossen ist, ist klar, dass dieser Prozess die Entstehung der kleinen und zerstreuten Bodennutzungen verursacht und somit große Schwierigkeiten für mögliche private und staatliche Investitionen in den ländlichen Gegenden, für die abgewogene ökonomische Entwicklung, Befriedigung der sozialen Bedürfnisse und für den Umweltschutz aufzeigt. Die Streulage der Grundstücke wirkte negativ auf die Erfordernisse der Landwirtschaft. Als Folge der Wiederherstellung der Eigentumsrechte der Bürger bildet oft die Bodennutzung der zusammengelegten kleinen Wirtschaften eine untaugliche Konfiguration für eine sinnvolle landwirtschaftliche Produktion auf den Grundstücken. Deshalb ist es für die Landwirte schwer, die neuen und in ständiger Fortentwicklung sich befindenden Geräte der landwirtschaftlichen Technik gezielt anzuwenden. Viele einzelne Land-

wirte sind gezwungen, die landwirtschaftliche Produktion für die Selbstversorgung zu betreiben; sie können die gewerbliche Produktion nicht aufnehmen. In solchen ländlichen Gegenden, besonders, wenn sie von den Absatzorten der landwirtschaftlichen Produkte sehr weit entfernt sind, wird der Boden nicht der landwirtschaftlichen Bestimmung gemäß bearbeitet bzw. genutzt. Mit Rücksicht darauf und nach der Untersuchung der gegenwärtigen Situation der Bodenordnung in Litauen und in einigen ausländischen Staaten schlage ich vor, zuerst die Wiederherstellung der gesetzlichen Grundlagen für Eigentumsrechte auf Grund und Boden, Wald und Gewässer durchzuführen.

2. In den Gegenden, wo die Wiederherstellung der Eigentumsrechte der Bürger auf den Boden beendet wird, ist von Beginn der Mitgliedschaft Litauens in der Europäischen Union die Flurbereinigung durchzuführen. Dieser Prozeß soll nach einem Sondergesetz durchgeführt werden, das die Umgestaltung der Grundstücke unverzüglich nach der Wiederherstellung der Eigentumsrechte regeln muss. Diese Etappe der Bodenordnungsarbeiten wird mit der Verwirklichung folgender Hauptziele verbunden:

- Verbesserung der landwirtschaftlichen Strukturen, in dem die im Laufe der Wiederherstellung der Eigentumsrechte der Bürger auf Boden geformten kleinen, für die konkurrenzfähige Wirtschaft untauglichen Bodennutzungen in rationellere, für die Konkurrenz unter Marktbedingungen einer sich vereinheitlichenden Europäischen Union tauglichere Grundstücksformen geschaffen werden;
- Verwirklichung der Forderungen des Umweltschutzes, indem die Bedingungen für die Entwicklung der umweltschonenden Bewirtschaftung und Erhaltung der Landschaft geschaffen werden;
- Entwicklung der Infrastruktur, indem die Nutzung des Bodens, der mit der Errichtung verschiedener Objekte der Infrastruktur verbunden ist, umgestaltet/neugeordnet wird.

3. Die Flurbereinigung in Litauen soll durch die Erstellung und Auflegung der Flurbereinigungsmaßnahmen durchgeführt werden.

4. Die Flurbereinigungsprojekte sind vorzubereiten, wenn dies freiwillig mehr als vier Bodeneigentümer wünschen. – Die Projekte sollen auf dem Territorium durchgeführt werden, wo es die Bodeneigentümer wünschen. Die Gesamtfläche darf dabei nicht weniger als 100 ha betragen. Die Projekte werden erstellt in den Versammlungen der Bodeneigentümer und Interessenten, wobei die Entscheidungen über die Notwendigkeit der Flurbereinigung getroffen, das Territorium des Projektes festgelegt und andere Fragen, die mit der Erstellung des Projektes verbunden sind, besprochen werden.

5. Zwecks Sicherung der Qualität der Flurbereinigungsprojekte und der Vermeidung von Fehlern sowie des Zuwiderlaufens von Rechtsakten ist es vorzusehen, dass die Projekte von einer staatlichen Institution vorbereitet werden.

6. Die Flurbereinigung ist ein langjähriger, kostenintensiver Prozess. Obwohl in vielen europäischen Staaten zur Finanzierung dieses Prozesses auch die Landwirte selbst und andere Bodeneigentümer beitragen, ist es jedoch verständlich, dass es für die litauischen Landwirte eine zu schwere Last wäre. Deshalb soll dieser Prozess in unserem Lande ausschließlich durch die Mittel der Europäischen Union und unseres Staates finanziert werden.

7. Eine richtig durchgeführte Flurbereinigung trägt der Vergrößerung der Erträglichkeit, Effektivität und Konkurrenzfähigkeit des landwirtschaftlichen Sektors bei. Sie sichert auch die Schaffung der Arbeitsplätze auf dem Lande, bildet bessere Bedingungen für die Arbeiten der Bodenordnung, hilft dem Umweltschutz und der Verwaltung der Naturressourcen. Jedoch ist in einigen Ländern Westeuropas bemerkt worden, dass die Flurbereinigung auch negative Folgen haben kann, wenn dabei die ökologischen und kulturellen Aspekte nicht berücksichtigt werden und nur nach der Vergrößerung der landwirtschaftlichen Produktivität gestrebt wird. Auf solche Weise entsteht eine

Gefahr für die Schonung oder Bewahrung der biologischen Vielfalt, entstehen die Voraussetzungen für die Erosion des Bodens und für die negativen Veränderungen der traditionellen Landschaft. Deshalb muss die Flurbereinigung ein Bestandteil sein der umfassenden Entwicklungsprogramme des Landes, einschließlich der regionalen Entwicklung, der Dorferneuerung und der Entwicklung der ländlichen Infrastruktur. Also sollte in Litauen, wie in anderen Ländern Europas, ein eigenes Modell der Flurbereinigung erstellt werden, das den eigenen Bedürfnissen gerecht wird und den größten Nutzen den Landwirten sowie den anderen Dorfbewohnern bringt.

8. Es bleibt zu hoffen, dass die Bodeneigentümer und die örtlichen Amtsstellen den wirtschaftlichen Nutzen dieses Prozesses verstehen werden und den Wunsch ausdrücken werden, dass möglichst zahlreiche solcher Projekte in ihren Gemeinden verwirklicht werden. Jedoch ist dafür in erster Linie eine schnellere Wiederherstellung der Eigentumsrechte notwendig.

Literaturverzeichnis

[1] Natkienė J., Pilypas J. Agrarinė ekonomika // Patirtis ir patarimai. Dotnuva – Akademija, 1995, 147 p.
[2] Republic of Lithuania Law on Agricultural and Rural development No IX – 987, 25, Vilnius, June 2002.
[3] Aleknavičius P. Žemės teisė. LŽŪU leidybinis centras, Kaunas, 2000, 110 p.
[4] Republic of Lithuania Law on Land No I – 446, 26 April, Vilnius.
[5] Aleknavičius P., Čaplinskienė R., Optažienė J. Žemėtvarka. Vilnius, 1999, 279 p.
[6] Jurginis J. Lietuvos valstie ūsĠistorija. Vilnius, 1978.
[7] Jučas M., Baudžiavos irimas Lietuvoje. Vilnius, 1972.
[8] Žemėtvarka ir hidrotechnika // Prisiminimai. Lietuvos žemėtvarkos ir hidrotechnikos inžinerių sąjungos žurnalas, Nr. 3, Vilnius, 2001.
[9] Republic of Lithuania Law on Land Reform, Vilnius, 1991.

Romualdas Survila, Rita Palciauskaite

Planning of Land Resources Use in Rural Areas of Lithuania

Introduction

Sustainable rural areas development is highly dependent on land policy rationality. Proper land policy is the fundamental of land degradation control caused by its improper use and the guarantee for effective use of land types according to productivity and restrictions. Such land policy would furnish necessary conditions for proper land management, in other words, land distribution should serve its best utilisation. Besides, there must be sufficient amount of land of different purpose, it must be expediently located and be of adequate structure. It should also meet environmental standards at reasonable costs and the title to the land should be clearly defined and reliable. Special attention should be paid to large areas of drained land and sensitive protected areas in Lithuania. All the work should be carried out at the scale and pace not to lack behind the growing and changing requirements of the developing free market economy. To achieve the above mentioned goals, farmers and decision makers need the soil and other land resource information. Therefore, it is necessary to take the following measures:

- to investigate in detail the existing land use situation and its best future use possibilities;
- to plan future land use;
- to implement land use plans;
- to control and monitor the planned land use in future;
- to evaluate and to develop land and water body protection policy.

These were the reasons that determined the demand for development of the national Land Resource Information System (LRIS), which would provide farmers and decision makers with precise, useful and timely information on soil, land and water resources. The system should contain all types of information, necessary for physical planning of rural areas (including land use planning) and programs on environment improvement. This would allow to focus on the areas, where economic duty and environmental situation is very complicated, as well as to use the existing scarce financial means in the best possible way. In this way, the local community could develop a positive and vital attitude towards realisation of an adequate, environmentally secure land use policy.

The purpose of planning

Not in all the areas of Lithuania agriculture and other economic duty is a profitable activity. In many areas, people are engaged in agriculture only because they do not have any other means of subsistence – no vacant jobs. Land of low productivity is not planted with forest as there is lack of cultivated land in densely populated areas. There is a pressing problem of drained land use and land improvement facility maintenance. The state has to respond to changing export and import conditions of agricultural products abroad. Land resources have to be preserved for future generations. Land has to be used for its most proper purpose (agriculture, water economy, constructions, recreation and etc.). Two conflicting goals to be reconciled: profitable, intensive farming and adequate environmental protection. Changes in farming technologies as well as the integration are stimulating changes in land use, farm size and crop structure. One tenth of the area of Lithuania is covered by protected territories (national and regional parks, reserves, strict reserves, cultural heritage entities and etc.), where specific farming measures are applied. About 200 special land and forest use restrictions are imposed on land users all over the country. The country, every county, district municipality and even every farmer must have long-term strategic plans and current work program – tactics for strategy implementation.

Land administration structure is changing-many functions are transferred to county administrations and district municipalities. Land must be distributed to enable the best use of it to meet the needs of inhabitants. All types of land must be expediently located and be of certain structure with regard to farming and environmental requirements.

There are three attitudes towards land use in our country, under the process of land ownership restoration to former owners and during the transition period to competitive market conditions.

They are as follows:

1) possible land use forms – private, state or long-term lease;
2) land parcels may be used for agricultural, forest growing, town construction, recreational and other purposes. For example, use purpose of an arable land parcel is changed by planting forest, in another case – exploited gravel-pit is re-cultivated and changed into arable land;
3) land use is designated for cultivation of certain crops, for example, with regard to soil, relief and other conditions, plant-growing is chosen with application of adequate plant rotation system. In case of soil erosion-anti-erosion measures are taken and etc.

The proper purpose of land use depends on both environmental protection regime and natural, economic and social factors (Figure 2).

Farming is the most successful in the parcels of respective purpose, provided that all the complex of environmental protection priorities as well as natural, social and economic factors are taken into consideration.

After GIS-based LRIS is developed, different models of data layers are established with regard to the most optimal land use alternatives. Development of data bases is based on the data base development scheme (Figure 3).

Planning Levels

With the purpose of land use planning, data of the Land Resource Information System are developed on three levels: national, district/regional and local (Figure 1).

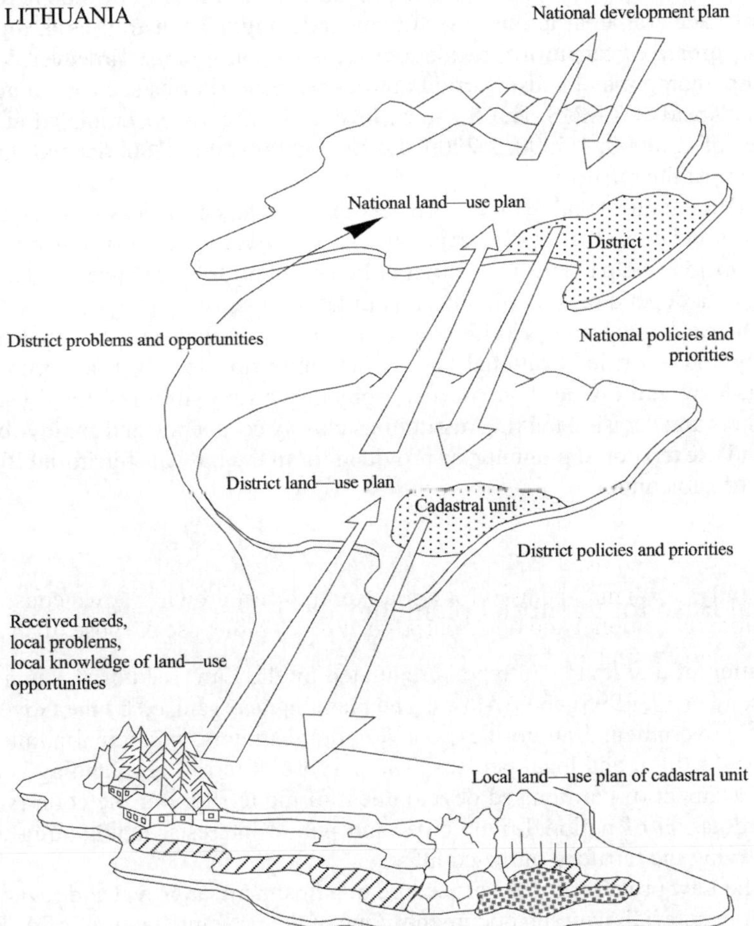

Figure 1. Two-ways links between Planning at different levels

On different levels, data of different precision and segmentation are accumulated. As Land Resource Information System (LRIS) is based on GIS, base maps at the scale of 1:200000, 1:300000 and 1:400000 are used for data accumulation on national level. LRIS data layers are developed at the scale of 1:300000.

Regional levels entities in Lithuania are territories of 51 district municipalities. In this case the scale of the base GIS map is 1:50000. Different territories of several districts or their parts may form the hot area. However, digital GIS data allow for easy solution concerning the use of regional and district data. In individual cases, regional data may be accumulated at the scale of 1:100000.

Cadastral areas are best suited for local level planning. Boundaries of 1400 cadastral areas in Lithuania were approved by the Decision of the Government. Cadastral areas are used for unique indexing of cadastral (land, forest, water, protected territories, residential areas and etc.) units. However, local level planning may be also carried out in smaller territories such as villages, farmer's and etc. LRIS data for local level planning are accumulated at the scales of 1:10000, 1:5000, 1:2000, 1:1000, 1:5000 and 1:100, depending on the size of the territory.

LRIS data for local planning are detailed and can be successfully used for farmers and co-operatives in decision-making activities. All forms of support and subsidies help to enforce district policies and priorities. However, district level data depend on specific interests of land users.

District policy and priorities must not be in conflict with national land policy and priorities. National land policy and priorities must take into account local and district level needs, problems and possibilities. Land users, planners and decision making institutions closely co-operate and are involved in land use territorial planning, as provided for in the Law on Territorial Planning of Lithuania.

Legal Base for territorial planning

Planning of any level and type is regulated by the Law Territorial Planning (December 12, 1995) and by-laws. The planning is organised by the Government, governmental authorities, county administrators, municipal administrations, natural and legal persons. The purpose of territorial planning is the establishment of harmonised development of the territory of the country by co-ordination of national, municipal and public interests, maintaining and preserving the territory of the country.

The Law provides for the procedure of adjustment, approval and registration of territorial planning documents. One of the principle features of democratic planning reflected in the Law is the publicity of territorial planning with the public involved in plan consideration. The State Inspectorate of Territorial

Planning and Construction is responsible for the control of the procedure of preparation, adjustment and public consideration of planning documents.

Principles of the territorial planning

Territorial planning system is established to regulate the activities of different kind in order to ensure a rational distribution of construction activities, formation of landscape and land use.

Territorial planning documents set conditions for farming activities that should be observed by the specialists who prepare detailed plans or concrete projects for the construction activities. Also they include land use regulations that should be observer by the land users. Suitably operating territorial planning system ensures:

- Balanced development of the territory in the country;
- Healthy and harmonic environment for life and activities of a person;
- Considered development of residential places and infrastructure;
- Rational use, protection and restoration of mineral resources, valuable objects of natural and cultural heritage and recreational resources;
- Maintenance of ecological equilibrium or its restoration;
- Co-ordination of the interests of the State, municipalities, legal and physical persons regarding possession of the territory, type of use and conditions of activity.

A system of territorial planning documents consists of general, special and detailed plans. Solutions in the plans belonging to the higher level are obligatory (as a condition for designing) for plans of lower level. General plans should be prepared for the territories of administrative units: territories of the Republic, country and municipality. Special plans are usually prepared for the territories that are connected by certain use conditions (for example national and regional parks, reservations, local forestry offices, protected landscape object, territories used for tourism and recreation, etc.) or for the territories of administrative units and land tenures.

Detailed plans are usually prepared for the territories where:

- Territory is defined planning city or town blocks and plots;
- A change in the land parcel layout is planned;
- Acquisition of land for public needs will take place;
- A change in the area, location and composition of lands (forests, water bodies, built-up territories, farming lands) is planned;
- Excavation of natural resources will take place;
- A change in the regime of territory building up and main objective land use is planned.

Depending on the importance of territorial planning documents, the Government, governing institutions. County Managers, Municipality Councils (Mayors) physical and legal persons may be the clients (initiators for preparation) for the territorial planning documents. Legal persons may prepare general and detailed plans if the laws provide for such activity, and suitable specialists do the work. The specialists having architectural, civil engineering, water management, land management, forest management, and geographical and other necessary education may prepare special plans.

Territorial planning documents upon the prescribed order should be discussed publicly, getting approbation from the society in this way. That ensures their obligatory nature for separate landowners and users.

Specialists and institutions authorised to undertake public control over territorial planning control the procedures of preparation, approval and public discussion of territorial planning documents and solutions in these documents.

Territorial planning documents should have a background defining the impact of the planned activity upon the environment and advantages. Plans that have been prepared, discussed with the interested institutions are approved following the procedure established by the Regulations for preparation of separate plans. The general plan of the Republic of Lithuania should be approved by the Seimas (Parliament), the general plan of a county – by the Government, and the general plan of a municipality or its part – by the Municipal Council. The Municipal Council usually approves detailed plans.

The following territorial planning documents are important for the regulation of land use in the Republic of Lithuanian.

General plan for the territory of the Republic of Lithuania. It will provide for the development of various economic branches by 2020. Approving the plan, this development will be expressed in concrete objects (motorways and railroads, large deposits of natural resources, territories necessary for the expansion of cities and industrial enterprises) or territories where priorities for the activities are set (e. g. urban, agricultural, forestry, recreational, etc.). General plan also will indicate areas where public support measures are necessary (for the improvement of social – economic conditions in not productive lands, for the preservation of cultural heritage, etc.).

Approved general plan for the territory of Lithuania will be a base for the preparation of general plans for the territories of counties and municipalities. At present, only land management schemes for the municipal territories in the districts (that execute the functions of general plans) and general plans of large cities are prepared.

Special plans that are prepared today include:

- Planning schemes for public parks (national and regional). Plans include zoning into the areas having special use and setting of regulations regarding land use and farming activities for every zone. After approving by the governmental decisions, these plans become obligatory and must be implemented.

- Forest management projects.
- Land management projects for farms. They solve the arrangement of farm territory, location of construction sites and roads, distribution of farming lands and crop areas. At present, rules for the preparation of projects and methods are revised taking into account provisions for new background and needs of farmer's farm.

Detailed plans that are prepared today include:

- Land management projects for taking land for public needs. They are prepared for those areas where it is necessary to build publicly important objects, roads and expand the deposits of natural resources. Projects provide for an obligatory purchase of land from the landowners or the termination of the right to use public land for other land users. For this purpose, land parcels are surveyed and evaluated.
- Land management projects for changing layout of the land parcels. Rules for the preparation of the project are developed anew taking into account specific character of private ownership and requirements for changing land cadastre information.
- Land parcel subdivision projects. They decide upon the subdivision of land parcels. During such projects, geodetic surveying is done and new land parcel pals are prepared. If there is a need, there are decisions made regarding the change of main objective land use, reorganisation of land cadastre indices and preparation of data for changes in the records of Real Property Register.

Environment Protection Strategy in land use planning

On September 25, 1996, the Seimas of the Republic of Lithuania, by the Decision No I-1550, approved the strategy of environmental protection of Lithuania.

The principal statements of the national environmental protection strategy were formulated with regard to the program document of 1995, consisting of three volumes.

Part 1 “Strategy Substantiation” provides for evaluation of environment situation, review on economy sectors, trends in changes of environmental protection as well as description of organizational, legal and economic system of environmental protection. Part 2 “Strategy Methodology” provides for the concept of the strategy based on the analysis of environment situation, methodology for evaluation of environmental problems, their acuity and complexity, as well as priority targets. Part 3 “Activity Program” provides for long-term strategy as well as short-and medium-term activity program, based on environmental components. Besides, the means of strategy enforcement, aspects of financing of environmental protection and etc are defined.

In 1983, in the national Integrated Nature Protection Scheme, the idea of Lithuanian's Nature Frame (ECONET) was raised and approved. Lithuania proposed the concept of Nature Frame, which became the concept and approach for the conservation and protection of Lithuanian's natural landscape (Table 1).

The Nature Frame, which offers a universal approach, was put forward and legally established under the relevant laws of the Republic of Lithuania on environmental protection and protected areas.

The Nature Frame links all natural protected areas with other ecologically valuable or relatively natural areas which underpin the general stability of landscape to form landscape system for natural buffering and connecting natural protected areas, but also at conversation of natural landscapes, biodiversity and natural recreational resources (Table 2, 3, 4). It does so by providing guidelines and conditions for the recovery of forests, optimizing the structure of agrarian landscape from the geoecological point of view, regulating development of agrarian activities, and defining sustainable urbanization. It is a concept based on catchment and biologically important areas. In effect, the principles of Nature Frame can be likened to the principles behind holistic watershed protection and management.

More specifically, the Nature Frame concept is based on analyses of migration processes in natural landscapes and the evaluation of the gravigenuous structure of natural landscapes and the evaluation of the gravigenuous structure of natural complexes and the conservation and enrichment of bio- and geo-ecological stabilizers in geosystems.

Nature Frame consist of the following:

- geoecological watersheds-belts which separate large geosystems and perform the function of ecological buffering between the systems;
- geosystem stabilization centers – areas which perform the function of ecological buffers in geosystems;
- migration corridors-valleys through which intensive geodynamic and bio-information circulation takes place.

The Nature Frame, however, is not a continuous network of green belts. Instead, it is an integrated process for all land use management protection.

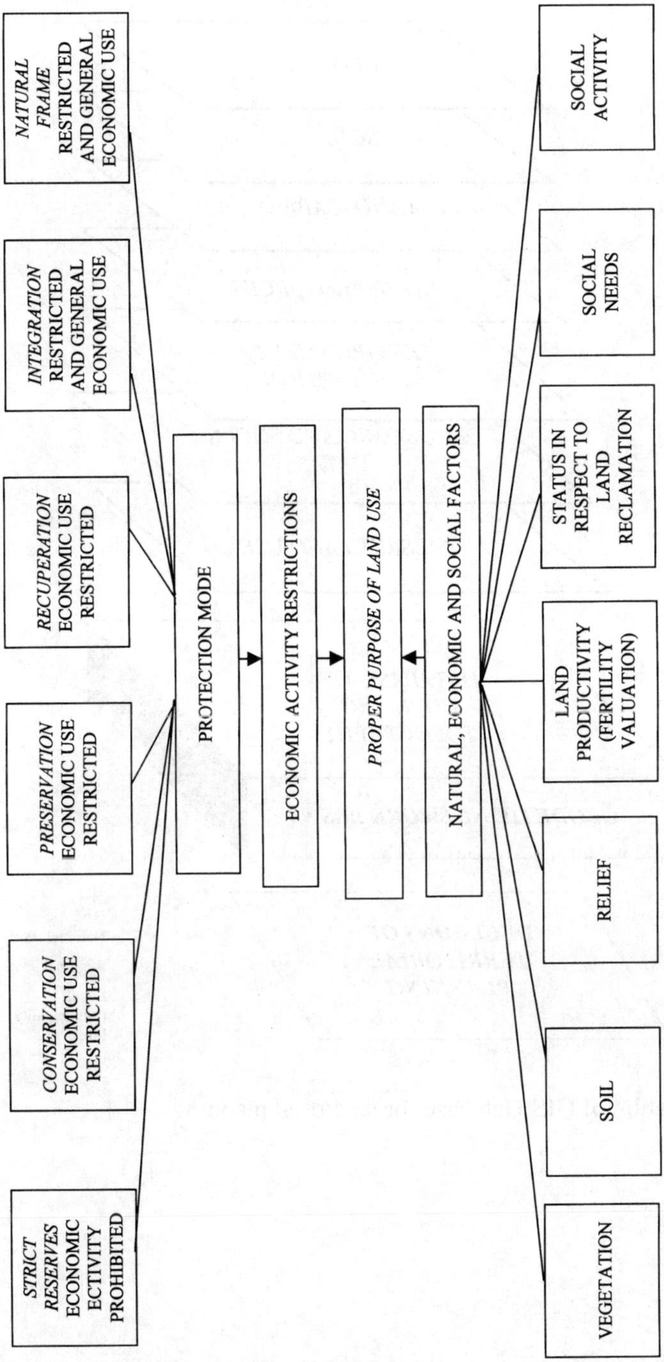

Figure 2. Establishment of proper purpose of land use

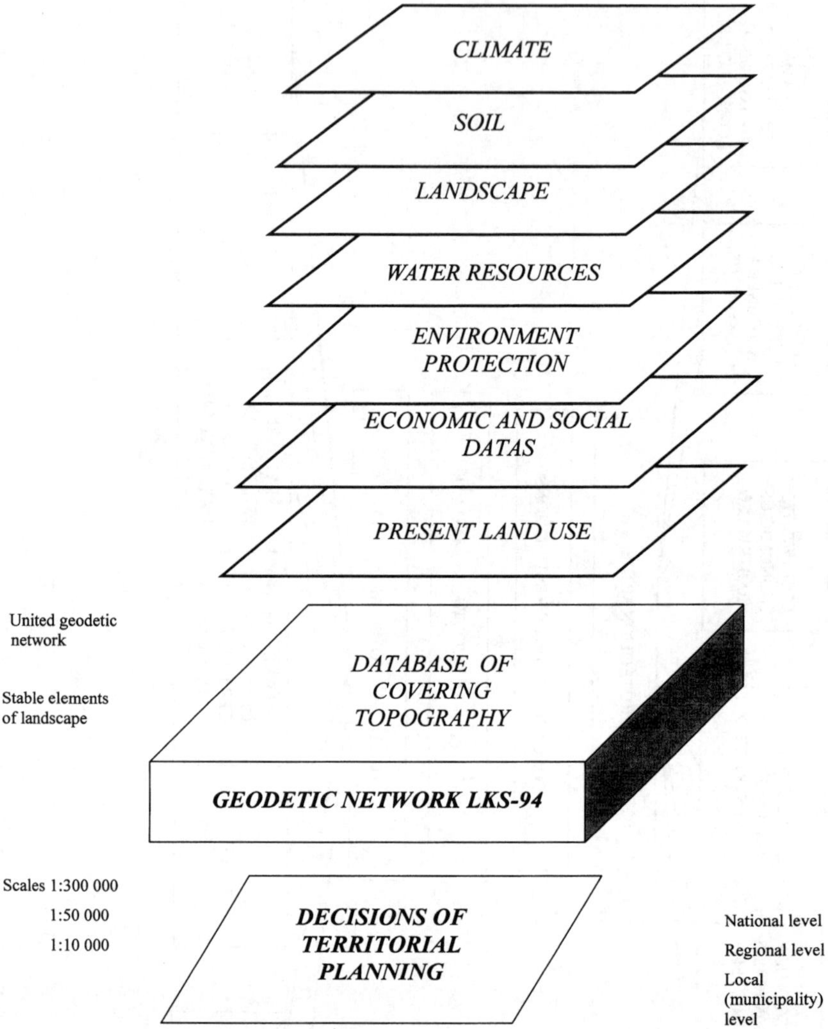

Figure 3. Formation of GIS Database for territorial planning

Table 1. The scheme of upper level nature frame element structure

	Regional geo-ecological watersheds and areas of inland stabilisation District geo-ecological watersheds and areas of inland stabilisation	Elements of territories of natural biocenosis prevail in the area of the natural frame (NF) NF territories usually cover the territories of the higher gravigenic position and in some cases-territories of middle landscape gravigenic position. In the upper level NF eco-functional compensational territories of water protection, marsh protection and forest protection prevail. Those territories are surrounding smaller areas of fields protection type. In flat areas of the country with intensively cultivated agrarian landscape areas eco-functional compensational territories of field protection prevail.
	Regional migration corridors Corridor axis (rivers, valleys, abounding in lakes and marshes and etc.) District migration corridors Corridor axis (rivers, valleys, abounding in lakes and marshes and etc.)	Migration corridor (MC) axes are riverbeds, valleys abounding in lakes and boggy valleys as well as other geo-dynamic and bio-informational circulation mains. Elements of MC nature frame are stretching and parallel to the axes. The main nature frame elements are forest plantations growing on slopes and grass vegetation in river creeks, around lakes and valleys.

Legend

	Bulky nature frame elements	Forests, marshes, areas abounding in lakes and other massives of natural biocenosis
	Group NF elements	Woods, various plantations, marshes, lakes and etc.
	Linear NF elements	Protective plantations of stretching shape
	Cultures territories	Ploughed fields, meadows, pastures, gardens, production centres, residential areas and etc.

The scheme of local level nature frame element structure

	Distribution of group NF elements Dominant in hilly landscape	The cultured landscape types are dominant in the rest of the territory – local level NF territory. Local level NF territories are mostly in the middle landscape gravigenic position. It is most problematic to produce a detailed description of NF located in agrarian landscape areas. All types of eco – functional compensational territories are there: territories of water protection, marsh protection, forest protection and field protection.
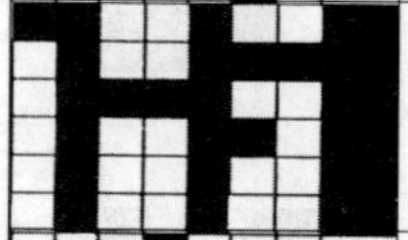	Distribution of linear NF elements Dominant in flat landscape areas	
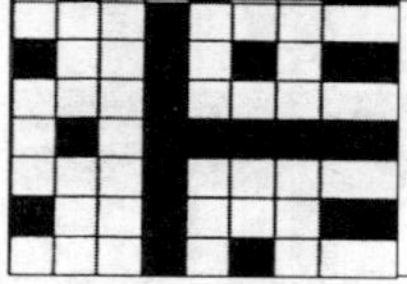	Distribution of NF elements of mixed type (group and linear) applied in different, mostly in highly cultured landscape areas	

Legend

	Group (compact form) NF elements	Little forests, shrubberies, plantation groups, little marshes and lakes, ponds and other compact of natural biocenoses.
	Linear (shape of belts) NF elements	Plantations of protective belts around fields and water bodies, insulating, of stretching shape.
	NF elements of mixed shape	Areas of natural and relatively natural biocenoses of compact and linear shape.
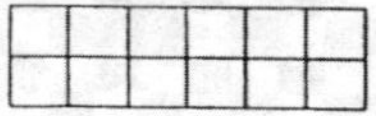	Cultures territories	Ploughed fields, meadows, pastures, gardens, production centres, residential areas and etc.

Table 2. Calculation of natural diversity indicator (I)

Area of the territory under investigation (S) …………….. ha

Item No	Names of relatively natural biocenoses	Ecological value coefficient	Perimeter In (m)	Area Sn (ha)
1	Forests of mixed and deciduous trees, protective belts with two and more tree and shrub lines, upper and middle-level marshes	1.0	I_1	S_1
2	Coniferous forests	0.9	I_2	S_2
3	Younger forests and plantations, lakes, natural rivers, areas abounding in springs and parks	0.8	I_3	S_3
4	Shrubbery	0.7	I_4	S_4
5	Lower marshes	0.6	I_5	S_5
6	Dams/ponds	0.5	I_5	S_6
7	Re-emerging marshes for agricultural lands	0.4	I_7	S_7
8	Natural meadows and pastures, positive relief forms, overgrown with perennial grass vegetation	0.3	I_8	S_8
9	One line tree belts, forest plant groups, gardens, garden plantations and drainage ditches	0.2	I_9	S_9
10	Cultured meadows	0.1	I_{10}	S_{10}

$$I = \frac{I_1 + I_2 \times 0.9 + I_3 \times 0.8 + I_4 \times 0.7 + I_5 \times 0.6 + I_6 \times 0.5 + I_7 \times 0.4 + I_8 \times 0.3 + I_9 \times 0.2 + I_{10} \times 0.1}{S - (S_1 + S_2 + S_3 + S_4 + S_5 + S_6 + S_7 + S_8 + S_9 + S_{10})}$$

Table 3. Reference indicators of biodiversity in Lithotops

Item No	Landscape	Name of soil texture	Recommended indicator of biodiversity of nature frame (m/ha)		
			Local level nature frame	District level nature frame	Regional level nature frame
1	Plain	Clay loam	25	30	35
		Light and medium loam	30	35	40
		Sandy loam	40	50	60
		Sand	60	70	80
		Peat mould	100	120	140
2	Hillplace	Light and medium loam	80	90	100
3	River valley	Light and medium loam	100	110	120
		Sandy loam	120	130	140
		Sand	150	165	180

Nature frame development in agrarian landscape areas

Nature frame		Recommended indicator of natural (ecological diversity)	Main elements and principles of nature frame development
Marked on land survey plans	Name		
1	2	3	4
	Regional geological watershed: – in plains – in hilly areas – in river valleys	 35-140 100-140 120-180	Areas abounding in lakes, marshy areas, sea litorale and area along the coast, territories of underground water replenishment and karstic areas, upper reaches of rivers located in the upper gravigenic position. Natural diversity in bigger part is determined by biocenoses of large forests, marshes and vegetation around water bodies. The missing relatively natural beogenoses may be compensated by protective forest plantations and re-naturalised parcels. Their shapes are similar to the natural forms.
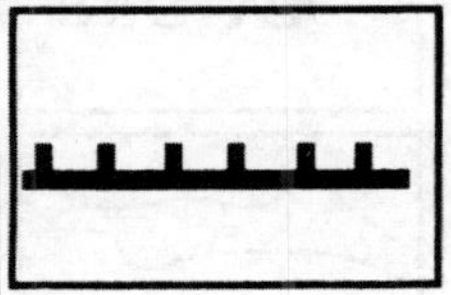	District geological watershed: – in plains – in hilly areas – in river valleys	 30-120 90-130 110-165	Natural and relatively natural vegetation growing in geo-ecological watersheds, separating different large geo-systems (districts). The missing natural diversity is compensated by protective forest plantations. All types of marshes, various relief forms overgrown with perennial grass vegetation, ponds and springs are preserved in the structure of land types. There is demand for cultured and natural meadows and pastures in the structure of agricultural lands. Private farms are established. Establishment of large, environmentally unfriendly production centres is not expedient. Forms of plantations and re-naturalised parcels are developed-similar to the natural forms.
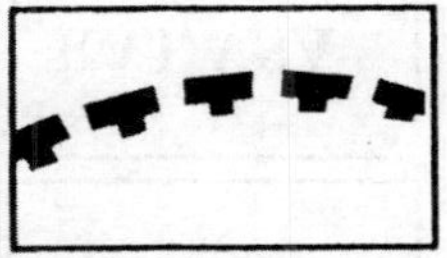 REG. VSA	Regional areas of land stabilisation: – in plains – in hilly areas – in river valleys	 35-140 100-140 120-180	Natural or relatively natural biocenoses located in the upper gravigenic position or in the territories of barrier type. They comprise the upper reaches of rivers. Plantation massives are distributed in such a way that they could perform side flow transformation functions. The developed environment should be similar to the natural in both its contents and form. Private farms are established. Establishment of large, environmentally unfriendly production centres is not expedient.

1	2	3	4
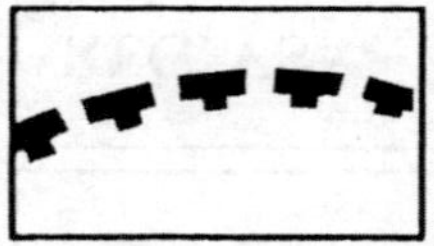 RAJ. VSA	District areas of land stabilisation: – in plains – in hilly areas – in river valleys	 30-120 90-130 110-165	The existing natural and developed relatively natural biocenoses located in separate large (district) geo-systems, in the upper gravigenic position or in territories of barrier type. The grater number of parcels overgrown with perennial grass and forest plantations is developed in agricultural lands. Private farms and agricultural companies with small environment friendly production centres are established.
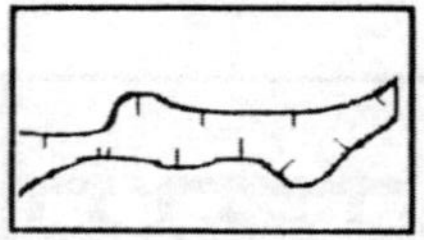 REG. MK	Regional migration corridor: – in plains – in hilly areas – in river valleys	 35-140 100-140 120-180	Lowland and valley mains of intensive geo-dynamic and bio-information circulation. Development of natural diversity is based on types of river valleys. Plantations (grass and forest plantations) of protective belts along riverbanks are developed. In protective zones of water bodies, water-protective and anti-erosion forests are planted. Restrictions of protective zones and belts of water bodies are applied for agricultural production. When the land of river valleys is privatised, land users have to observe economic activity restrictions and have to take obligations of nature frame formation. Valleys abounding in lakes as well as river valleys are applied the requirements for protective zones of rivers. In marshy valleys, natural diversity is increased by planting protective plantations on slopes and by separate parcel (located in over-cultured areas) re-naturalisation. In valleys where natural diversity is sufficient, cultured meadows may be established. There is demand for larger areas of perennial grass in migration corridors in different land types.
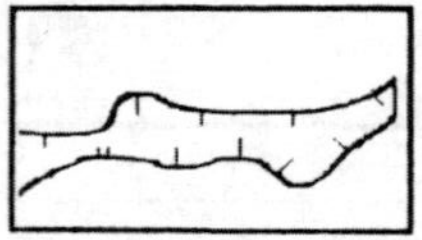 RAJ. MK	District migration corridor: – in plains – in hilly areas – in river valleys	 30-120 90-130 110-165	Elements and principles of development are analogous as for the regional migration corridor, however, natural diversity is not that well developed.

1	2	3	4
	Nature frame of local level: – in plains – in hilly areas – in river valleys	 25-100 80-120 100-150	In the drained areas of high natural diversity, the recommended natural diversity is maintained by regulation of land type transformation into agricultural lands. Natural diversity of highly cultured areas and drained territories is increased by planting protective forests and grass plantations. To increase natural diversity, green areas are developed in settlements and around production centres. Natural diversity is increased by cultured pastures, meadows and gardens. Forest plantations are developed: 1) in loamy plains – distance between them must not exceed 1000 m. 2) in other landscape areas – distance between them must be 500 - 800 m. Before land reform, it is necessary to prepare nature frame projects for individual farms. The projects should provide for the minimal area of plantations and re-naturalised parcels around the stable landscape objects (rivers, lakes, ditches, constructions and etc.) that would insure the recommended natural diversity. The established private farms would contribute to the minimal natural diversity, provided for in the projects, by farmstead plantations, gardens and small crop areas. In the extremely cultured drained land areas, it is expedient to develop natural diversity according to land reclamation reconstruction projects.

Pranas Aleknavičius, Marius Aleknavičius

Influence of the Environmental Protection Factors on the Formation of Agricultural Structures in Lithuania

Introduction

In the course of land reform in Lithuania, essential reorganization is going on in the structure of land ownership and land use. Laws that regulate privatization of land give a priority for the restitution of land to its former owners. It means, that those people eligible for restitution, who owned land before 1940 and who are willing either to get it back in the former existing place (where laws allow) in kind or to get an equivalent area of land in another place, have a priority right against the current users of this land. Agricultural and forestry land is restored without imposing any restrictions on land disposition and without any compulsory requirements for new owner to have a qualification of readiness for farming. Moreover, there are no requirements to compensate the state or former land users for the improvements made to that land at their expense. Legal acts provide only for general land use conditions that are mostly related to environmental protection.

In the second stage of the land reform, the state will pay more attention to the intentional regulation of formation of *agricultural structures* in regions. By *agricultural structures* we mean the units of economic activity established in the manner prescribed by laws, i. e. farms, agricultural companies, family farms (personal farms). Subjects of agricultural activity that may get the state support are farmers and legal entities growing agricultural production for trade. In the meantime, they use only ½ of the total area of agricultural land. The rest of land is used by local people for their self-sustenance needs (personal farming) or for short-term lease. Besides that, a part of land is desolate and forms wasteland. During the land reform the main objective of agrarian policy for the state is to support and promote the development of existing agricultural structures, so they can increase their possessions up to the *rational size*, to perform consolidation of land parcels possessed by those structures and to support the establishment of new farms by privatizing unoccupied state-owned land. This process may be successful only in a presence of certain conditions. Analysis of these conditions lets us identify the main factors that determine the progress in formation of agricultural structures in different parts of Lithuania.

Research Methods

Methods of statistical sampling, correlation analysis and descriptive analysis were applied in this research. The following material was used for analysis: cumulative data from the Center of agricultural information and rural business *(Zemes ukio informacijos ir kaimo verslo centras)* on the declared farmland and crops in 2002 and 2003, the cumulative data of state statistics on land prepared by the State Enterprise Center of Registers, the data about protected areas prepared by the Service of State Protected Areas under the Ministry of Environment and the data prepared by the National Land Service under the Ministry of Agriculture describing the indicators of land reform.

Results and discussion of the research

As a unit of agricultural activity land tenure of a farm is considered. It consists of agricultural land parcels related by common economic activity and held by the same physical person (when a farmer possesses a farm) or legal entity (for other types of farms). State regulation measures focus on ensuring rational land tenures of perspective farms. *Rational land tenure* means privately owned land possession that creates favorable conditions for efficient farming and enables application of modern production technologies in the compliance with environmental protection requirements by its size and by dislocation of separate fields and parcels in relation to the road network and to the farm center. The Minister of Agriculture of Lithuania has approved minimum size of farm to be 100 ha of farmland for farms specialized in crop growing and pig breeding and 40 ha for farms specialized in cow and other animals breeding to be considered as rational land tenures. Assuming that about 2/3 of farms will specialize in stockbreeding the average size of a farm in Lithuania should be at least 60 ha of farmland. Area of rational land tenure recommended for each separate farm can be identified from business plan for agricultural activity, i.e. from the corresponding expected scope of production and from the expected demand for farmland. The maximum allowable size of land tenures of a farm can be determined by the dislocation of land plots and fields used in a farm in relation to each other and to the place, where agriculture machinery and livestock is held or warehouses of agricultural production are located. At present time (in 2003) the average size of a farm which has declared its farmland and crops, is 29.3 ha including leased land or twice as less as the recommended average size of farm. Out of this only 12.6 ha is actually owned by the possessor. In addition to that, 17% of farms do not own land at all. Agricultural structures in Lithuania are only in the stage of formation, which can be illustrated by these figures: 30,000 persons declared land (totally 1,111,800 ha) in 2002 and 53,800 persons (totally 1,572,400 ha) in

2003. The *degree of formed agricultural structures* is expressed as a percentage of the declared land in total area of private land in municipality. It is used as a dependent variable (y) in the analysis. As of July 2003, this indicator makes 53.4% in average varying from 4.0 up to 74.0% in different municipal areas. Another indicator that describes the formation of farms in the area is the average area of land owned by a single registered farmer. On 1 August 2003, this indicator was 10.4 ha in average, varying from 5.5 ha up to 15.6 ha in different municipalities. On the other hand, the average area of land granted for one owner in Lithuania in the process of restitution was only about 6 ha. This fact clearly indicates that farmers have significantly increased their holdings since that time by buying private land.

State regulation measures aim on the following objectives: privatization of the remaining state-owned land, creating favorable conditions for the perspective agricultural structures to acquire currently rented private or state-owned land and, if necessary, initiating the consolidation of land parcels used by farms. These measures can be implemented and promoted by giving a priority when buying state-owned land and by providing economic support for those taking credits to buy private land.

New amendments to the Land Law set for the procedure of preparation of land consolidation plans. Following this procedure, land consolidation plans will be prepared at the expense of the state. They will allow the reorganization of large farms to form more compact land tenures. As the scientific research shows, the compactness of land tenure of a farm is expressed by a ratio $K = l/\sqrt{P}$, where l is an average distance from a farm center to the fields; P is an area of land tenure in ha. For the rational land tenure in Lithuania this ratio equals K = 0.09-0.25. However, only about 39% of farms fall within this range. For the majority of large farms, the compactness ratio equals K = 0.45 and the average distance to the fields makes 3-10 km and even more. Land consolidation will allow reducing transportation costs after the exchange of land parcels. Standard calculations of hypothetical annual production costs for field works show that the decrease of average distance to access the fields by one kilometer reduces costs by 1-1.5%. Moreover, land consolidation allows to increase total farming area of fields and to use agricultural machinery in a more efficient way.

While analyzing the possibilities to increase land tenures of farms we have identified reasons why the degree of agricultural structures differ throughout Lithuania. The main reasons are the different pace of land reform, different economical value of land and the restrictions on land use.

Indicator of the progress of land reform may be expressed as a ratio of privately owned agricultural land and the total area of agricultural land that can be privatized and used for farming. This ratio makes 60.2% in average and varies from 19.7% up to 77.4% in different municipalities of Lithuania. Each year the area of private land increases by 200,000 ha or by about 5% of total area of agricultural land. Correlation analysis revealed the dependency between the degree of formed agricultural structures (y) and the indicator of progress of land reform (x_1) with a correlation coefficient $r_1 = 0.52$.

Economic value of land is expressed by the *soil productivity grade* of farmland, which is equivalent to the value of agricultural production (feeding units) gained from an area unit. Soil productivity grade depends on natural features of land (type of soil, granuliometeric structure, agro-chemical characteristics, and amount of humus) and man-made land improvements such as installed drainage systems and other land improving measures. Due to a variety in patterns of these conditions, the average productivity grade in different municipalities varies from 30.4 up to 50.0 and makes 39.3 in average. Correlation analysis revealed the dependency between the degree of formed agricultural structures (y) and average productivity grade of soil (x_2) with correlation coefficient $r_2 = 0.70$.

Influence of restrictions on land use is evaluated through estimation of the distribution of protected areas and regulations on their use and management. The system of protected areas in Lithuania consists of reservoirs, sanctuaries, national parks, regional parks and ecological protection zones. In some cases agricultural land may be located inside some sanctuaries and national parks. Then, the following activities are restricted within these areas: exploitation of natural resources (minerals and peat-bogs), regulation of streams, drainage of larger swamps and construction of buildings. These restrictions of economic activity in certain areas may slow down in a smaller or larger extent the development of intensive agriculture. Zones of environmental protection are divided into protection zones of water bodies and water reservoirs, protection zones of cities and resorts, field protection zones, buffer protection zones of national parks and sanctuaries, physical protection zones of roads, electricity lines and other linear installation, sanitary protection zones of industrial and utility objects, and protection zones of perspective mineral areas. Depending on type of zone, cultivation of land, fertilization and construction of buildings may be restricted. In some zones, setting out the requirements for crop structure regulates the growing of agricultural production. The presence of ecological protection zones has smaller impact on agricultural production than the regulation of activity in other protected zones. Therefore, *the indicator of land use restrictions* (x_3) is expressed by a percentage of area used by reservoirs, sanctuaries and national parks in the total area of municipality. The above-mentioned protected areas in Lithuania (excluding *Kursiu Marios* lagoon and territorial sea area) covers a total area of about 748,000 ha or almost 11.5% of the total area of the country. The correlation analysis revealed the dependency between the degree of formed agricultural structures (y) and the indicator of land use restrictions (x_3) with coefficient of correlation $r_3 = 0.54$.

Analysis resulted in the following multiregression equation:

$Y_K = 0.3821\ x_1 + 0.0129\ x_2 + 0.3638\ x_3 - 0.2098$

Coefficient of correlation $R = 0.7787$; coefficient of determination $R^2 = 0.6064$.

We can conclude from the analysis, that the most important factors on which the formation of agricultural structures depends are the economic value of land and land use restrictions. Municipalities with the highest grades of soil productivity as a rule have less of protected areas than anywhere else. According to zoning of Lithuania by natural-ecological features, the most sensitive is the eastern part, which covers about 1/3 of the total territory. Comparing to the rest of the country, the average indicators here are far less favorable for the establishment of competitive farms:

Indicators	In eastern part of Lithuania	In other places in Lithuania	Totally
1. Number of administrative units	16	35	51
2. Area, in thousand ha	2277	4253	6530
In percent	34.9	65.1	100
3. Area of reservoirs, state sanctuaries and state parks, in thousand ha	4081	3399	7480
Percentage of the total area	17.9	8.0	11.5
4. Agricultural land that may be privatised, in thousand ha	1189.6	2729.9	3919.5
Percentage of the total area	52.2	64.2	60.0
5. Privately owned land, as for 1 July 2003, in thousand ha	612.1	1745.9	2358.0
In percent	51.5	64.0	60.2
6. Land owned by farmer, out of private land, in thousand ha	154.3	457.4	611.7
In percent	25.2	26.2	25.9
7. Own and leased private land declared in 2003 by agricultural structures (the degree of formed agricultural structures), in percent	18.6	65.6	53.4

Thus, the development of eastern part of Lithuania should be more oriented towards the activity other than agriculture, i. e. the alternative activity such as rural tourism, forestry and commercial fishery. It should be also mentioned, that the existing farms in this zone are relatively smaller than in other places (on 1 January 2003, the average area of farmer's holdings was 16.9 ha there and 23.9 ha elsewhere), and the amount of desolate state and privately owned land is larger.

According to the multi-regression equation it can be estimated that even in case of entire privatization of agricultural land, the perspective agricultural structures in eastern Lithuania would make only 54% of area, while in the rest of the country – almost 68 %. Agricultural land if not used by agricultural structures will be used by villagers for self-sufficiency needs (not for trade) or it will be used for forest growing and for other non-agricultural activity. In order to fulfill the territorial differentiation of economic activity, regional development programs are necessary and municipal territory planning documents must be prepared. According to the strategy on developing the system of protected areas in Lithuania prepared in 2002, it is intended to identify the critical anthropogenic parameters within protected areas, to apply measures for the implementation of ecological farming and to turn low productivity land into the natural areas. These measures performed in eastern part of Lithuania where protected areas cover about 1/5 of the area will have certain impact on agricultural development, as well.

In the rest of the country – in central and western Lithuania – the formation of agricultural structures is based on intensive agricultural production. However, it should be regulated in such a way as to ensure protection of water bodies against chemical pollution and protection of soil against deterioration. Protection of water bodies is realized by identifying and installing the shore protection strips and protection zones and ensuring their proper exploitation. Therefore, more important issue is to protect soils against the too intensive cultivation, when the agro technical requirements for the crop rotation and for accumulation of organic substances in the cultivated land are violated. It is estimated that the amount of humus decreases by 0.8-2 t/ha per year (1 t/ha in average) in soil where crops, rapes, flax and cumulative plants grow and it increases by 1-2 t/ha where perennial grass grows. In areas with low level of humus (not higher than 2-2.5 %) the land cultivation can improve soil quality and increase the level of humus in soil. For now, the current amount of livestock provides with such amount of organic fertilizers, which can restore not more than 0.3 t of humus per 1 ha of arable land. Considering this, the average area of crops and technical and cultivated plants in municipality should not exceed 68 % of arable land. In farms with no livestock and where shredded straw is used as organic fertilizers (after the harvesting) the maximum allowable area should not exceed 62 %. Requirements on land use are set in more detail in the *Good Agricultural Code*. The current situation in Lithuania shows, that according to the data of farmland and crop declaration in 2003, this area under the crops and technical and cultivated plants made 58.9% of arable land in average for the involved in declaration farms. However, in 12 municipalities out of 35 in Middle and West Lithuania such land made 68-82% of arable land or 74% in average. It means, that now it is already needed to regulate the farm specialization, the number of livestock and the crop structure. This would ensure the harmonization of perspective production with rational land use and avoiding soil deterioration.

Conclusions

1. Different pace in forming system of perspective farms is determined by variety of natural conditions in the territory of Lithuania and by the processes of land reform.
2. By more than 60% the degree of formed agricultural structures is influenced by qualitative factors – productivity of farmlands and land use restrictions.
3. Land use restrictions have the greatest impact in the eastern part of Lithuania, where the percentage of protected areas is significantly higher than in other places.
4. According to the current trends in establishing of farms, the regulation of agricultural development must be differentiated. It is of crucial importance to ensure intentional planning of measures on agricultural production and on land use in compliance with the rational land use and environmental protection requirements. In the eastern part of Lithuania there is a need to promote the development of alternative activities on land. In the future, farms producing the agricultural production for trade will use not more than 54% of the area of former farmland in these territories. Agricultural activity in the middle and western part of Lithuania must be more active, however it may not exceed the maximum allowable size limits for areas on which plants that exhaust soil grow.

References

1. Law on Protected Areas of the Republic of Lithuania, No. 1-301, 1993 (Official Gazette, 1993, No. 63-1188; 2001, No. 108-3902).
2. Territory Planning Law of the Republic of Lithuania, No. 1-1120, 1995 (Official Gazette, 1995, No. 107-2391).
3. Land Law of the Republic of Lithuania, No. I-446, 1994 (Official Gazette, 1994, No. 34-620).
4. Agricultural and Rural Development Law of the Republic of Lithuania, No. IX-3987, 2002 (Valstybes Zinios, 2002, No. 72-3009).
5. Strategy on Development of Protected Areas System in Lithuania. Ministry of Environment and Energy of Denmark and Ministry of Environment of Lithuania. Vilnius, 2000, p. 82
6. Rules and guidelines for modern farming. Ministry of Agriculture of the Republic of Lithuania, Ministry of Environment of the Republic of Lithuania. Kedainiai, Vilainiai, 2000, p. 63.

JURATE KUMPIENE, ČESLOVAS AKSAMITAUSKAS

Soil Remediation – Do we need it?

Abstract

Intensive agricultural and industrial activities caused large areas of land to become hazardous for both wildlife and humans. The risk assessment of contaminated soil to the human health is complicated and soil remediation is often excluded from the priority list of the environmental quality protection programs. Humans can be exposed to contaminated soil trough three main routes: ingestion, inhalation and through dermal contact. Soil should be remediated in order to reduce the pollutant exposure. Despite the large variety of available techniques developed for soil remediation, traditional methods as landfilling and encapsulation are still dominating. Soil remediation strategies are mainly determined by economic factors, aiming at the lowest prices and the fastest solutions. This often leads to non-sustainable choices when the problem is not solved but only relocated passing the responsibility to e. g. landfill owners. A high cost and non-sustainability associated with such remediation method has been encouraging to apply more innovative technologies. Biological soil remediation, including phytoremediation, is an environmentally friendly solution for remediation of contaminated soils. However, improvements of these technologies are still needed for the full-scale applications of soil decontamination. Examples of the environmental and health problems caused by soil contamination over the world indicate that the problem should not be ignored.

Introduction

The problem of land and groundwater contamination originates from industrial and agricultural activities as well as from improper handling and disposal of hazardous materials. However, not all the countries acknowledge the soil contamination as an actual environmental problem. Soil remediation is often excluded from the priority list of the environmental quality protection programs. In some countries, e. g. Lithuania, the negative impact of industrial and agricultural activities on the soil and even on the surface and ground water, were not taken into account until recent years. Soil monitoring lately has been noticeably extended with more attention being paid to the processes connected to agricultural activities – the leaching of biogenic and toxic materials into surface and ground water. For 12 of EU countries, the estimated number of potentially contaminated sites is 1,500,000, of which more than 300,000 have been identified. According to European Environment Agency (1999), this number of sites is not expected to increase, due to already established national policies to follow the precautionary principle. But, the large number of existing contaminated sites requires the development of innovative remediation technologies as well as adjustment of financial and legislative means.

Contaminant exposure to humans

The risk of contaminated soil to the human health is often complicated to estimate. Models of human health risk assessment are based on the contaminant exposure to humans through a selected set of pathways (SNV, 1996). Soil and dust are important carriers of contaminants that can be exposed to humans through ingestion, dermal contact and inhalation (Health Canada, 1995). Bare ground, contaminated soil blown as dust, contaminated sediments in river and lake bottoms, vegetables grown on the contaminated soil, are all possible sources of contaminants. Soil may also act as a contaminant pool, by slow leaching and volatilisation polluting groundwater, air or agricultural products. Determination of the generic values for contaminated soil is mainly based on the toxicological tests, i. e. information on the dose-effect or dose-response relationship of contaminant for humans (SNV, 1996).

Remedial policies

The policy of environmental protection agencies regarding remediation (from Greek remedium = "correct evil") is to choose long-term solutions that, if possible, solve the problem once and for all (U. S. EPA, 1999). The highest priority is given to the methods that destroy the contaminant (e. g. biodegradation or combustion) and enable land and soil to be re-used. When this is not possible, as in the case of metals, methods that concentrate the contaminant, which then can be separated from the soil and treated separately, should be used (e. g. soil-washing, soil-venting and thermal desorption). Only in the last case should methods such as containment, immobilisation and landfilling of untreated soil, be selected. Selection of a remediation strategy is usually determined by the intended future use of a site (U. S. EPA, 1998).

The Remediation Technologies Screening Matrix and Reference Guide (2002) lists 28 soil treatment technologies. Despite the large variety of available techniques developed for soil remediation, traditional methods as landfilling and encapsulation are still the dominant (U.S. EPA, 1999). Soil remediation strategy is mainly determined by economic factors, aiming at the lowest prices and the fastest solutions.

Such approach is described in the Baltic Marine Environment Protection Commission – also known as HELCOM – Recommendation 12/3 as a best available technology (BAT), when the most appropriate solution, considering the time frame, technical and economic feasibility of a particular measure, is made to take care of the discharges (SNV, 1995). From environmental point of view, the reduction of discharges, resulting from the use of BAT, as defined in the Recommendation, does not necessarily lead to environmentally acceptable results. On the contrary, it often leads to non-sustainable choices when problem is not solved but only relocated passing the responsibility to e. g. landfill owners, despite that the application of the BAT, according to the Recommendation, should not result in any increase in pollution in other parts of the environment or any increased risk to the human health or living resources. This is however a contradictive statement, when the target contaminants are heavy metals. The concentration of trace elements in the earth crust is constant and it is therefore impossible to reduce the metal concentration in one place without increasing it at the other (e. g. at landfill sites).

Remediation of metal contaminated soil possesses different problems compared with most organic pollutants. Metals cannot be degraded but are only transformed from one oxidation state or organic complex to another (Garbisu and Alkorta, 2001). Some of the metals are micronutrients (trace elements) necessary for organism growth, such as Zn, Cu, Mn, Ni, Co, Mo, i. e. they provide essential cofactors for metalloproteins and enzymes; while others have no known biological function, such as Cd, Pb and Hg (Garbisu and Alkorta, 2001; Leyval, 1997). At high concentrations, metals can act harmfully by blocking essential functional groups, displacing other metal ions, or modifying the active conformation of biological molecules (Collins and Stotzki, 1989).

In most of the European countries the soil contamination level is determined according to the total metal concentration of soil assuming that all forms of a metal have an equal impact on the environment. The question about which metal fraction should be used to determine the contamination level is still under discussion. Metal toxicity depends on its bioavailability, hence determination of potentially bioavailable fraction of contaminant suppose to give a better prediction of contaminant behaviour. This question is particularly important for the establishment of such soil remediation methods as in situ (on site) stabilisation of metal contaminated soil. In this case, the total metal content of soil remains the same, while leaching, bioavailability and toxicity of contaminant is significantly reduced.

Soil remediation requites high financial inputs. Clean up of the 67700 hectares of soil contaminated with oil products, rocket fuel, radioactive wastes and explosive materials, that were left by Soviet Union military units in Lithuanian territory was estimated to be 733 million US dollars. Paradoxical, but "dig-and-dump" strategy is not the cheapest measure to take care of contaminated site. The cost to remove, transport and place contaminated masses to the landfill for hazardous waste can be prohibitively expensive. For example, to remediate 10 h site highly contaminated with wood impregnation chemicals (chromated copper arsenate) in Sweden was estimated to be about 7 mln Euros.

A high cost and environmental disturbances associated with such remediation method has been encouraging to apply more innovative technologies. Biological soil remediation, including phytoremediation, is believed to be an environmental sustainable solution for remediation of contaminated soils. Some sites, such as contaminated with organic pollutants could be successfully cleaned up employing bioremediation methods (using microorganisms to degrade organic contaminants). Some companies, e. g. the Lithuanian enterprise "Naftos terminalas", are already using a bioremediation method for the clean up of the oil-polluted soil (Pilipaviūus, 2002). However, legally permitted landfilling of contaminated soil in most European countries is pointed as the most important barrier for the development of the innovative soil treatment methods (Terra Nova, 2002).

Development of in situ soil treatment methods e.g. stabilization using natural and synthetic soil amendments are less disruptive to the natural ecosystems techniques and less expensive compared with conventional soil remediation methods (Knox et al., 2001). Soil additives, e. g. organic matter, aluminosilicates, Fe and Mn (hydro)oxides, etc., could facilitate the stabilization of metals and reduce their bioavailability as well as prevent their transport to the surface and groundwater through the formation of insoluble contaminant species and without drastically altering the physical or chemical properties of the soil (Knox et al., 2001; Berti and Cunningham, 2000).

The significance of natural tools (e. g. bacteria, fungus, plants) is lately brought into active consideration. Plants as well as micro-organisms can be employed for in situ immobilization of metals in soil. A method that uses plants and metal immobilizing soil additives to reduce the bioavailability of pollutants in the environment is called phytostabilisation. The role of plants in phytostabilisation is a physical stabilization of the soil with dense root system to prevent erosion, minimize water runoff and percolation through the soil (Berti and Cunningham, 2000). However, improvements of biological soil remediation technologies are still needed for large scale applications of contaminated soil remediation.

The USA, one of the leading countries in soil investigations and remediation, was one of the first to experience the catastrophic consequences of an unacceptable hazardous waste disposal. Environmental and health problems caused by waste dump in Love Canal in New York State initiated a release of a new law in the USA. The Comprehensive Environmental Response, Compensation, and Liability Act (CERCLA), commonly known as Superfund, became an effective legislative and financial tool. It created a tax on the chemical and petroleum industries and provided broad Federal authority to respond directly to threatened and actual releases of hazardous substances that may endanger human health or the environment. Due to this law $ 1.6 billion was collected over five years and went to a trust fund for cleaning up uncontrolled hazardous waste sites (U. S. EPA, 2003).

Examples of the environmental and health problems caused by soil contamination over the world indicate that the problem should not be ignored. Although the risk for the health of wildlife and humans might be complex and not so clearly understandable, the importance of soil as being a junction of all environmental spheres should not be underestimated.

References

Berti, R., Cunningham, S. D. (2000) Phytostabilisation of metals. In: Raskin, I., Ensley, B. D. (Eds.) Phytoremediation of toxic metals: using plants to clean up the environment. John Wiley & Sons Inc. NY. 304 p.

Collins, Y. E., Stotzky, G. (1989) Factors affecting toxicity of heavy metals to microbes. In: Beverige, T. J., Doyle, R. J. (Eds.), Metal Ions and Bacteria. Wiley, Toronto, Canada, pp. 31-90.

European Environment Agency (EEA), (1999) Environment in the European Union at the turn of the century. Environmental assessment report No 2, Copenhagen, Denmark.

Garbisu, C., Alkorta, I. (2001) Phytoextraction: a cost-effective plant-based technology for the removal of metals from the environment. Bioresource Technology 77: 229-236.

Health Canada (1995) Investigating Human Exposure to Contaminants in the Environment: A Handbook for Exposure Calculations. Ottawa.

Helsinki Commission, Baltic Marine Environment Protection Commission (2002) Helsinki, Finland.

Leyval, C., Turnnau, K., Haselwandter, K. (1997) Effect of heavy metal pollution on mycorrhizal colonization and function: physiological, ecological and applied aspects. Mycorrhiza 7(3): 139-153.

Pilipavičius, V. (2002) General ecological problems in Lithuania. In: Technological and Economic Aspects of Soil Bio/Phyto Remediation. Country reports and selected papers from the Training Workshops of International Centre for Science and High Technology and the United Nations Industrial Development Organization (ICS-UNIDO).

Swedish Environmental Protection agency (SNV), (1995) Action program for the treatment. Report no. 4454. Stockholm (in Swedish). p. 125.

Swedish Environmental Protection agency (SNV), (1996) Development of generic guideline values. Model and data used for generic values for contaminated soils in Sweden. Report no. 4639. Stockholm. p. 49.

Terra Nova (2002) Summary Report on Technical and Legal Barriers for the Introduction of Ex-Situ Biological Soil Treatment in the EU Member States. Commission Directorate General Research, Holte, Denmark.

U. S. Environmental Protection Agency (1998) Technology Innovation Office, NATO/ CCMS Pilot Study: Evaluation of Demonstrated and Emerging Technologies for Treatment and Clean Up of Contaminated Land and Groundwater, Phase II Overview Report, EPA/542/R-98/001b, p.16.

U. S. Environmental Protection Agency (1999) NATO/CCMS Pilot Study. Evaluation of Demonstrated and Emerging Technologies for the Treatment and Clean Up of Contaminated Land and Groundwater (Phase III, Annual Report). EPA/542/R-99/007. p.192.

U. S. Environmental Protection Agency, (2003) CERCLA Overview. http://www.epa.gov/superfund/action/law/cercla.htm

Van Deuren, J., Loyd, T., Chetry, S., Liou, R., Peck, J. (2002) Remediation Technologies Screening Matrix and Reference Guide, 4th edition. U. S. Army Environmental Centre, Platinum International Inc., PII-0111641.

Steponas Deveikis

Le patrimoine des anciens domaines/manoirs dans un processus de la planification territoriale en Lituanie

Résumé

Le patrimoine culturel a toujours sa valeur. Il faut garder cette valeur, il faut créer la valeur et la force économique dans un processus de la planification territoriale. Les anciens manoirs présentent une structure vivante dans le paysage lituanien. Ça doit être utilisé pour le développement durable du pays.

La pratique de la sauvegarde du patrimoine des anciens domaines ou manoirs en Lituanie est médiocre en ce moment. Il faut souhaiter de plus d'intégration dans toutes les phases des décisions de la mise en valeur de ce patrimoine dans les espaces ruraux, plus d'intégration dans les documents de la planification territoriale.

Les anciens domaines ont une longue histoire des reformes agraires et d'aménagement foncier ainsi que des infortunes à partir du XVIe siècle. Le réseau des manoirs est bien connu pour les experts fonciers et les experts du patrimoine. Il y a toujours «deux choses dans ces édifices – son usage et sa beauté. Son usage appartient au propriétaire, sa beauté à tout le monde» (Victor Hugo).

Aperçu historique

Je voudrais commencer mon aperçu par les anciens documents cartographiques[1] du XVIe siècle. Ces documents gardent les traces des anciens domaines. Le Grand Duché de Lituanie en cet époque a finit déjà sa première reforme agraire ou Reforme foncière (Reforme Valakas, 1557). Cette reforme a désigné l'arpentage et le bornage des terres du XVIe siècle ainsi qu'a donné un aperçu général de la propriété, de la planification territoriale, du régime

1 Joachim Lelewel et A. E. Nordenskiold sont les deux personnalités les plus célèbres qui ont beaucoup travaillé sur ce sujet et qui ont gardé les oeuvres cartographiques pour l'Europe, l'un pour l'université de Vilnius et un autre pour Helsinki et les pays scandinaves. Il y a beaucoup des anciennes cartes dans les bibliothèques de l'Europe, en particulier dans la Bibliothèque de l'Université de Vilnius.

Plan du domaine Zagaré (dressé en 1989)

foncier. L'arpenteur juré qui avait travaillé dans cet époque a créé l'aménagement foncier du pays. Ça donnait le système des domaines qui quadrillaient notre pays.

Les levées de reforme Valakas ont été présentées dans un oeuvre très important Tabula Magni Ducatus Lithuaniae désigné sous la direction de duc M.K. Radvila (Radziwil) et imprimé à Amsterdam en 1613. Nous trouvons ici dans le territoire de notre actuel pays 60 rivières et 150 centres d'habitation. Nous avons sur cette carte des domaines nobles. Nous pouvons suivre sur les autres anciennes cartes ainsi que dans les inventaires les étapes du développement des domaines privés et d'Etat.

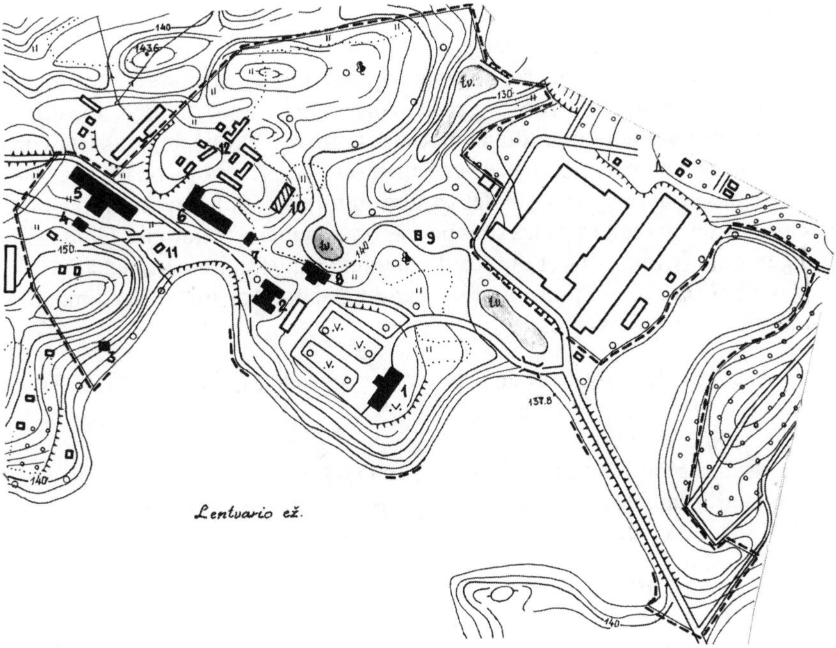

Plan du domaine Lentvaris (dressé en 1989)

En bref, nous pouvons marquer les étapes de l'évolution des domaines en Lituanie suivantes :

1) Jusqu'au XVIe siècle, la période de la naissance et de la croissance des domaines féodaux.
2) Entre les XVIe et XIXe siècles (1557-1861), les années de la prospérité des domaines féodaux.
3) Les années 1861-1922, années des reformes agraires et d'économie rurale.
4) Depuis 1922, cette période est la période des reformes politiques et économiques qui ont dirigé vers la dévastation permanente du patrimoine des anciens manoirs en Lituanie.

Une base juridique et la pratique actuelle de la sauvegarde des anciens manoirs

Il y a 15 ans en Lituanie que nous avons dressé l'inventaire des anciens domaines. C'était un travail immense fait par les organismes des Ministères de l'Agriculture, de la Culture – Institut de la Réglementation Foncière, Institut de la Restauration du Patrimoine Culturel, Centre du Patrimoine Culturel ainsi le Département de l'Environnement.

Nous avons reçu les chiffres des anciens domaines ou manoirs assez impressionnants. En époque de 18e-19e siècles dans le territoire du pays nous trouvons de 3500 des domaines développés. A la fin de 20e siècle nous trouvons seulement 800 objets matériels de ce patrimoine – les manoirs ou les sites culturels. Nous pouvons y trouver des trésors culturels pour la préservation, pour la sauvegarde ou seulement les sites pour le développement durable. Cinq cents d'objets sont inventoriés et ont les dossiers de l'inventaire complets.

Une base juridique pour la préservation des anciens domaines / manoirs comporte les lois suivantes:

- Loi sur le rétablissement du droit de la propriété privée (1991; 1997)
- Loi sur la reforme foncière (1991; 1997)
- Loi foncière (1994)
- Loi sur la planification territoriale (1995)
- Loi sur la sauvegarde des trésors culturels immobiliers (1992)
- Loi sur les territoires protégés (1993; 2002)
- etc. ainsi que les arrêts de Gouvernement et ministériels (1992-2003).

L'arrêt de Gouvernement No. 256 du 7 avril 1992 sur les principes de la sauvegarde des anciens domaines – des monuments culturels en réalisant la reforme agraire et la privatisation a confirmé les principes essentiels: le terrain autour de la ferme centrale, un propriétaire pour toute la complexité du manoir, etc. L'arrêt gouvernemental No 369 du 27 mai 1993 «Sur la protection des anciens domaines et manoirs» a mis en vigueur «Le règlement général des anciens domaines monumentaux». Mais ça reste les actes dormants. Pas d'action commune, pas de contrôle stricte, pas de moyens économiques...

Le programme de la mise en valeur du patrimoine des anciens domaines ou manoirs

Faut-il concevoir les manoirs dans le paysage du XXI siècle? Comment nous devons posséder ce patrimoine? Seulement cent sites des anciens domaines sont inscrits dans la liste des monuments culturels. Parmi eux les domaines des comtes Tyszkewicz à Palanga et à Lentvaris, Traku Voké et Uzutrakis, pas loin de Vilnius avec les châteaux, avec les parcs dessinés par les architectes français père et fils Edouard et René André, ainsi que le manoir de Zagare avec le châteaux, les bâtiment d'exploitation, le parc dessiné par paysagiste G. F. Kuphaldt (1858-1938).

Schémas directeur du territoire de Lituanie, Bendrasis Planas, voté par le Parlement de la République de Litanie en novembre 2002 nous montre les directions générales de développement durable du pays. Schémas directeur du territoire de Lituanie est un document principal, un document – clé pour la planification territoriale où nous pouvons trouver la base des décisions économiques et sociales pour les anciens domaines aussi.

L'idée même de stabilité paysagère est absurde – dit Jean-Robert Pitte, spécialiste français de l'histoire du paysage. Il faut protéger et développer les sites culturels. La Convention européenne du paysage (2000), les Conventions de l'UNESCO et les autres documents internationaux nous présentent les solutions dans ce domaine. Dans les années 2002 et 2003 la Lituanie a mis en valeur la Conception de la sauvegarde et de l'intégration dans la vie sociale du patrimoine des anciens domaines, adoptée en 2002 par la Commission du Patrimoine culturel ainsi que le Programme de sauvegarde des anciens domaines adopté par le Département des trésors culturels et par le Gouvernement en 2003.

Nous pouvons travailler et concevoir la mise en place de ces documents légaux . Nous avons assez de projets dans les dossiers, pas de projets mises en place… On manque des documents de la planification urbaine et rurale dans les municipalités et des moyens économiques. Restent beaucoup de choses à faire dans le champ d'action sur le système du patrimoine des anciens domaines.

Quelques conclusions sur la pratique actuelle et à l'avenir

Une première conclusion faite déjà par Victor Hugo, l'homme des lettres français, comme je viens de dire – «il y a deux choses dans un édifice (manoir): son usage et sa beauté. Son usage appartient au propriétaire, sa beauté à tout le monde». Il faut trouver un partenariat dans la mise en valeur du patrimoine culturel.

La deuxième conclusion – la structure des anciens manoirs ou domaines est le paysage culturel utilement planifié. Ça doit être bien utilisé dans toutes les phases de la planification territoriale, de l'aménagement foncier pour le développement durable du pays.

La troisième conclusion – l'utilité des manoirs doit être spécifiée dans tous les documents de la planification territoriale ainsi que de la reforme foncière. Les plans et les projets architecturaux et paysager doivent conclure les données de l'inventaire des domaines. La Lituanie possède l'inventaire assez riche.

La quatrième conclusion sur la base juridique. Les chartes et les lois sur le paysage de l'Europe ainsi que de Lituanie doivent être bien lues et bien compris dans le milieu des experts fonciers ou de planification rurale ainsi qu'urbaine. Il faut concevoir une loi d'orientation sur les anciens manoirs – le patrimoine culturel.

La cinquième conclusion – nulle manoir sans propriétaire. L'Etat, le secteur publique ou privé doit mettre en valeur ce patrimoine du pays. Le *marketing* du patrimoine culturel doit se concentrer et embaucher sur la complexité des domaines. L'actualité et l'avenir a pour l'objet de trouver des solutions de la complexité économique, écologique, esthétique ainsi que le développement durable de ce patrimoine.

JURGITA ANTUCHEVIČIENĖ, EDMUNDAS KAZIMIERAS ZAVADSKAS

Problems of Regeneration of Mismanaged Rural Buildings and Built Environment in Lithuania

1. Introduction: An unique set of environmental problems in the newly enlarged European Union

Since the beginning of restoration of independence in Lithuania the attention has been focused on integration into European State and political structures. Lithuanian politicians made every effort for this intention. In the near future Lithuania will be initiated to European Union. Enlargement of the European Union from 15 to 28 or more countries will bring with it at least a further 170 million inhabitants, a 58 % increase in land area and a unique set of environmental problems and assets[1]. There are specific problem areas and objects in these countries, such as heavily polluted ex-military sites, industrial centres or large agriculture complexes that do not satisfy present technological, economic and ecological requirements and located in the former Soviet Union countries. The question of rational management of such objects and their territories must be solved immediately, because their negative influence on the landscape, environment and economy is increasing on and on.

The problem of derelict and mismanaged rural buildings and built environment in Lithuania is analysed in this paper. The question of decline and deterioration of property is also relevant to other countries. Most countries have areas of land, within city environments or on a regional scale, that are deteriorating or deteriorated and that exhibit characteristics such as areas with a large proportion of empty industrial or commercial buildings, potentially derelict and based on a declining or vanishing industry[2]. But the solution of the problem regarding rural property is a new and having no analogies process in the developed European countries. The scientific research is performed being conscious of an importance of rural buildings in the structure of the economic potential of the state and being aware of natural and social environmental damage. The structural analysis and the spatial distribution pecu-

1 Commission of the European Communities (2001). *Communication from the Commission to the Council, the European Parliament, the Economic and Social Committee and the Committee of the Regions*, On the six environmental action programme of the European Community 'Environment 2010: Our future, Our choice', Brussels.

2 International Council for Research and Innovation in Building and Construction (2002). Regeneration of the built environment. *Information*, No 2, p. 5.

liarities of mismanaged Lithuanian rural buildings observed with the help of mathematical statistical analysis were carried out in the paper, having the aim of suggesting principal trends for the establishment of revitalization priorities of derelict property according to the general principles of sustainable development.

In this paper regeneration of objects and their environment is analysed according to the aspect of sustainable development for several reasons. Derelict and mismanaged buildings in former Soviet countries merged on the ground of political, economic and social changes. These objects were built without proper environmental legislation. They are today, threatening the natural environment and people's safety as the buildings decay. The proper and sustainable management of these buildings and their territories could help decrease the rural, social and economic crisis as well as prevent environmental and landscape degradation. Moreover, sustainable development is greatly appreciated while planning the distribution and use of various national, local funds and financial funds of Europe Union. It focuses on integration of the environment into economic and social areas. Environment and economic development are not mutually exclusive neither is the welfare of people, according to the point of view of the European Union.

2. Sustainable development trends in EU and specific features of countries in transition

Sustainable development continues to receive increasing international recognition and it has become a key guiding principle for the global society at the start of the new millennium[3]. The European Union pays great attention to attitudes towards sustainable development. It has the most comprehensive and advanced environmental legislation in the world. That is the reason for making demands of a clean and healthy environment for EU candidate countries from Central and Eastern Europe.

When analysing sustainable development, two categories of countries are usually distinguished – developed countries and developing countries. Problems of poverty, equity, education and health care services are typical of developing countries. The developed countries face problems of an excessive consumption of natural resources and environmental pollution. Despite the mentioned differences, both groups of countries have a common feature, because their development takes place in the way of natural evolution. At the

---

3 National Research Council Board on sustainable Development, Policy Division (1999). *Our Common Journey: A Transition Towards Sustainability*, National Academy Press, Washington, DC.

beginning of the last decade, after the former Soviet Union collapsed and the Soviet block was broken down, a third group of countries with especially specific features of their development were formed. The main feature of these post–communist countries is that they were deprived of the possibility of natural evolution[4]. The Soviet dictatorship showed respect neither for nature nor for humankind. Both were subordinated to the main goal of expanding material production, without regard for the well being of the population, their quality of life, or for the natural environment. Environmental legislation meant little in a nation that was not based on constitutional principles. Soviet legislation consisted of grand pronouncements and largely unenforceable standards[5]. Soviet agriculture was environmentally destructive and wasteful of natural resources. Land was impoverished and eroded by unscientific attempts to extract higher yields. Chemical fertilizers and pesticides were overused on collective and state farms. Unique to the Soviet Union was the environmental impact of the militarised Soviet economy. Military bases contaminated the environment with fuel, lubricants, dumped garbage and equipment, and polluted groundwater supplies. Troop exercises destroyed forests and ruined agricultural land. Also, pollution was generated by the production and storage of chemical and nuclear weapons[6, 7]. The mentioned problems were typical of the Baltic States because the Soviet government treated the non-Russian republics like colonies, exploiting their natural resources, degrading the land and locating polluting industries on their territory.

Lithuania started to implement its political, social and economic reforms a decade ago. Since the beginning of the restoration of independence in Lithuania attention has been focussed on targets of a sustainability policy. During the last ten years Lithuania had the possibility to choose a way in which it was going to be developed. Lithuania has chosen the way based on the market economy and private property. In June 1992 the representatives of Lithuania took part in the United Nations conference on Environment and Development and they declared support of the way of sustainable development. Our country has joined into a new era of economic, social and environmental sus-

4 Juknys, R. (Ed.) (2002) *National Report on Sustainable Development from Rio to Johannesburg. The Republic of Lithuania,* Vilnius, Lithuania.

5 Ziegler, C. E. (2001) Soviet studies: environment. *International Encyclopaedia of the Social & Behavioural Sciences,* Elsevier Science Ltd, pp. 14723-14728.

6 Baltrėnas, P., Ignatavičius, G., Vaišis, V. (2001) Investigation of soil pollution with heavy metals in the Pabrade central military ground. *Environmental Engineering (Aplinkos inžinerija)*, Vol. 9, No 1, pp. 3-8 (in Lithuanian).

7 Baltrėnas, P., Oškinis, V., Ignatavičius, G., Kumpienė, J. (2001) Soil transgression and possibilities of improving environmental protection in the tank field of the Lithuanian central military ground. *Environmental Engineering (Aplinkos inžinerija)*, Vol. 9, No 2, pp. 103-109 (in Lithuanian).

tainable growth. Now Lithuania is in the process of transition to a market economy with all the difficulties of the transition period[8].

On the basis of a new national economic development policy and taking into account new environmental protection problems and objectives, the Parliament of the country approved the Environmental Protection Strategy of Lithuania in 1996. The Government adopted an Action Program that aimed at a clean and healthy environment, biological diversity and landscape conservation and an effective consumption of natural resources. Then, for the first time the term "sustainable development" was officially mentioned in our country[4]. The environmental protection strategy was based on principles laid down in the Rio de Janeiro Declaration.

The Master Plan of the Republic of Lithuania is prepared till 2020 with correct realization principles for sustainable development. The basic principle of the Master Plan is to ensure sustainable development of the country's territory, designating the best possible way to use the territory without harm to landscape and without violating the interests of the present and future generations. The Lithuanian Environmental Strategy and Action Programs are also based on these principles. In 1996-1998 the National Strategy for the Implementation of the Framework Convention on Climate Change, the National Strategy and an Action Plan for Biodiversity Conservation, Ecological Education Strategy and Action Plan were prepared and adopted. In 2000 Lithuania signed the Convention on European Landscape. This allows the country to develop protection and management of the landscape more intensively within the context of the system of European law and to formulate a more accurate state policy in this sphere.

Essential changes to the political and economic system determined great changes in different fields, which should be taken into account when analysing the experiences and prospects of development of the country.

3. Mismanaged rural buildings and their environment as the object of the research

A dominating element of the total area of Lithuania is constituted of rural areas. It is this part of the territory which is situated beyond the city borders. According to the data from the Main Statistical Office so defined rural areas cover 97.4 % of Lithuania's area, inhabited by about 32 % of the total popula-

---

8 Pakalnis, R. (2001) Sustainable development is the current modern way without alternatives in W. Leal Filho, R. Pakalnis, L. Sakalauskas (Eds.) *Proceedings of the International Conference Sustainable Development in the Information Society*, Institute of Mathematics and Informatics, Vilnius, Lithuania, pp. 10-13.

tion. Lithuania is more "rural and agricultural" than the majority of European countries. Over 18% of employed persons still work in agriculture (in the EU countries this index is 5.5% on average). Agriculture generates 6% GDP (in the EU about 2%). The share of agriculture in export is 11.6%, in import 10.2%. Accordingly rural property constitutes an important part of the economic potential in Lithuania. Buildings used for farming and as part the rural infrastructure constitute a great part of this type of property. Most of these objects were built during the Socialist years under social economic conditions. Political and economic changes were followed by an unsuccessful reorganisation of the agricultural sector in Lithuania.

The past decade has been a period of cardinal changes for the Lithuanian villages. Within the system of collective farms deteriorating and rapidly forming private farms, not only agricultural production but also the entire life of the rural population changed in essence. After the re-establishment of Lithuania's independence in 1990, the land reform legitimated private land ownership. In 1992, after the properties of collective and state farms had been privatised, people's farms, agricultural partnerships and other agricultural enterprises were established. However, most agricultural partnerships collapsed in a short time. In 2000, the areas of farm crops belonging to farmers and other inhabitants constituted over 87% of the total area of farm crops. This is characteristic in that private farms are small in Lithuania. In 2000, the average farm was 12.6 hectares and over 80% of the farms were smaller than 10 hectares[4]. Also, a majority of agricultural buildings are private property, but they are not used and have almost been destroyed. Many rural properties, due to their large parameters, energy susceptibility, technological and external (economic) depreciation do not meet the contemporary production requirements. Small farmers are not capable of using or holding large complexes in proper condition. These buildings do not satisfy the present technological, economic and ecological requirements. Much investment is required for the purpose of using these objects. Furthermore, quantities of agriculture production are declining and a part of the agricultural buildings will become derelict unless they will be redeveloped and adapted for other uses. Animal production has decreased rapidly in the transitional period. As compared with 1990, only about 40% of meat and 60% of milk and eggs are produced. The main reasons for such decrease are the restructuring of the agricultural sector and essential changes in the domestic and foreign markets. During the years of the collective farming in seeking to provide the deficit market of the former Soviet Union with agricultural products, Lithuanian agriculture was thrown off balance and the production sector has been developed hyper-trophically.

As a result of the mentioned transformations, there are many derelict and mismanaged buildings in the rural areas now. According to data from the Lithuanian State Territorial Planning and Construction Inspectorate, also from Lithuanian Institute of Agrarian Economics and data on the carried out questionnaire by previous researchers, there are nearly 6300 derelict buildings and nearly 1100 mismanaged buildings in the countryside, that is 7400 unexploited objects in total. Many of them are semi-destroyed and are falling into decay. Such a phenomenon is negatively influencing the economy of the country. It is hard to revitalize these buildings because of technological and economic changes during the last decade. They are also negatively influencing the environment and landscape, threatening people's safety and wasting the real estate while decaying irreversibly.

Generalized findings on investigative property on the base of mentioned data are carried out. The place of unused rural buildings in the total structure of Lithuanian real estate is presented in Table 1. These objects occupy about 962 ha. This constitutes 0.52 percent of the total building area in Lithuania.

According to classification of rural property in conformity with Soviet catalogues of former kolkhoz building is established that rural property dominated by farming buildings (Table 2). While in the whole structure of Lithuanian real estate dwelling houses constitute the largest part of buildings, namely 75 %. Industrial buildings constitute 6 % of total stock of buildings. This fact expresses the importance of redevelopment of rural buildings. Deteriorated farming buildings amount even 3 % of total stock of Lithuanian industrial buildings. It is a national asset of the Republic of Lithuania and it must be protected and effectively used.

It is important to find out how many rural buildings are fit for reconstruction subject to their depreciation. It is considered that only objects with depreciation no higher than 30 percent are suitable for repair and future use. S. Lunkevicius et al. estimated, that some 60 percent of buildings became of poor state since 1991 [9]. They have already decayed or are quickly decaying and are not proper for reconstruction. According to data from the Lithuanian State Territorial Planning and Construction Inspectorate, there are some 70 percent of derelict and mismanaged buildings of poor state that need to be demolished. Determination of the number and the building area of objects designated for demolition are presented in Table 3. This is important estimating their environmental impact.

Forasmuch as reuse and recycle of building materials is important requirement for sustainable construction, quantity of building materials and their decomposition were estimated (Table 4). The estimation is based on data from the Lithuanian State Territorial Planning and Construction Inspectorate, also from Lithuanian Institute of Agrarian Economics and from the research of S. Lunkevicius et al.[9]. It is estimated that the quantity of scrap of buildings to be demolished is some 2870 thousands of cubic meters.

Some 30 percent of rural buildings are suitable for repair according to their physical state. It is important to choose the most effective strategy of revival of these objects. The strategy ought to be based not only on physical characteristics of an object, but also on regional peculiarities, on environmental performance, and ought to meet the requirements of local community, that is ought to be based on principals of sustainable development.

Table 1. Building area of derelict and mismanaged rural property

Factor	*Value*
Total building area in the territory of Lithuania, ha	185098 ha
Building area of unexploited rural buildings, ha	962 ha
Building area of other real estate, ha	184136 ha
Building area of unexploited rural buildings in proportion to building area of other real estate, %	0,52 %
Density of unexploited rural buildings (average number of buildings, situated in 1,000 ha area of farming land), unites / 1 000 ha	1,88

Table 2. Distribution of derelict rural property according to purpose of use

Purpose of use	*A part of property*
Industrial buildings	83 %
Social buildings	2 %
Dwelling-houses	1 %
Other buildings	14 %

Table 3. The number and the building area of objects to be demolished

Factor	Derelict and mismanaged rural buildings	Buildings, suitable for future use	Buildings, designated for demolition
The number of objects, unites	7401	2105	5296
Percent of objects, percents	100 %	28 %	72 %
Building area of objects, ha	962 ha	269 ha	693 ha

Table 4. Decomposition of rural buildings to be demolished

Material	Quantity, percents
Bricks, ferroconcrete, asbestos	65 %
Wood	20 %
Metals	10 %
Other material	5 %

4. Analysing spatial distribution peculiarities of mismanaged rural buildings in search of their sustainable regeneration

The revival of Lithuanian rural property as a process of investment and substantiation of financing method has been analysed in several scientific papers[9-11]. Though various activities of construction and environmental sector have to be regarded and analysed when considering sustainable development. Sustainability has become an important issue mainly in countries in which construction is in decline[12]. The prevention of urban decline and urban sprawl is concern expressed by most countries. In the Western and Eastern European countries conservation of open space and safeguarding the structure of rural settlements is a major issue. This concern was consolidated by the two main challenges and future trends to construction industry thrown down by Kohler[13] and Rees[14], namely that countries must reduce the environmental impact of their built environment tenfold by 2040 and that they must stop constructing additional new buildings, instead to improving their existing stock. These worldwide attitudes amplify the importance of redevelopment of derelict buildings in rural areas of Lithuania.

Rational use of derelict buildings is analysed from the aspect of sustainable development for the reason mentioned above and depending on local peculiarities of a ground of the explored phenomenon and subsequence of a possible decision. No priorities are given to the economic benefits of revitalization. A great attention is paid to natural and social environmental potential, because within the system of collective farms deteriorating and rapidly forming private farms, not only agricultural production but also the entire life of the rural population changed in essence. Many rural people are unemployed due to a rapid and not very successful reorganization of the agricultural sector, during the transition period. A lot of social and cultural properties have

---

9 Lunkevičius, S., Ustinovičius, L., Zavadskas E. K. (2001) Possibilities of revitalization of unused rural property. *Journal of Civil Engineering and Management (Statyba)*, Vol. 7, No 1, pp. 44-55 (in Lithuanian).

10 Lunkevičius S., Ustinovičius, L., Zavadskas, E. K. (2001) Substantiation of financing approach for rural property. *Journal of Civil Engineering and Management (Statyba)*, Vol. 7, No 2, pp. 148-157 (in Lithuanian).

11 Lunkevičius, S., Ustinovičius, L., Zavadskas, E. K. (2001) Ranking efficiency of rural property investment projects using multicriteria decision methods. *Journal of Civil Engineering and Management*, Vol. 7, No 3, pp. 238-246 (in Lithuanian).

12 Bon, R.; Hutchinson, K. (2000) Sustainable construction: some economic challenges. *Building Research & Information*, Vol. 28, No 5/6, pp. 310-314.

13 Kohler, N. (1999) The relevance of the Green Building Challenge: An observer's perspective. *Building Research & Information*, Vol. 27, No 4/5, pp. 309-320.

14 Rees, W. (1999) The built environment and the ecosphere: a global perspective. *Building Research & Information*, Vol. 27, No 4/5, pp. 206-220.

also been damaged. This phenomenon exacerbates the social and moral crisis of the rural population in our country. Also, the territories of these buildings are not ecologically stable due to weak environmental legislation in the socialist years. Devastated objects are ruining the landscape and negatively influencing the environment.

There are several possible ways of arranging mismanaged buildings and their territory. Revitalization variants should be selected according to technical conditions, social interests and environmental possibilities. It is necessary to set new ecologically motivated priorities in areas where huge anthropogenic intervention was made. Derelict buildings in State parks and other protected areas need to be demolished in order of the Environmental Ministry. But this process is still going very slowly. Buildings that are less depreciated should be renovated and used for proper purposes, especially since in European Union and in other countries attention to urban renovation is given as opposed to the development on new sites. Revitalization variants should be selected according to technical conditions, social interests and environmental possibilities. Buildings with great depreciation and those not fit for renovation must be dismantled instead of falling into decay or simply demolition. Dismantling instead of demolition helps to separate different building materials and to reuse and recycle materials giving superior utilization options. Recovering and reusing building materials can partially reduce the environmental impacts of the construction industry by decreasing the volume of construction and the demolition waste stream, conserve natural primary resource stocks and landfill space and generate a new source of building materials that are less energy and resource intensive[15, 16].

The development policy and priorities must be based on reliable scientific information and knowledge. In order to manage derelict buildings, firstly the existing situation needs to be explored[17]. Therefore, scientific research of the present state has been completed in this paper. Observations of these buildings' territorial distribution and their peculiarities were made in the context of the conception of the country's economic, social and ecological sustainable development. Economic, social and environmental factors determining

15 Peng, Ch.; Scorpio, D.E.; Kibert, Ch.J. (1997) Strategies for successful construction and demolition waste recycling operations. *Construction Management and Economics*, No 15, pp. 49-58.

16 Formosso, C.T.; Masce, L.S; De Cesare, C.; Isatto, E.L. (2002) Material waste in building industry: main causes and prevention. *Journal of Construction Engineering and Management*, Vol. 128, No 4, pp. 316-325.

17 Bossel, H. (Ed.) (1999) *Indicators for Sustainable Development: Theory, Method, and Applications.* A Report to the Balaton Group, International Institute for Sustainable Development, Canada.

the distribution peculiarities of unused buildings and their environment were established by using methods of mathematical statistics[18].

The data were grouped into several regions. The main distribution was made into three regions according to the concept of the country's spatial development[19]. The concept of the country's spatial development is based on tendencies in the industrialization of the country's economy, influences from the internal and external markets, the processes of the internal economy of the country, the economic, social, environmental quality, the system of settlements as well as on the inertia of the country's development. Lithuania's territory is divided into three main types of areas: areas of active development, areas of regressing development and "buffer" areas. The largest amount of facilities held and the greatest variety of activity and maximum internal as well as foreign investment is characteristic to areas of active development. The main industries, science, culture and facilities centres and major highways are located in these territories, in contradistinction to areas of regressing development. Areas of active development cover main cities and zones of their influence. The economic base of areas of regressing development is composed of agriculture, forestry and recreation. They cover northeastern and southern parts of Lithuania. "Buffer" areas take a middle place according to the character of activity, geographical and environmental situation and singularities of the local population. They are situated in territories of a slight influence of largest cities, farther from them. These differences must be taken into consideration when finding the most rational ways of derelict buildings' management. The priorities of disposals should differ according to local singularities of intention and of finding a sustainable solution and meeting the environmental, social and economic needs and restrictions.

Distribution according to territorial singularities was performed also. This distribution mostly depends on land fertility and other differences of agricultural conditions. The Central – Northwestern and the Southeastern parts of Lithuania's territory were divided for the purpose of the analysis.

One more distribution is made into two types of regions: surroundings of main cities and zones of their influence, and rural areas, situated farther from larger towns.

Sets of mismanaged agricultural industrial buildings and rural buildings of other functions (housing and facilities) have also been analysed.

A number of derelict and mismanaged buildings in 1,000 ha of territory in an administrative region were used as a dependant variable in the analysis.

---

18 Aivazian S. A.; Mxistarian V. S. (1998) Applied Statistics and essentials of econometrics (Прикладная статистика и основы эконометрики). Moscow: Junity (in Russian).

19 Juskevičius, P. (1999) The concept of the country's spatial development. *Town Planning and Architecture (Urbanistika ir architektūra),* Vol. 13, No 2, p. 49-55 (in Lithuanian).

Indexes used by the authors as independent (factorial) variables are as follows: farming land productivity grade, farming land percent rented by farming communities, parameters of life quality, population activity indices, indices of farming and corn agriculture territorial concentration in the years 1990 and 1997.

Several correlation matrixes according to regional peculiarities and the use of buildings have been calculated and statistical connections have been estimated.

Multiple Regression Analysis of Forward and Backward selection was applied to determine a causative relationship between dependent variable and independent factors. Method of Principal Components was performed for all mentioned groups of variables.

Some conclusions can be drawn according to the results of this analysis. It is estimated that a connection between derelict and mismanaged rural buildings and the regional concentration of economic, social and agricultural indices are different in zones of active and regressing development and in "buffer" areas. In areas of active development, the distribution peculiarities are influenced mostly by the population's activity indices and life quality parameters. These connections are reverse. There are less mismanaged buildings, rural property is renovated and used for various kind of activity in regions, were inhabitant activity is higher (a higher percent of adults are employed) and the life quality is better. This indicates better possibilities of local inhabitants for investment and business development as well as attractiveness of a region for foreign investors.

In areas of regressing development, indices of farming territorial concentration are statistically most significant. The connection is reverse. This point can be explained by the fact that stronger collective farms existed in districts with better agriculture conditions. Stronger collective farms were less deteriorated and more farming communities were founded here. Farming communities use buildings for farming purposes and for their agricultural activities. So there are less derelict and mismanaged buildings in these regions. A part of the analyzed buildings are private property, but they are not used and almost destroyed, because small farmers are not capable of using large complexes and maintaining them in proper conditions.

"Buffer" areas take an intermediate place according to their concept as well as to the results of the mathematical statistical analysis[20-22]. The mostly

20 Antuchevičienė J. (2002) Territorial distribution peculiarities of derelict rural buildings in Lithuania according to the conception of the country's economic, social and ecological sustainable development. *Environmental engineering (Aplinkos inžinerija)*, Vol. 10, No 2, pp. 93-101 (in Lithuanian).

21 Antuchevičienė J. (2002) Relation between socio-economic development and mismanagement of buildings in rural areas. *Geography (Geografija)*, Vol. 38, No 2, pp. 74-80 (in Lithuanian).

22 Antuchevičienė J. (2003) Differential renovation of soviet agro-industrial rural landscape in the context of country sustainable development. *Technological and eco-*

statistically significant relation established between mismanaged buildings and inhabitant activity index.

In rural areas situated far from towns most significant statistical connections established between density of derelict buildings and farming land productivity grade as well as indices of farming and corn agriculture territorial concentration.

In areas under the influence of large towns distribution peculiarities of derelict and mismanaged rural buildings statistically depend on population activity indices and parameters of life quality. The influence of farming factors is insignificant in areas of such type.

Analysis of mismanaged buildings used for farming showed that the distribution peculiarities of dependant variables are influenced mostly by indices of farming territorial concentration, farming land productivity and farming land percents that are rented by farming communities. Distribution peculiarities of mismanaged buildings of other uses in rural areas are mostly influenced by the population's activity indices and life quality parameters[21, 23]. There are many renovated buildings used for commercial purposes or manufacture in areas of higher activity or those with a higher life quality.

Matrixes of various analysed groups of variables were compared having the purpose of proving that their structure is statistically different. The appropriate index, namely statistics M was counted according to the methodology presented in the respective literature[18]. The established indexes were compared to a radical of Pirson's distribution. The fact was proved that correlation structures of derelict and mismanaged buildings of groups mentioned above are different to the credibility $p=0.95$. Consequently, scientific recommendations for a rational regeneration of buildings and their surroundings could be prepared on the base of performed explorations. The research demonstrates that the same solution concerning management of abandoned real estate is inexpedient for application within the whole territory of a country. Possibilities of rational revitalisation of unused rural buildings in Lithuania mostly depend on features of locality. One ought to make renovations and use buildings not for farming in the areas of active development. In regressing or other localities these buildings can be effectively used for farming purposes and there are less possibilities of changing their functions successfully. But the quantities of agricultural production are not increasing in Lithuania. So, damaged and not useful farming objects ought to be dismantled or renovated and adapted for other activities according to local possibilities.

---

nomic development of economy (Ūkio technologinis ir ekonominis vystymas), Vol. IX, No 1, pp. 27-33 (in Lithuanian).

23 Antuchevičienė, J. (2001) Relationship between social economic conditions and distribution of unused rural buildings in Lithuania. *Social Sciences (Socialiniai mokslai)*, Vol. 32, No 6, pp. 49-55 (in Lithuanian).

5. Conclusions

According to the assessment of existent situation, derelict and mismanaged rural buildings make a considerable amount of total building stock in Lithuania. It is important to solve the question of their rational revitalization.

Derelict and mismanaged rural buildings cause a negative influence on the economy and on the natural as well as social environment of the country. Consequently, revitalization of unused rural buildings must be analysed in a context of sustainable development, combining the economic benefit of redevelopment of buildings with the environmental and social interest.

Economical and social factors determining distribution and revitalization perspectives of unused buildings were established by using methods of mathematical statistics. It was estimated that the peculiarities of derelict and mismanaged rural buildings territorial distribution are statistically different in various zones of development activity as presented in the conception of Lithuania's spatial development and they differ according to the purpose of buildings.

Revitalization of rural property includes restoration, redevelopment and/or dismantling of buildings. The research indicated that the priority of revitalization alternatives of buildings and their territories depends on local features and peculiarities of the object. The same solution is inexpedient for application to any object and within the whole territory of a country.

On a base of this research, scientific recommendations for a rational revitalization of buildings and built environment could be prepared, ensuring the sustainability of the solution.

ZYDRUNAS DĖNAS, ALGIRDAS KUMETAITIS, SAULIUS SLIAUPA, ALGIMANTAS ZAKAREVIČIUS

Application of GIS for Surface Erosion Model of Lithuania

Abstract

Surface runoff and soil erosion are major threats to sustainable agriculture, therefore, mapping regional erosion risk is increasingly needed for land use management. Erosion results from the interaction of many different parameters that vary in space and time. The aim of this work is to develop a methodology based on application of GIS tools for the evaluation of erosion risk at national scale and incorporating different types of data. The erosion factors have been graded by previous studies based on field experiments. Soil lithology is considered as a key factor in runoff and erosion risk on cultivated soils. It was combined to data on land use CORINE Land Cover. Digital Elevation Model has been created at the scale of 50*50m pixels that was incorporated in calculations. The modeling indicates that the main areas affected by erosion risk are related to topographic highs of the western and eastern Lithuania.

Introduction

Assessment of surface soil erosion is needed by national environment agencies. Erosion results from the interaction of many parameters, no simple model can take into account all relevant factors. The previous models of Lithuania were based on semi-quantitative approach (cf. Cesnulevicius, 2001). Quantitative information on relief is often used in soil studies including the modeling (Pennock et al., 1987; Moore et al., 1993; Beven et al., 1995; Le Bissonnais et al., 2001; Florinski et al., 2001). The aim of this work is to develop a methodology of complex assessment of different factors contributing to erosion risk in Lithuania. The various erosion factors have been graded for different geographical situations and erosion mechanisms according to previous field and laboratory experiments, the various erosion types observed in Lithuania were classified (Racinskas, 1990; Jankauskas, 1996). Sub-soil lithology, land cover and relief are considered as the key factors in runoff and erosion risk on cultivated soils. A model of erosion risk has been developed within a Geographical Information System GIS. The model uses expert rules to combine data on land use CORINE Land Cover database, sub-soil lithology model provided by Geological Survey of Lithuania and relief Digital Elevation Model that was considerably improved.

Erosion Equation

Erosion rates of Lithuanian soil were estimated using different approaches, including field experiments and observations of natural processes, and laboratory studies (Racinskas, 1990; Jankauskas, 2000). Different factors contribute to the erosion rate; their impact is quite variable. One of the main factors is topography. The equation by (Racinskas, 1990) was applied to model the contribution of this factor to the erosion rate:

$$ETOP = 0.09*l + 1.62*i + 0.18*e - 1.46,$$

where E is the erosion potential (m^3/ha), l is the slope length (m), i is the slope steepness (deg) and e is the slope exposure (32 for the eastern slopes, 34 for the western slopes, 32-i for the northern slopes, and 32+i for the southern slopes).

Besides to topography, the sub-soil lithology and land cover type are the other considered factors. They were graded using the same approached as presented by (Racinskas, 1990):

$$ELC = B * [1-(C-0.23)],$$

where B is the lithology factor of sub-soil and C is the land cover factor. These factors are discussed below.

The total erosion rate was calculated as a sum of afore described parameters:

$$ETOTAL = ETOP * ELC.$$

The climate factor was assumed constant (K = 1) over the Lithuanian territory.

Methods

Calculation of the Erosion Potential was performed using MapInfo Vertical Mapper software. In the Vertical Mapper all datasets to be used for calculations must be in the raster form, common Coordinate System and the same grid cell size. Therefore, the all required digital data of the Lithuanian territory were converted to the required form first.

The relief data are available in raster form in Arc/Info format. Using Arc/Info software the relief data were exported to ASCII format and imported to

Vertical Mapper using Import Grid command. In our relief model, the grid cell size is 50 meters. This accuracy was kept the same in all calculations of the Erosion Potential.

Other datasets are in the vector form. After importing to MapInfo format, these datasets are converted to raster form with grid cell size 50 × 50 meters. Transformation of data is performed using Region to Grid function in Vertical Mapper.

Calculation of the Erosion Potential was performed in two steps. The first step is calculating the Erosion Potential accounted for the Relief factor. Slope steepness and slope aspect were calculated from digital relief model using Vertical Mapper command Create Slope & Aspect. Slope steepness is directly used in calculations of Erosion Potential. To calculate the slope exposure factor, both slope aspect and slope steepness are necessary. Different dependences of the slope steepness and exposure are for the north-, south-, west- and east-facing slopes. Using Grid Query function slope aspect layer is separated into four subsets. After calculations of the slope exposure factor for each subset, they are aggregated using Vertical Mapper Splicer command. Approximated slope length values of Lithuania were incorporated according to (Cesnulevicius et al., 1984). The Erosion Potential is calculated using Grid Calculator function.

The Erosion Risk related to relief factor was multiplied by the erosion potential values related to the Sub-soil lithology and Land cover factors. For that reason, the adjusting weights of each factor are calculated. Using Grid Query function the grid cells that have common adjusting weight are selected. For each selected subset adjusting weight is estimated. After applying required weights to the grid subsets, they are aggregated to one raster layer using Vertical Mapper Splicer command. Final adjusting of Erosion potential by Sub soil lithology and Land cover factors are performed using Vertical Mapper Grid Calculator function.

Main Factors

Relief

The relief is an important factor controlling erosion rates (Dokuchaev, 1892; Jenny, 1961; Huggett, 1975; Gerrard, 1981). The topography of Lithuania changes from 0m at the seaside to almost 300m in the southeast. Two prominent topography highs are located in the eastern and western Lithuania attributed respectively to the Baltija Upland and the Zemaitija Upland (Fig. 1). They are separated by the Middle Lithuanian Lowland characterised by low topography.

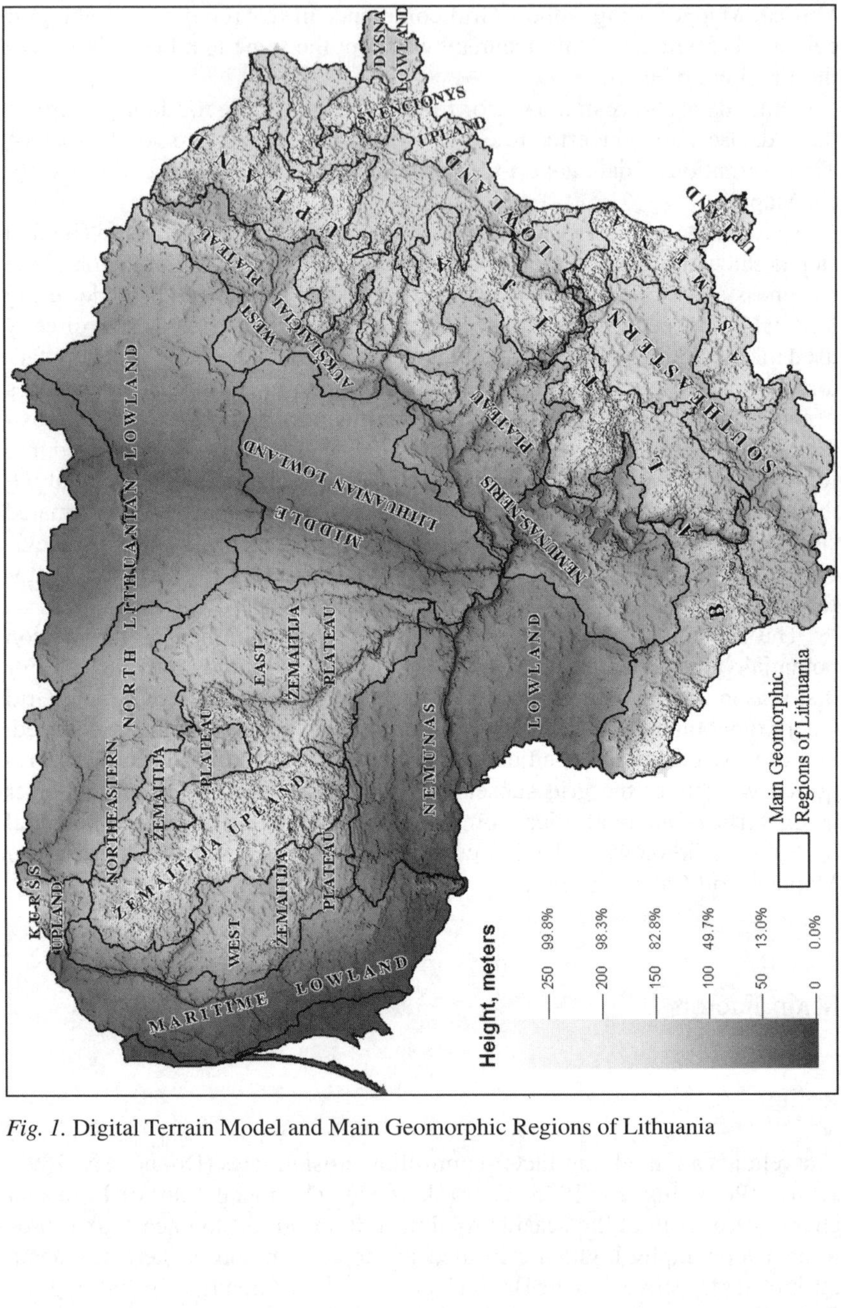

Fig. 1. Digital Terrain Model and Main Geomorphic Regions of Lithuania

For calculation of the Erosion Potential, the input data were used from the Digital Terrain Model (DTM) of Lithuania. It was still necessary to considerably improve the available version distributed by “GIS-Centras” in 1996 that has insufficient accuracy, essentially as regards the hydrographical polygons. Elevation points and contour lines digitized from M 1:50000 topographical maps were used in previous version.

Accuracy of the DTM was analyzed using topography points digitized from 1:10000 scale cartographical maps. Density of points is 1 point per square kilometer. More then 64000 points were tested all over the territory of Lithuania (Table 1).

Table

SEQ Table * ARABIC \s 1 . Statistical parameters of DTM	Statistical data		
	DTM Year 1996	DTM Year 1999	DTM Year 2000
Number of points	64451	64451	64450
Mean	(0,8	0,0	0.5
Dispersion	9,0	9,1	8.8
Standard deviation	3,0	3,0	3.0
Min	(27,0	(27,0	-29.0
Max	29,0	30,0	30.0
Asymmetry	0,0	(0,1	0.1
Excess	5,6	4,5	4.9

In 1999 the DTM was improved by adding hydrographical data. Polygons of lakes were assumed as plane surfaces, and contours of hydro-objects were incorporated as elevation contour lines. The Anudem method, distributed by Australian National University, was used for interpolation. Anudem method uses Kriging interpolation and incorporates hydrographical lines and polygons. As a result, the accuracy of this DTM was improved. However, problematic areas remained close to the river valleys with steep slopes. In 2000 more than 1600 points were digitalized along valleys of the major rivers of Lithuania. A new version of the DTM was calculated that incorporates digitized points, elevation contour lines and points and hydrographical polygons. The last version of DTM has the best accuracy (Table 1).

Three parameters of the relief were used to calculate the Erosion Potential, i. e. the slope steepness, the slope exposure and the slope length.

Sub-Soil Lithology

The sub-soil lithology considerably controls the Erosion Potential, as the different types of sediments have different resistance to erosion. The Sub-soil lithology factor was calculated based on the Geological Map of the Quaternary of Lithuania at scale of 1: 200 000 (author R. Guobyte, 2002). These data were released at Lithuanian Geological Survey in 1999 and are updated all the time.

Lithology factor was assessed by grouping different lithology values into four classes: sandy deposits, sandy loam, clayey deposits and peat and assigning required factor weight values for them. For sandy deposits and clayey deposits the sub-soil lithology factor values depend on the slope steepness. For sandy deposits the highest values (1.5) are attributed to the steepest slopes (>15 degrees), and the lowest values (0.7) characterise the plain areas (<3 degrees). For clayey deposits the dependence is different, as the highest values (1.4) are documented from the plain areas and the lowest values (1.1) are reported from the steep slopes. For sandy loam we assign the weight value of 1. For this type lithology the slope steepness have no influence in factor weight. Peat areas were eliminated from the Map of Erosion Potential because in these areas sedimentation processes dominate erosion processes.

Land Cover

The land cover type is an important factor as the vegetation reduces Erosion Potential. Land cover factor was calculated from CORINE Land Cover of Lithuania (scale 1: 100000) dataset. All different land cover records were grouped into four classes: forests and grasslands, gardens, agriculture areas and areas without vegetation. Different values were assigned to each class, i. e. 1 for forests and grasslands, 0.5 for gardens, 0.25 for agriculture areas and 0.1 for areas devoid of vegetation (pits, fired areas, sea-banks). Urbanized areas were omitted as natural processes are changed by human activities and erosion in those areas is governed by other factors.

Modelling Results

As mentioned above, the modelling was performed in two steps. The relief parameters were calculated first. The steepness of the slopes is mainly in the range of 0-1 degrees (Fig. 2). The high values attaining 19 degrees are confined to the river valleys and uplands that show dissected relief. Accordingly, the higher potential of erosion is predicted in the western and eastern Lithuania.

The exposure factor varies from 15 to 45 (Fig. 3). The dominating values are about 32-33. Increased values are reported from the West Zemaitija Plateau and east of the Middle Lithuanian Plain.

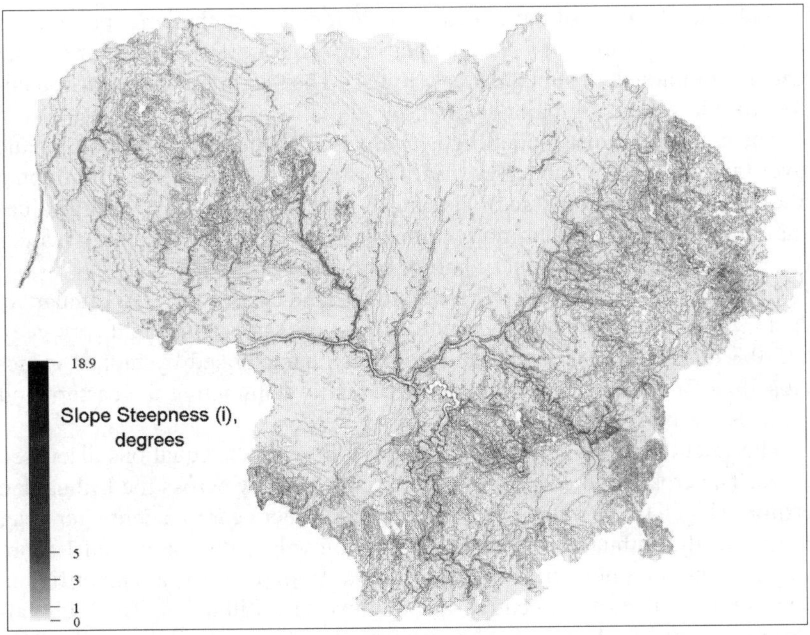

Fig. 2. Slope Steepness factor

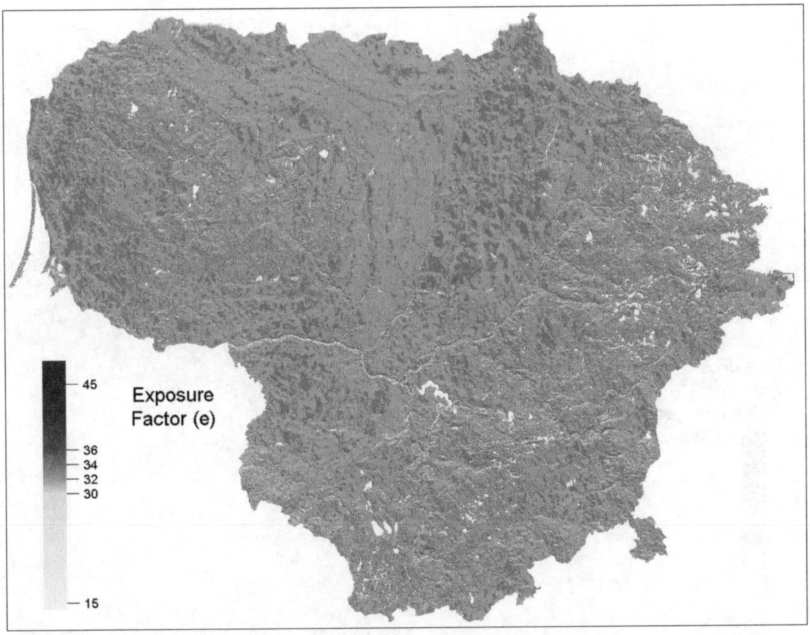

Fig. 3. Slope Exposure factor

The calculated values of the erosion potential related to the relief parameters are mainly in the range of 7-9m^3/ha per year. High values of 10-15m^3/ha per year are common in the upland areas (Fig. 4). The maximum values are calculated for the valleys of major rivers.

The next step in the modelling is evaluation of the sub-soil lithology and cover factors. The soil factor (B) varies drastically depending on outcropping lithologies. The lithology factor also much depends on the relief. High values attaining 1.5 are somewhat more common in the north, whereas low values (0.7-0.8) dominate the southern half of Lithuania (Fig. 5).

Land cover factor is directly related to the land cover type. The Lithuanian territory is dominated by rather low Land Cover factor values in the range of 0.25-0.3 due to intense land use. The areas characterised by higher values (close to 1.0) are more rare, they are essentially dominating the eastern and southeastern parts of Lithuania (Fig. 6).

The calculated factors were combined, following the equations discussed above. The erosion potential varies quite considerably across the Lithuanian territory (Fig. 7). The values vary from 0.12 mm per year to 3.5 mm per year. In most of the Lithuanian territory the erosion values do not exceed 1.5 mm per year. The minimum erosion rates (bellow 1mm per year) are identified in the eastern, southeastern, central and southwestern Lithuania. The high values are common in the uplands and essentially in the river valleys (Fig. 8).

Fig. 4. Erosion Potential related to Topography factor

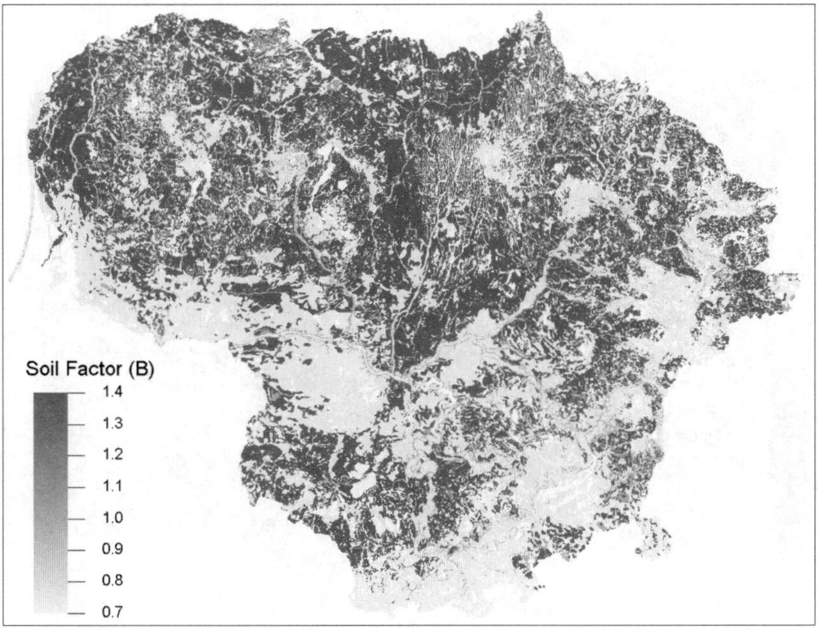

Fig. 5. Sub-Soil Lithology factor

Fig. 6. Land Cover factor

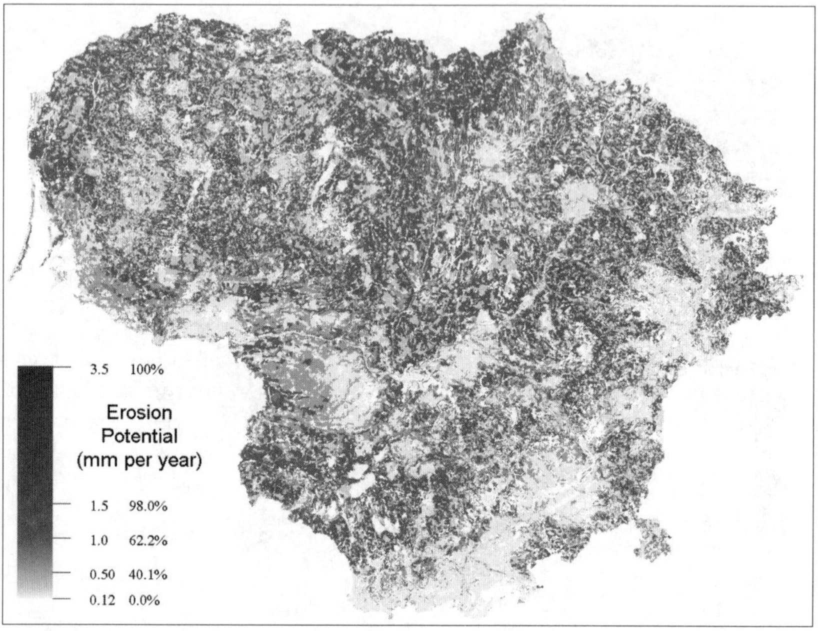

Fig. 7. Erosion Potential of Lithuania

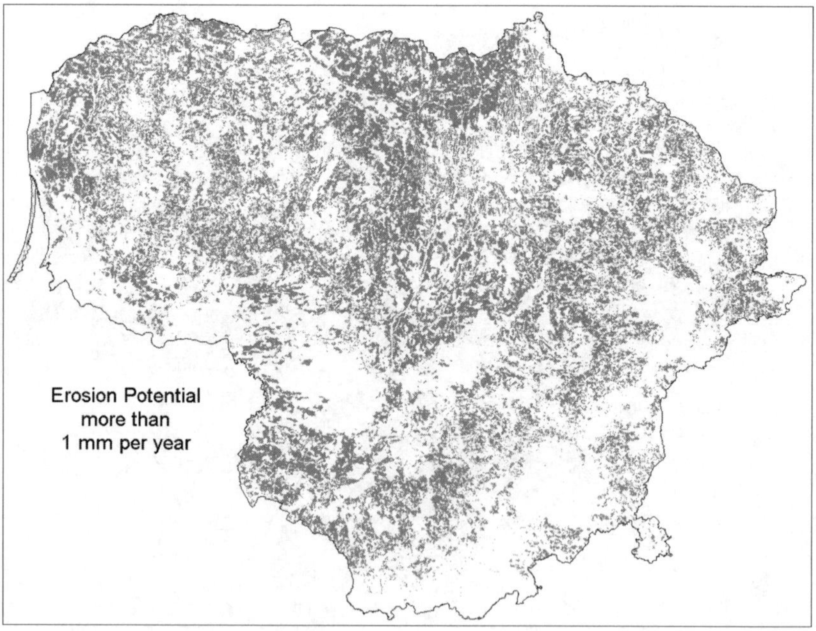

Fig. 8. Erosion Potential higher than 1 mm per year

Conclusions

The application of GIS tools is highly efficient in the modelling of the erosion potential of different regions. The main advantage is high accuracy of results not limited by largeness of the territory and amount of data, as well as the number of factors. The accuracy is depended on the input data only and is not lost regardless to the territory scale. The factors are limited to the mapping of the factors as well as parameterisation of those factors as regards the impact on erosion, which is obtained via laboratory and field experiments and observations. Therefore the model can be easily updated, as more information is available.

The erosion model shows distinct variations across Lithuania. The erosion rate ranges from <0.1 mm/year to 3.5 mm/year. The minimum erosion rates (bellow 1 mm per year) are identified in the eastern, southeastern, central and southwestern Lithuania. The high values are common in the uplands and essentially in the river valleys.

Acknowledgment

The work of D. Z. is a part of the Ph. D. at the Vilnius University, as well as K. A. studies are related to Ph. D. at the Vilnius Gediminas Technical University. S. S. performed this work within the frames of the National Scientific program "Lithosphere".

References

Beven, K. J., Lamb, R., Quinn, P., Romanowicz, R., Freer, J., 1995. Topmodel. In: Singh, V. P. (Ed.), Computer Models of Watershed Hydrology. Water Resources Publications, Colorado.

Cesnulevicius A., 1998. Intensity of the geodynamic processes affecting the Lithuanian relief. Litosfera, Vilnius, No. 2, 157-163.

Cesnulevicius A., Minkeviscius V., Mikalauskas A., Biaconis M., Mikutiene L., 1984. Morphometrics of Lithuanian SSR (on example of typical administrative regions). Geographia Lithuanica, Vilnius, 36-42 (in Russian).

CORINE, see EEC.

Dokuchaev, V. V., 1892. Our Steppes – At One Time and Now. Yevdokimoff Press, St. Petersburg (in Russian).

EEC, 1992. CORINE soil erosion risk and important land resources in the southern regions of the European Communities. Brussels, 97 pp.

EEC, 1993. CORINE Land Cover. Guide technique, Brussels, 144 pp.

Florinsky I. V., Eilers R. G., Manning G. R., Fuller L. G., 2002. Prediction of soil properties by digital terrain modeling. Environmental Modelling & Software 17, 295-311.

Gerrard, A. J., 1981. Soils and Landforms. An Integration of Geomorphology and Pedology. George Allen and Unwin, London.

Gessler, P. E., Moore, I. D., McKenzie, N. J., Ryan, P. J., 1995. Soil-landscape modelling and spatial prediction of soil attributes. Inter-national Journal of Geographical Information Systems 9 (4), 421-432.

Guobyte R., 1999. Quaternary Geological Map of Lithuania. Geologijos Akiraciai, Vilnius, No. 1/99, 7-16.

Huggett, R. J., 1975. Soil landscape systems: a model of soil genesis. Geoderma 13 (1), 1-22.

Jankauskas B., 1996. Soil erosion. Vilnius "Margi rastai", 167 pp.

Jenny, H., 1961. Derivation of state factor equations of soils and eco-systems. Soil Science Society of America Proceedings 25 (5), 385-388.

Le Bissonnais Y., Montier C., Jamagne M., Daroussin J., King D., 2001. Mapping erosion risk for cultivated soil in France. Catena 46, 207-220.

King, D., Fox, D., Daroussin, J., Le Bissonnais, Y., Daneels, V., 1998. Upscaling a simple erosion model from small area to a large region. Nutr. Cycling Agroecosyst. 50, 143-149.

Moore, I. D., Gessler, P. E., Nielsen, G. A., Peterson, G. A., 1993. Soil attribute prediction using terrain analysis. Soil Science Society of America Journal 57 (2), 443-452.

Pennock, D. J., Zebarth, B. J., De Jong, E., 1987. Landform classification and soil distribution in hummocky terrain, Saskatchewan, Canada. Geoderma 40 (3-4), 297-315.

Racinskas A., 1990. Soil erosion. Vilnius "Mokslas", 135 pp.

SAULIUS SLIAUPA, ALBERTAS BITINAS, ALGIMANTAS ZAKAREVIČIUS

Predictive Model of the Vertical Movements of the Earth's Surface: Implications for the Land Use of the Lithuanian Coastal Area

Abstract

The coastal areas are sensitive to the vertical movements of the earth's surface as it leads to either drowning of the mainland territories that are often densely populated, or conversely to shallowing of the harbor areas. The latter phenomenon is well known in the Scandinavian coastal area of the Baltic Sea that is accounted to the isostatic rebound of the lithosphere after the glaciation. As for the eastern coast of the Baltic Sea, no consistent data were provided on the tectonic activity of the earth's crust. The predictive model of the vertical movements of the earth's surface of the Lithuanian coast is presented based on detailed study of the Quaternary, Holocene and Recent tectonic activity. Different approaches were employed. The model shows vulnerability of the southern part of the Lithuanian coastal zone to drowning that may lead to significant environmental and economic problems in 100-1000 years AP. The central part is rather stable, whereas the northern area is subjected to moderate subsidence that can associate with coastal erosion and sediments accumulation in the near-shore zone.

Introduction

The coastal processes are influenced by different factors, listing the sea currency direction and velocity, wind direction, morphology and lithology of the coast, land cover, etc. The vertical movements of the earth's surface are accounted for considerable environmental and economic problems in some areas, like the sea advancement due to surface subsidence in the Netherlands or sea recession in lithosphere resulted in a high-rate uplift of the earth's surface.

So far, the costal processes of Lithuania were considered in terms of the interaction of exogenic (surface) factors, wile the tectonics was only scarcely addressed (Gudelis, 1955). The geological studies and geodetic measurements indicate certain activity of the vertical movements of the earth's crust during the Neotectonic stage (Sliaupa et al., 1995; Zakarevičius, 1994). The tectonic trend was not stable due to changing tectonic mechanism and glacial loading-unloading. Glacial isostasy and related processes are widely reported from the formerly glaciated territories (cf. DeGeer, 1963). Against the isostatic trend, a distinct tectonic differentiation is identified in the Baltic region (Sliaupa et al., 1995). Due to superposition of different mechanisms, it is difficult to construct short- and long-termed prognostic models of the vertical movements in the Baltic area. It can only be achieved by detailed reconstruction of the vertical movements of the earth's crust for various time slices and separation of different tectonic mechanisms.

Methods and data

The presented study consists of the three stages. As the first, the total magnitudes of the vertical movements of the earth's crust during the Neotecotnic stage (several dozens of million years) were identified from the deformations of the sub-Quaternary surface that was regarded as the main reference level, as discussed by (Mozhaev, 1973; Sliaupa et al., 1995; 1996). A detailed map of the relief of the sub-Quaternary surface was recently compiled for the Lithuanian territory based on inventory of more than 10,000 wells (Sliaupa et al., 2001). The relief is a result of superposition of both the exogenic and endogenic processes. The formers list the glacial exaration, melt-water erosion, the latter is accounted for the tectonic activity, also the glacial isostasy is an important factor. As a result, the relief is highly dissected. The tectonic factor was separated using morphoisoline method (summit altitudes), whereas the sharp negative forms are considered as the representatives of the exogenic processes (Sliaupa, 2001). The palaeogeographic situation of the Neogene was taken into consideration to correct the Neotectonic magnitudes.

The vertical movements of the earth's surface during the Holocene time (past 10,000 years) were identified from the deformations of the sea terraces established during the Baltic Ice Lake, Ancylus, Litorina and Postlitorina stages of the Baltic Sea development. A study of terraces was a part of the integrated geological mapping of the Lithuanian coastal area held in 1998-2002. The particular terraces were traced based on lithological and morphological features and OSL dating (Bitinas et al., 2001; 2002). It should be noted, that the first evidences on the neotectonic movements of the Baltic Region were documented from the deformation of the Baltic Sea shorelines in 19th century. Assuming no vertical movements took place after the terrace establishment, each terrace should be at the same level, whereas the deformations imply vertical movements of the earth's surface. The differences between particular terraces are explained in terms of the sea level changes and the tectonic movements.

The map of the recent vertical movements of the earth's crust was compiled based on data collected from the seven monitoring sites and the western leveling line of the Lithuanian geodynamic polygon. The vertical movements are documented since the beginning of the twentieth century (Pobedonoscev, 1973). Major trends and superimposing short-termed fluctuations are identified. The geodynamic polygon of Lithuania was measured several times during past century. A comparison of the leveling data reveals the vertical movements of the earth's surface at different sites (Zakarevičius, 1994).

A comparison of the Neotectonic, Holocene and Recent vertical movements enabled construction of the long-termed predictive model of the surface movements in the coastal area. These data were incorporated into the Elevation model of Lithuania (Denas et al., this volume). The topography was calculated for different time slices (100, 500 and 1000 years after present (AP). It enables the prediction of the location of the seashore line and prognoses of related surface geodynamic processes. Additionally, the predicted global sea level rise was included in calculation of the "pessimistic" model.

Results

1. Neotectonic vertical movements

The Neotectonic movements are quite differentiated in the Baltic area. The sub-Quaternary surface is generally tilted to the west attaining –250 m in the central part of the Baltic Sea, while exceeding +100 m in the eastern Lithuania.

Against this general trend, the smaller scale Neotectonic structures are identified. In the Lithuanian coastal area the depth of the sub-Quaternary surface is in the range of –20 m to –60 m (Fig. 1). The northern half of the coastal zone is dominated by altitude of –40m, it is dissected by distinct west-east trending high of relative amplitude of about 10 m. This high is bounded by deep-seated Karkle (south) and Telsiai (north) faults. A distinct depression is mapped south of the Klaipeda City, the depth of the sub-Quaternary surface attaining –60 m. It has a complex geometry. The depression is traced into the mainland (about 20 km) and is bound by west-east tectonic terrace in the south of about 10m amplitude. This structural terrace is further mapped in the Kursiu Marios Lagoon, the orientation changes to the north-south. The southernmost part of the coastal area (Ventes Horn, Nemunas Delta) is confined to the uplifted area, where the sub-Quaternay surface is documented at the depth of –30 to –20 m.

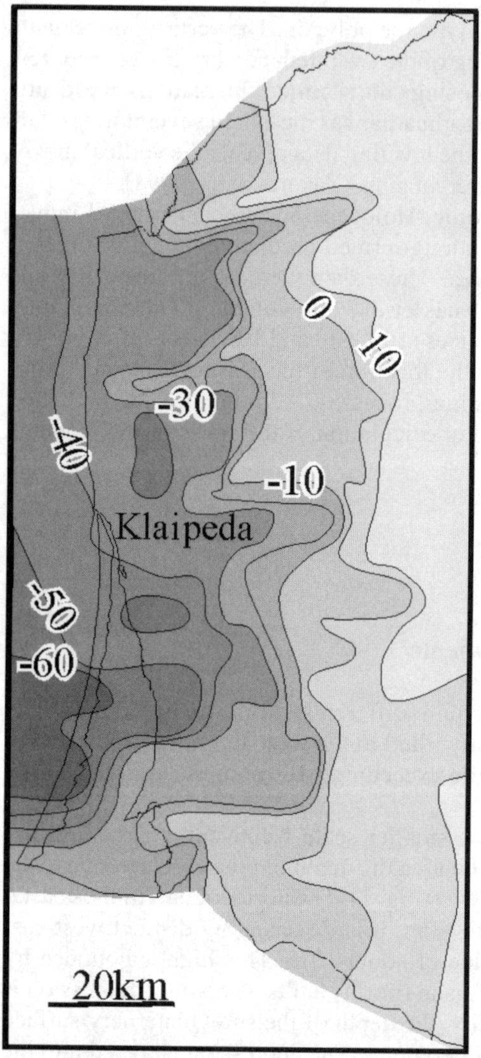

Fig. 1. Structural map (m) of the sub-Quaternary surface that reflects total amplitudes of the Neotectonic vertical movements of the western Lithuania

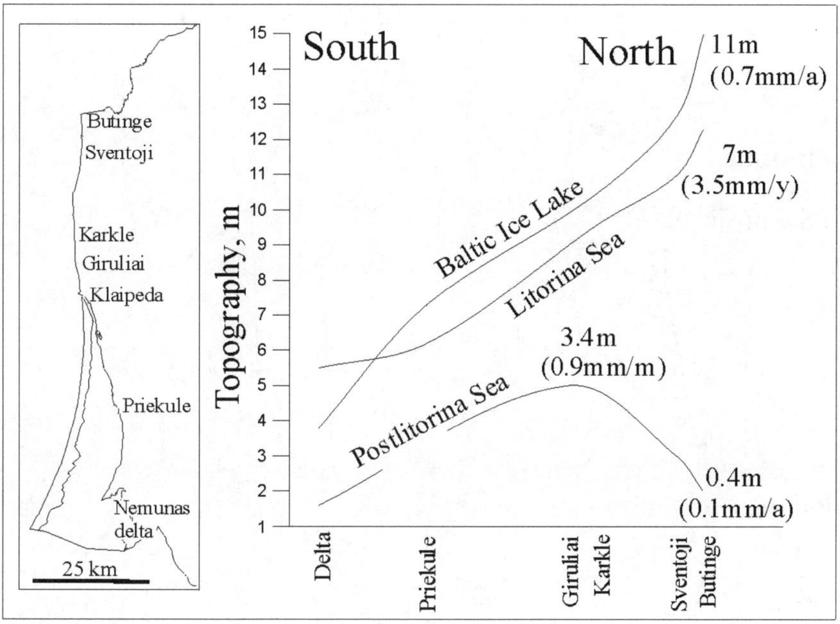

Fig. 2. Present topography of the Baltic Ice Lake, Litorina Sea and Postlitorina Sea terraces in western Lithuania. Numbers indicate calculated relative uplift and its rate of the northern part of Lithuania with respect to the southern Lithuania (for Postlitorina terrace calculation is done for the central part).

2. Holocene vertical movements

The Late Veichselian ice sheet abandoned the Baltic area about 13,000 years ago. The shape of the Baltic Sea was not stable, relating to global sea level fluctuations, climate variations, etc. The Baltic Ice Lake, Ancylus, Litorina and Postlitorina stages of the sea evolution are defined. The changes of the basin shape are manifested in the system of the sea terraces that are mapped onshore all along the present sea shoreline (Fig. 3). The Baltic Ice Lake and Litorina stages are characterized by maximum transgression, accordingly, these terraces are the best preserved, whereas the Ancylus Lake and Postlitorina Sea terraces are someplace obliterated. The type of the terraces varies from erosional to accumulation in nature, depending on the topography. West of each terrace the open shelf sediments were accumulated, mainly of sandy composition. In the south of the western Lithuania, the lagoonal deposits are distributed due to partial confinement of the basin by marginal moraine inliers of belt geometry. It associates with domination of the accumulative nature of the sea terraces, whereas the accumulation alternated with erosion in the north.

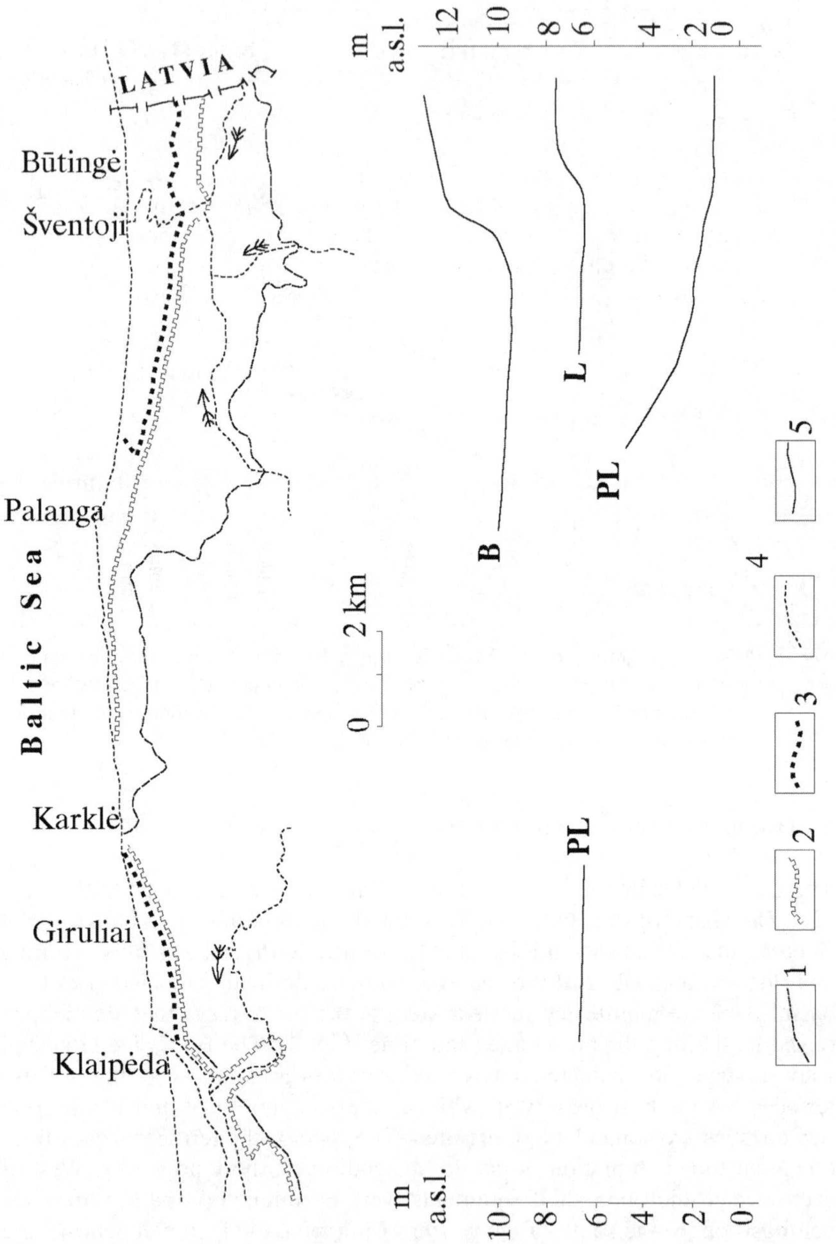

Fig. 3. Deformation and topography of terraces of different basins of the Baltic Sea in the northern half of the western Lithuania. Shorelines of basins: 1 – Baltic Ice Lake (B), 2 – Litorina Sea (L), 3 – Postlitorina Sea (PL). 4 – the present day coastline and main rivers; 5 – longitudinal profile of terrace.

The present topography of the Baltic Ice Lake, Litorina and Postlitorina Sea terraces was plotted as a function of distance from the south to the north (Fig. 2). Due to considerable sea level changes (Lampe, 2002) it is difficult to reconstruct the absolute values of the vertical movements from the terrace topographies, however they are effective in identification of the relative vertical movements. In the northern extreme of the profile, the present altitude of the Baltic Ice Lake terrace is of +15 m, while it is of +3.7 m in the south (i. e. 11.3 m relative uplift of the northern area). The southernmost part of the profile indicates the maximum gradient, as well as the northern extreme. The Litorina Sea terrace has very similar deformation pattern and is tilted from +12.3 m in the north to +5.5 m in the south (6.8 m relative uplift). Some differences are discernable though, as the deceleration of the subsidence of the southernmost part of the studied profile and miserable tendency of the uplift in the center of the profile (about 0.3 m). The Postlitorina Sea terrace shows very different deformation pattern. The difference between the southern and northern extremes is only of 0.4 m (altitude +1.6m and +2.0 m), with prominent uplift in the central part of the profile of +3.4 m. The northern extreme indicates the maximum subsidence gradient, by contrast to preceding tectonic trend.

Based on the presented results of the deformations of the sea terraces we imply two different stages in evolution of the vertical movements of the earth's crust. A prominent tilting to the south of the Baltic Ice Lake and Litorina Sea terraces is related to the isostatic rebound of the lithosphere after recession of the Late Veichselian ice sheet. This mechanism ceased before the Postlitorina stage (around 4,000 years BP), when the tectonic mechanism has driven the deformations of the lithosphere. Accordingly, the Baltic Ice Lake – Litorina stage movements should not be incorporated in the predictive model of the vertical movements of the Lithuanian coastal area.

The measured altitudes of the terraces enable calculation of the rate of the vertical movements of the earth's surface. The Baltic Ice Lake terraces is assumed as old as 12,000-11,000 years, while the Litorina and Postlitorina Sea terraces are dated respectively about 6,000 years and 4,000 years. Subtracting the Litorina terrace altitudes from the Baltic Ice Lake terrace topography, the rate of the isostatic rebound during the former part of the Holocene time is estimated as high as 0.7 mm/a. The rate of the uplift of the northern part of the Lithuanian coastal zone considerably increased during the middle Holocene, attaining 3.5 mm/a.

The rates of the Postlitorina stage movements are of the same order, though the pattern of deformations considerably changed. The southern and northern parts of the Lithuanian coastal area subsided with respect to the central part (Klaipeda block) at the rate of 0.8-0.9 mm/a, the southern part showing somewhat more intense (for about 0.1 mm/a) downwarping than does the northern area.

It is notably that only miserable uplift of the central (Klaipeda) area is discernable in the Litorina Sea terrace, and no evidence of uplift of this area is noticed in the Baltic Ice Lake terrace, though the younger Postlitorina Sea terrace is considerably uplifted. This points to inversion of the vertical movements, i.e. the Klaipeda block relatively subsided in order of 3-4 m during the early-middle Holocene time, while it warped up during the late Holocene.

Against the regional trend of the vertical movements, activity of some large-scale faults is identified. A sharp offset of sea terraces is mapped above the Early Palaeozoic Akmene fault in the northern part of the coastal area (Fig. 3). The Baltic Ice Lake terrace shows 3 m subsidence of the northern flank of the fault, while the Litorina Sea terrace is set off for 1.5 m. It implies rather stable subsidence of the northern flank at the rate 0.4-0.5 mm/a during the early-middle Holocene. The Postlitorina Sea terrace shows different trend, the northern flank is relatively uplifted about 0.5 m. It points to tectonic inversion of the fault, as it was before stated for the regional tectonic pattern.

3. Recent vertical movements

The rates of the Recent vertical movements range from –1 mm/y to +2.5 mm/a in Lithuania (Zakarevicius, 1994). The subsidence trend is documented in the central Lithuania, southern Lithuania and coastal area. The latter still shows distinct differentiation. The analysis of the seven monitoring sites indicates the rate variations from 0 mm/a to -4 mm/a (Fig. 4). The subsidence (about -1 mm/a) is reported in the south, and stable regime (about 0 mm/y) is identified in the Klaipeda block. A maximum subsidence of -2.4 mm/a and -4.9 mm/a is documented in respectively the Dreverna and the Nida sites.

The leveling of the western line of the Lithuanian geodynamic polygon comes in agreement to the monitoring data (Fig. 5). A few very sharp changes in the rate of the vertical movements of the earth's surface are recognized implying fault activity. Four blocks, bound by faults, are readily distinguished. The most prominent discontinuity is related to the well-known Silute fault striking W-E in the south (Sliaupa, 1996). The southern flank of the fault is relatively uplifted. The rate of he fault movement is estimated as high as 0.5 mm/a. The Silute and Mickai faults bound the most actively subsiding Nida-Dreverna block. The block is tilted to the north. The activity of the Mickai fault is calculated as high as 0.7mm/a. The Klaipeda block is relatively uplifted and is bound by Mickai fault in the south and Karkle fault in the north. The latter is the most active feature indicating the rate of the vertical movements 0.9 mm/a.

Based on collected data, a detailed map of the resent vertical movements of the earth's surface was compiled for the coastal area of Lithuania. As mentioned above, the southern area shows the highest rate subsidence in the range of -1.5-1 mm/a, while the northern part subsiding –1 to 0.5 mm/a. A stable block is defined in the Klaipeda area (center).

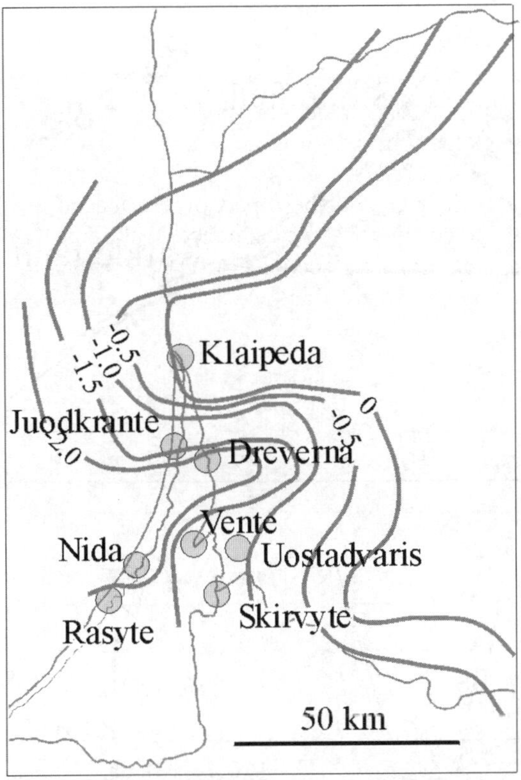

Fig. 4. Rates of the Recent vertical movements of the earth's surface of the western Lithuania. Locations of monitoring sites of the vertical movements are indicated.

The pattern of the vertical movements of the Postlitorina and Recent stages show striking similarity in terms of both the areal distribution and intensity. It is therefore concluded that the tectonic mechanism is rather persistent for the past 4.000 years. It enables the calculation of the reliable predictive model. Actually, the rate and pattern of the vertical movements defined for the Recent stage can be extrapolated with great level of confidence for at least the next 1,000 years.

A very good correlation of the Neotectonic map and Recent movements should be noted. The Nida-Dreverna depression is the most distinct feature on the both maps, showing the same shape in a plan view. The Silute fault also shows very distinct tendency of the uplift of the southern flank, as well as the Mickai fault indicating the uplift of the northern flank on the both maps. The Karkle fault is punctuated by distinct west-east trending Neotectonic structure. All this implies inheritance of the tectonic mechanism that was evidently disturbed during the glaciation leading to tectonic inversion.

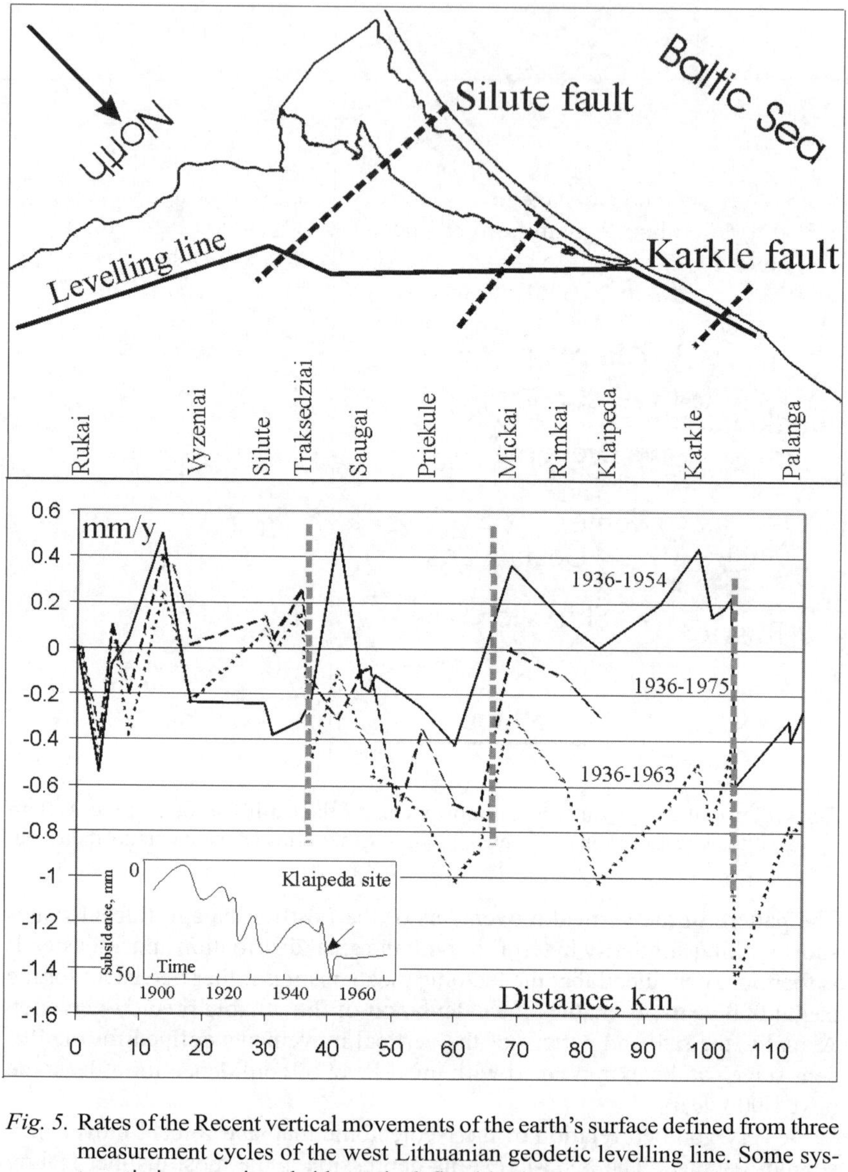

Fig. 5. Rates of the Recent vertical movements of the earth's surface defined from three measurement cycles of the west Lithuanian geodetic levelling line. Some systematic increase in subsidence is defined in the north, that possibly relates to short-termed subsidence event as monitored in the Klaipeda monitoring site (inset). Major faults are indicated by hatched lines.

Present topography

The present topography of the coastal area is rather dissected (Fig.6). The southern part is characterised by low topography that is complicated by distinct roughly north-south trending high of marginal moraine. The southernmost part of the Lithuanian coastal area is dominated by the wide Nemunas Delta. The mainland is separated from the open shelf by Neringa Spit striking NNE-SSW for about 100km from Klaipeda to the Sambian Peninsula.

The northern part of the coastal area of Lithuania shows higher topography, essentially in the central part (Klaipeda block). This topography pattern is well reflected in the configuration of the coastline, the mainland slightly protruding into the sea in the central part, while the sea invades to a maximum extent in the south. North of the Neringa spit, the coast is embroidered by foredune, while the mainland coast of the Kursiu Marios Lagoon represents the flat plain. Numerous wetlands and lakes are scattered in this area, essential in the Nemunas Delta region.

The configuration of the Baltic Sea shoreline is intimately related to the pattern of the tectonic vertical movements as discussed above. The southernmost subsiding area indicates the maximum sea transgression, while the stable Klaipeda block is protruding. The Kursiu Nerija Spit is of the greatest width within the limits of the Neringa-Dreverna depression that suggest increase in sediments accumulation rates within the subsiding areas, essentially in the Kursiu Marios Lagoon side protected from the marine abrasion processes. It implies important role of the tectonic factor in shaping the coastal area that, in turn, governs the surface geodynamic processes.

Predictive topographical model

The predictive model of the relief was calculated for the time slices 100 years, 500 years and 1000 years by superposing the digital model of the present topography and defined tectonic trends (Fig. 6). As it was shown before, the vertical movements of the earth's crust are characterized by persistent trends during the Neotectonic stage in terms of both the pattern and intensity, therefore they can be predicted for at least a few thousand years with great confidence.

The calculations indicate rather distinct change in the topography of the coastal zone in 1,000 years AP. It mainly concerns the southern part of the studied area. The advancement of the sea for about 10 km is predicted, leading to drowning of large areas. The high belt composed of the marginal moraine represents the isolated island, while the Nemunas Delta area is broadly flooded. The sediment accumulation potential of the Nemunas River should be however taken into consideration that can partially compete the sea invasion.

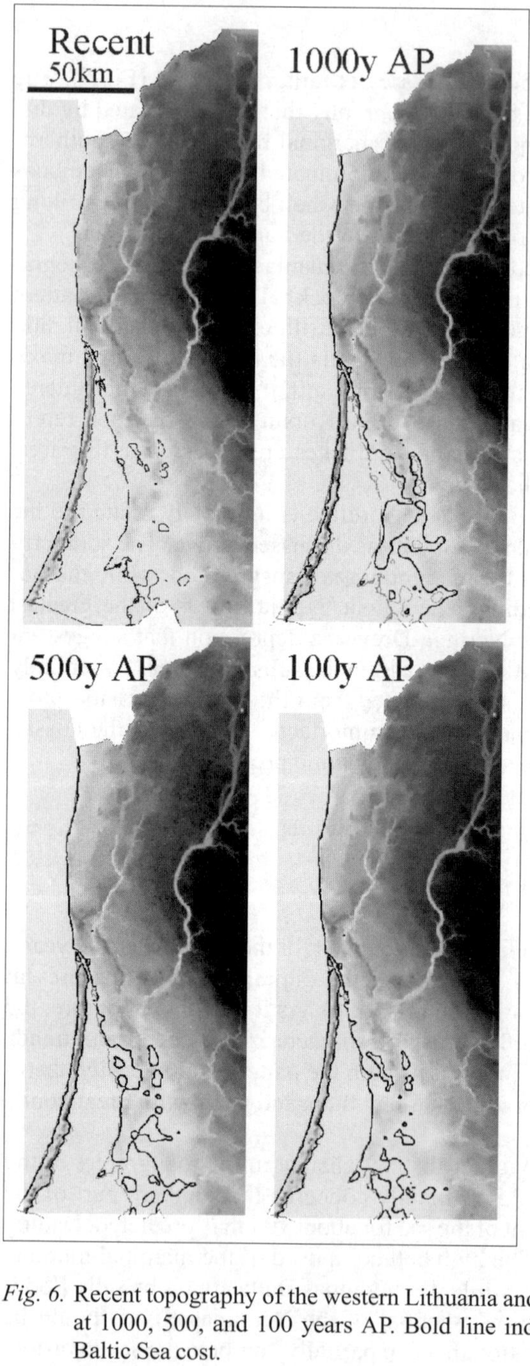

Fig. 6. Recent topography of the western Lithuania and calculated topography models at 1000, 500, and 100 years AP. Bold line indicate position of the modelled Baltic Sea cost.

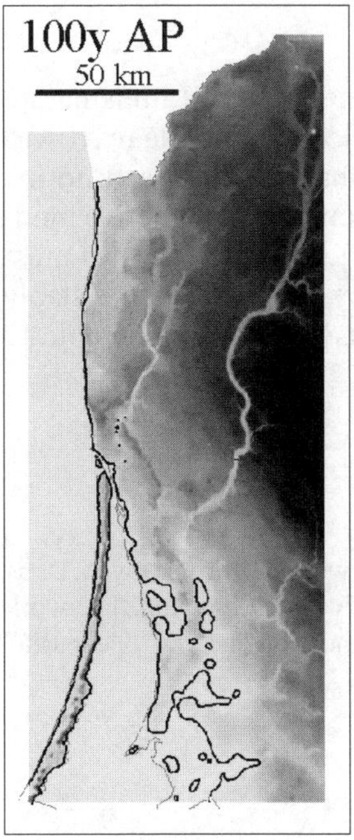

Fig. 7. Calculated topography model at 100 years AP with incorporated global sea level rise of 75 cm.

The western part of the Kursiu Nerija Spit is also subjected to cannibalization within the Nida-Dreverna depression that likely will lead to some migration of the spit to the east. The Klaipeda block actually does not show any significant modifications, as the rate of the vertical movements is close to 0mm/a. Some sea invasion is foreseen north of the Palanga resort, though the flooding scale is rather miserable, though increasing to the north.

The shorter-time models are of value for planning the near-future land use activities. For this, the model was constructed for 100 years AP. The changes of the coastal area are much less as compared to 1,000 years AP model. Still, rather important changes are foreseen in the south. The wetland and lake areas seems to progress, while the abrasion will increasingly affect the shore zone. The model of 500 years AP presents the transition model between the short-time and long-time models discussed. Also, predictive topographical model was calculated at 100y AP with incorporated global sea level rise of 75 cm. Large areas are thus invaded by the sea in the southern part of the western Lithuania (Fig.7).

Conclusions

The vertical movements of the earth's surface of the western Lithuania are characterized by persistent trends throughout the Neotectonic Stage. It was somewhat disturbed by glacial isostasy processes that ceased after Litorina stage. This persistence enables construction of the consistent predictive model of the topography. The southern part of western Lithuania indicates most intense subsidence, with minor subsidence tendency in the north and stable block in the center. Accordingly, the southern areas are most vulnerable to costal processes and sea invasion.

Acknowledgment

A detailed study of sea terraces of the western Lithuania were performed by A. B. as a part of integrated geological mapping by Geological Survey of Lithuania. The work by S. S. was partially financed by Lithuanian Science Foundation and AB "Geonafta" (Integrated Basin Analysis of the Western Lithuania).

References

Amantov, A., Spiridonov, M., Moskalenko, P. (1988) Geology of Pre-Quaternary formations along the margin of the Baltic Shield as an indicator of the evolution of the Gulf of Finland up to the most recent stage. – Geol. Surv. Finland, sp. pap. No 6, pp. 87-93.

Bitinas, A., Damušytė, A., Hütt, G., Jaek, I., Kabailienė, M. (2001) Application of the OSL dating to stratigraphic correlation of Late Wechselian and Holocene Sediments in the Lithuanian Maritime Region. Quaternary Science Review, 21. pp.767-772.

Bitinas, A., Damušytė, A., Stančikaitė, M., Aleksa, P. (2002) Geological development of Nemunas River Delta and adjacent areas, West Lithuania. Geological Quarterly, 46 (4), pp. 375-389.

De Geer E. H. (1963) De Geer's part in exploration of the history of the Baltic Sea. Baltika, 1. Vilnius. 15-33.

Galkus, A. (1998) The changes of the parameters of Baltic Sea geosystem by the input of sedimentary matter from the land. The Geographical Yearbook 31. Vilnius: 145-157

Gudelis, V. K. (1955) Neotectonic activity of the Lithuanian coastal area. Tr. AN Lit.SSR 5(3): 81-98

Janukonis, Z. (2000) Morpholithodynamic investigations of the coastal zone of South-eastern Baltic in 1970-1990. The Geographical Yearbook 33. Vilnius: 152-166.

Jarmalavicius, D., Zilinskas, G. (2001) Dependence of wash away during hurricanes on the coastal morpholithological indices. The Geographical Yearbook 34. Vilnius: 88-94.

Lampe, R. (2002) The post-glacial water-level variability along the southern Baltic coast – a short review. Greifswalder Geographischer Arbeinten 27: 13-19.

Mozhaev, B. N. (1973) Recent tectonic on the north-western part of the Russian plain (in Russian). – Leningrad, Nedra. 232 pp

Pobedonoscev, S. V. (1973) Vertical movements of the Riga and Kurshiai lagoons of the Baltic Sea. Recent movements of the earths crust. Tartu. 314-322.

Saarse, L., Heinsalu, A., Karhu, J., Vassiljev, J., Veski, S. (2000) Holocene shoreline displacement and palaeogeography of the Kõpu Peninsula, Hiiumaa Island, Estonia Mälkki. Baltica 13. Vilnius: 15-23

Sliaupa, A., Gelumabauskaite, Z., Straume, J., Sliaupa, S. (1996) Neotectonic Map of the Baltic region. Geologija 20: 67-72 (in Lithuanian, with English summary)

Sliaupa, A., Gelumbauskaite, Z., Straume, J., Sliaupa, S. (1995) Neotectonic structure of eastern part of Baltic Sea and adjacent land area. Technika Poszukiwan Geologicznych. Krakow: 63-65

Sliaupa, A., Straume, J., Sliaupa, S. (1995) Peculiarities of neotectonic investigation in Baltic // Technika Poszukiwan Geologicznych. Krakow: 67-73

Sliaupa, S. (1996) Major neotectonicaly active fault zones in Lithuania. Lithosphere N5. Minsk: 108-115

Sliaupa, S. (2002) Influence of the last glaciation to stress field and tectonic activity of faults of the Baltic region. Geologija 39. Vilnius: 12-24

Sliaupa, S., Cyziene, J., Lazauskiene, J. (2002) Stratigraphy of Sub-Quaternary surface: new edition of the Pre-Quarternary Map of Lithuania at a scale of 1 : 200 000. 5th Baltic Stratigraphic Conference. Vilnius: 31-33.

Trimonis, E., Gulbinskas, S. (1999). Lithodynamical processes in the Klaipeda Strait. The Geographical Yearbook 32. Vilnius: 93-102

Winterhalter, B., Flodén, T., Ignatius, H., Axberg, S., Niemistö L. (1981) Geology of the Baltic Sea. – The Baltic Sea. In: Voipio A.(ed.). Amsterdam: Elsevier. P.1-121.

Zakarevičius, A. (2002) Research into relations between the Earth's crust present vertical movements and geologic structure in Southern Lithuania // Geodesy and Cartography, Vilnius: Technika, Vol XXVIII, No 2, pp. 29-34. (In Lithuanian, with English summary).

Zakarevičius, A. (1994) The investigation of present vertical Earth's crust movements in the territory of Lithuania. Vilnius: Technika, 276 p. (In Lithuanian, with English summary).

Zakarevičius, A. (1998) Horizontal deformations of the Lithuanian geodetic network // Geodesy and Cartography, Vilnius: Technika, Vol XXIV, No 4, pp. 169-178. (In Lithuanian, with English summary).

Zakarevičius, A. (1996) Coordinate systems of Lithuanian geodetic networks and their connection. Vilnius: Technika, 200 p. (In Lithuanian, with English summary).

Zakarevičius, A., Stanionis, A. (2002) The dispersion structure of horizontal deformations of Lithuanian geodetic networks // Geodesy and Cartography, Vilnius: Technika, Vol XXVII, No 4, pp. 117-124. (In Lithuanian, with English summary).

Česlovas Aksamitauskas, Petras Petroškevičius,
Algimantas Zakarevičius, Arminas Stanionis

Geodetic Measurements for the Estimation of Eco-Geodynamic Risks in the Area of the Ignalina Nuclear Power Plant

Summary

The investigation has been performed in order to ascertain the existing tectonic activity of the region where Ignalina Nuclear Power Plant is situated; this activity is one of the indicators for an assessment of ecodynamic risk.

The investigation was carried out according to the precision levelling method. The measurements were made in 1989, 1990, 1991, 1992 and 1994. It has been determined that Earth's crust movement in the surroundings of the Ignalina Nuclear Power Plant are block structure. The blocks are related to the main breaks of crystalline foundation. Vertical movement of the Earth's crust are obviously distributed along the east-west direction, crossing Lake Drûkðiai. The northern block is ascending relatively to the southern one. The relative shift during the investigations was about 5-6 mm.

The special ground for measuring the horizontal deformations of the earth-crust had been arranged in 1998. It includes 10 points in the depth of 2.5 m with forced centration of the devices. The points are located taking into account the tectonic structure of the territory. Measurements with GPS devices were carried out in 1998 and 1999. The average quadratic errors of establishment of the coordinates of the points do not exceed 1.2 mm.

The horizontal deformations of the earth-crust were examined according to the results of the measurement applying the method of tensoric analysis. The maxinum and minimum relative lengthening in the territory under investigation and their directions as well as the relative dilatation of the territory have been calculated. The parameters of the deformations have been assessed using the method of finite elements. The wide range of changes of the indicators of earth-crust deformations is caused by peculiarities of situation of the principal tectonic breaks.

The conclusion was made: breaks of crystalline foundation in the surroundings of the Ignalina Nuclear Power Plant are active.

1. Introduction

The Ignalina Nuclear Power Plant is located in the northeastern part of Lithuania, close to the borders of Belarus and Latvia. The plant is built on the southern shores of Lake Drūkšiai, 39 km from Ignalina. The nearest towns to the plant are Vilnius at 130 km, and Daugavpils in Latvia at 30 km. Six km from the plant, the town Visaginas is situated. In the vicinity of the nuclear power plant there are lakes Drūkšiai, Visaginas, Apyvardė, Alksnas, Dysnai, Smalvės and the River Daugavam, which passes 30 km north of the plant.

The seismic activity of the territory that is closely related to the contemporary tectonic activity of the territory is an important aspect in the estimation of risk and the forecasting of the safe working of the nuclear power plant.

Ignalina Nuclear Power Plant was built in a region with complicated tectonic conditions. It is situated at the junction of three large-scale tectonic structures. The large-scale junction at the intersection of breaks of different range is situated in the zone of the lake Drūkšiai.

Many breaks intersect the sedimentary layers. Seismic processes of this region, fixed in the XIX century and the beginning of the XX century are related to active breaks. Epicentres of earthquakes are related to the main tectonic breaks.

In 1988, the State committee for the study of building conditions of the third block of the nuclear power plant stated that the possible seismic activity of the region reaches 7 points. Furthermore, the same committee stated that insufficient investigations of the contemporary tectonic activity of the earth-crust have been performed and emphasized the necessity to establish a special geodynamic polygon.

The geodynamic polygon for the investigation of contemporary vertical movements of the earth-crust in the region of the Ignalina Nuclear Power Plant was arranged in 1989.

It was established on the basis of the results of high-scale survey of the polygon in 1989, 1990, 1991, 1992, 1994 and 1998 that contemporary vertical movements of the earth-crust within the limits of the geodynamic polygon that appear to be of a block character. Furthermore, they are related with especially large-scale tectonic breaks of the crystal foundation [1], [2], [3], [4], [5], [6], [7]. The most active breaks cross the lake Drūkšiai are from the west to the east and pass to the southwest direction in the center of the lake (Drūkšiai bend). The northern block rises with respect to the southern block with an intensity of 1-2 mm/year. Besides the sharp-distinct northern and southern blocks, some smaller areas related with blocks limited by breaks have been distinguished.

A conclusion was drawn on the basis of the investigation of the contemporary vertical movements of the earth-crust, which says that the main tectonic breaks in the region of Ignalina Nuclear Power Plant continue to be active at present. The more detailed information about the contemporary tectonic activity of the earth-crust and fields with geotectonic tension is available from the results of an investigation of horizontal movements of the earth-crust. With that end in view, the special geodynamic polygon was established in the region of Ignalina Nuclear Power Plant in 1998. We can affirm that the geodynamic polygon provides valuable information on recent tectonic movements in the area of the nuclear power plant and hence on the possibility of danger of seismic processes. The polygon can be used as a part of a control system monitoring the seismic conditions in the Ignalina Nuclear Power Plant area.

2. Investigation of Vertical Movements of the Earth-Crust in the Region of Nuclear Power Plant

The Ignalina Nuclear Power Plant (INPP) is located in the area of the East-European platform at the intersection of three large tectonic structures: the Baltic Syneklise, the Masurian-Belarus Anteclise and the Latvian Soddle. The crystalline basement lies at more than 700 m depth. The crystalline foundation and sediment case are separated by a series of tectonic breaks.

The main section of breaks from different ages and of various activities is close to Lake Drūkšiai. The seismic processes could be related to active breaks. Not only seismic processes, but also intensive ground movements can have a negative influence on the power plant maintenance.

The geodynamic polygon was created in the vicinity of the nuclear power plant in order to investigate vertical movements of the Earth's crust. It was established in 1989. The polygon leveling network perimeter is 66 km (with 114 bench marks). Levelling results of 1989, 1990, 1991, 1992 and 1994 have been analyzed. The southern part of the polygon has been levelled four times, the northern one twice. The accuracy of levelling is 0.30 mm/km.

Different variants of recent vertical movements of the Earth's crust in the geodynamic polygon of the Ignalina Nuclear Power Plant have also been analyzed in [1, 2, 3, 4, 5, 6, 7]. The present summary of the investigations is based on the above-mentioned publications.

The main purpose of the investigations was to determine whether the Earth's crust breaks are active at present in the nuclear power plant region. Apart from the main task, the investigation results help to evaluate for better general tectonic activities in the territory of Lithuania and to interpret the results of investigations performed in other territories.

The determined vertical movements of benchmarks are shown in Figs. 1 and 2.

The measurements show that vertical ground movements are differentiated according to the crust blocks that are divided by tectonic breaks. Clear by divided separate blocks are almost perpendicular to the meridian and extend to the eastern turn of the polygon, crystalline basement break at Lake Drūkšiai. The shift amplitude of the northern and southern blocks during the period of investigation was 5-6 mm. The northern block is coming up in relation to the southern block.

Between the distinct northern and southern blocks, the third block in the south-east of the site can be seen; and is limited by the intersection of breaks along the meridian and the parallel. The earth's crust movement inversion was present during the period of investigations. This is evidently seen in the south-east of the polygon.

These preliminary conclusions, i. e. preliminary model of the Earth's crust (ground) vertical movements in the territory of Ignalina Nuclear Power Plant geodynamic polygon, were checked by mathematical statistical analysis of the vertical ground movements measured [3, 4, 5, 6, 7].

For checking the block structure model of the Earth's crust vertical movements and identification of probably active breaks, the proposed methods of disperse analysis and cumulative sum analysis were used [3, 4].

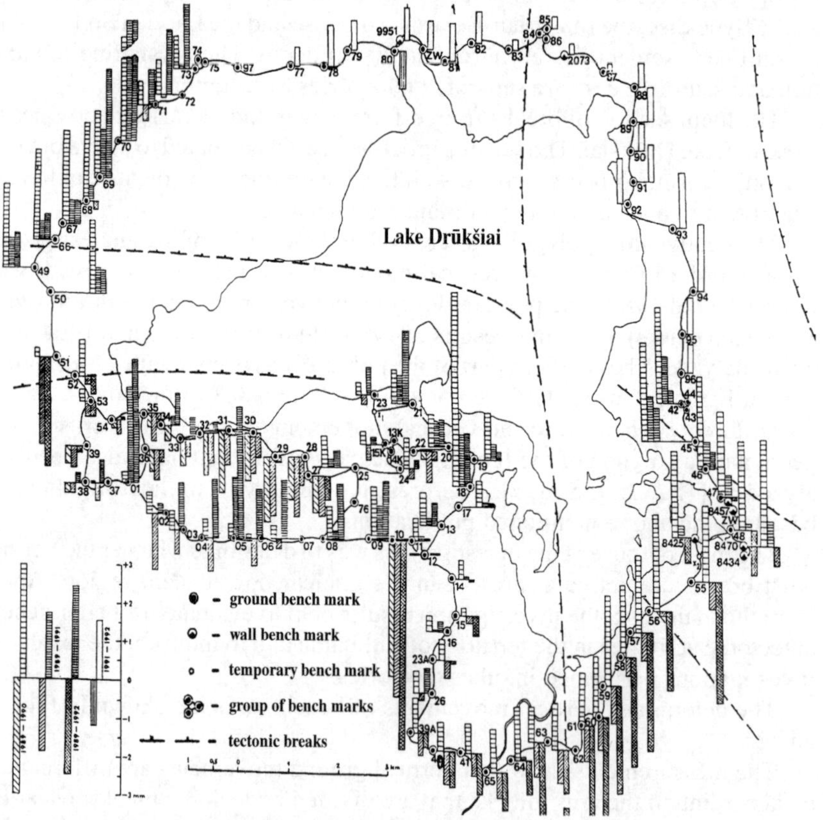

Figure 1: Earth's crust vertical movements in the territory of the Ignalina Nuclear Power Plant geodynamic polygon

The results of disperse analysis of 0.99 probability confirms the preliminary model hypothesis that is related to the crystalline basement tectonic breaks. Methods of cumulative sum analysis were used for the identification of active crystalline basement breaks according to the results of the measurements. The most reliable sites of possible active breaks are determined by the calculated parameters [4, 7].

$$D_t = \min\left(D_{t-1}; a\right) + \left(\frac{a}{b} + \delta h_t\right), \quad (1)$$

$$R_t = \min\left(R_{t-1}; a\right) + \left(\frac{a}{b} - \delta h_t\right), \quad (2)$$

where $\frac{a}{b}$ is the value of limit allowed of the Earth's crust vertical movements gradient change value, dependent on measuring error; δh is deviation of the observed data from the movement background values. In our case, taking into consideration the errors of measurements, with a probability of 0.9, coefficient $a \approx 0.5$. When $D_t < a$, the Earth's crust vertical movements can change into negative direction, while at $R_t < a$, the change takes positive direction.

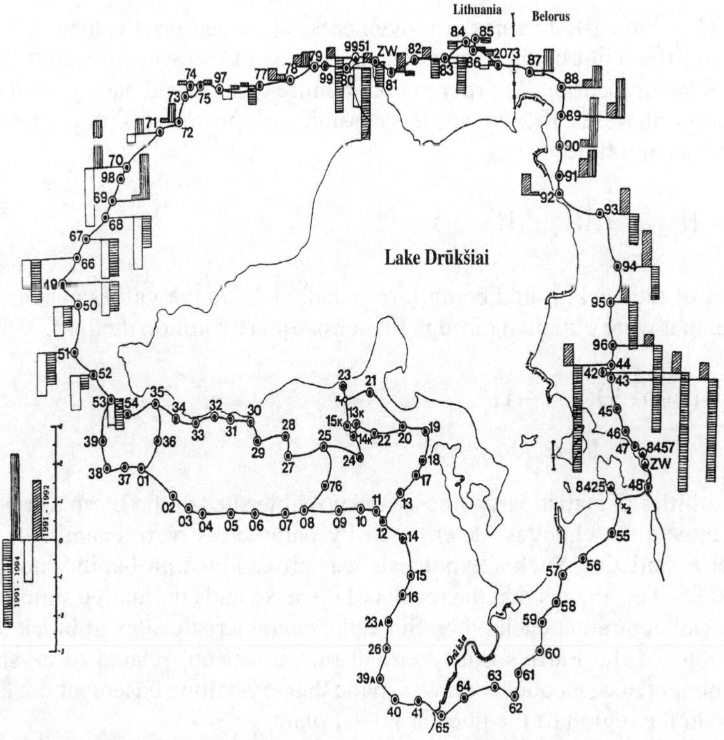

Figure 2: Earth's crust vertical movements in the northern part of the Ignalina Nuclear Power Plant geodynamic polygon

The analysis of cumulative sums of all geodetic observation results of the geodynamic polygon does not contradict the conclusions of the disperse analysis and the preliminary model of the movements.

Vertical earth surface movements are often of complex structure unstationary processes. Regional background of movements sums up with differentiated movements of periodical multiple nature and with random instability results of benchmarks. Then the direct application of cumulative sums' method for the determination of property changes' zones of the earth's surface movements is not always possible. In such cases cumulative sums' analysis methods are applied for the investigation of the remaining quantities that are obtained by eliminating regional background of movements. The application methods of adaptive indexes prognosis model, for the determination of the regional background of movements worked have been made out. Forecasted values of the regional background movements, according to this model, are the following

$$\overline{\mathrm{H}}_t = \overline{\mathrm{H}}_{t-1} + \alpha\left(\mathrm{H}_t - \overline{\mathrm{H}}_{t-1}\right), \quad (3)$$

here $\overline{\mathrm{H}}$ – forecasted values of movements, H – measured values, t – coordinate of the point being forecasted, $0 < \alpha < 1$. On a value depends the degree of generalization. In order to obtain more generalized background of the vertical earth's crust movements it is possible to apply indexes prognosis model for the second time:

$$\overline{\overline{\mathrm{H}}}_t = \overline{\overline{\mathrm{H}}}_{t-1} + \alpha\left(\overline{\mathrm{H}}_t - \overline{\overline{\mathrm{H}}}_{t-1}\right), \quad (4)$$

Values of the background being forecasted, close to the values of regressive model that were obtained through the least-square solution method, would be

$$\tilde{\mathrm{H}}_t = \frac{(2-\alpha)\overline{\mathrm{H}}_{t-1} - \overline{\overline{\mathrm{H}}}_{t-1}}{1-\alpha}. \quad (5)$$

Probabilities of statistic hypothesis and possible sites of the Earth's crust vertical movement changes, determined by parameters were examined, using Fisher F statistics. Such a hypothesis was proved by a probability not lower than 0.95. This means that the results of disperse and cumulative sums analysis do not contradict each other. Since the characteristic sites of block differentiations of the Earth's crust vertical movement are related to crystalline basement breaks, a conclusion was made that crystalline basement breaks are active in the region of the nuclear power plant.

To find out whether common vertical ground movement characteristics, except the intensity of movements, are similar on the whole to a polygon, e. g.

whether active movement inversion in the south-east of the polygon is spreading to other parts of the polygon, correlation analysis of measured vertical ground movements was performed [7]. Two groups of benchmarks were taken for an analysis of the correlation: the western and south-eastern groups. The calculation results are shown in Tables 1 and 2.

Table 1: Correlation matrix of the western group benchmarks

Leveling	1 – 2	1 – 3	1 – 4	3 – 4	2 – 4	2 – 3
1 – 2	1	0.73	0.37	–0.97	–0.69	–0.71
1 – 3		1	0.86	–0.08	–0.05	–0.14
1 – 4			1	0.40	0.41	0.21
3 – 4				1	0.97	0.76
2 – 4					1	0.86
2 – 3						1

Table 2: Correlation matrix of the south-eastern group benchmarks

Leveling	1 – 2	1 – 3	1 – 4	3 – 4	2 – 4	2 – 3
1 – 2	1	0.63	0.49	0.52	0.14	–0.71
1 – 3		1	0.95	0.78	0.82	0.72
1 – 4			1	0.98	0.93	0.33
3 – 4				1	0.89	0.69
2 – 4					1	0.92
2 – 3						1

It has been established that matrixes of correlation in both groups of benchmarks have the same regularities. Therefore, it is possible to affirm that the movement inversion observed in the south-eastern part of the polygon was noticed in other parts of polygon too, but it was less distinct.

Recommendations for future investigations of geodynamic processes in the geodynamic polygon of the Ignalina Nuclear Power Plant are presented in [7]. An expansion of the geodynamic polygon is foreseen. Investigations in the geodynamic polygon and measurements of deformations of are suggested to be combined. Investigations and methods of both horizontal and vertical movements are foreseen. Investigations into local vertical and horizontal Earth's crust movements in the geodynamic polygon are to be included into regional investigations. The geodynamic polygon is connected with Lithuania's first order of GPS geodetic network and Baltic zero order geodetic network as well as with observations performed according to the Baltic Sea Level Project. Therefore, for future observations according to the presented recommendations, of local the Ignalina Nuclear Power Plant region Earth's crust deformations could be related to regional and global Earth's crust blocks movements. Following the suggested conception for future investigations,

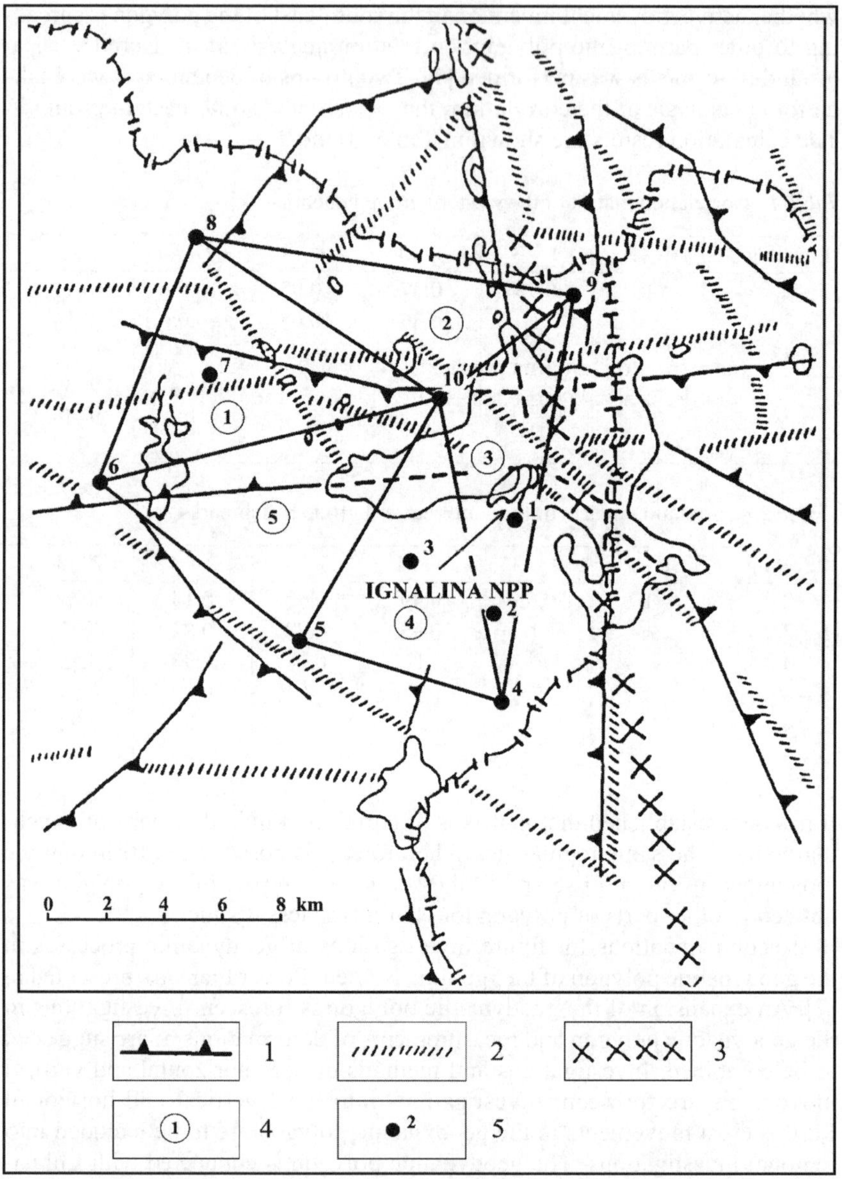

Figure 3: The scheme of the locations of the tectonic breaks at Ignalina NPP (according to I. Juknelis, V. Marcinkevičius, I. Šeštokas and A. Šliaupa): tectonic breaks discovered by: 1 – seismic survey, 2 – aeromagnetic survey, 3 – morfostructural analyse; 4 – number of triangle, 5 – GPS points

research at the Ignalina Nuclear Power Plant geodynamic polygon would not be autonomous, but part of a complex research and one of the seismic monitoring of Lithuania.

3. Investigation of Horizontal Movements of the Earth-Crust in the Region of Nuclear Power Plant

A special polygon of 10 ground points deepened up to 2.5 m with forced centering was established for the measurement of horizontal movements of the earth-crust in the region of Ignalina Nuclear Power Plant (Fig. 3).

Four points (No 1-4) were distributed close to the nuclear power plant in order to determine the present tectonic activity. The other six points were established around INPP at a distance of some kilometers. The points No 1, 2, 3 are very important for the investigation of tectonic stability. A neotectonic linear zone extends in an east-west direction and is very close to the INPP. Point No. 1 is in a zone of tectonic breaks according to seismic and geologic data. Points No 2, 3 and points No 4, 5 are also separated by the east west direction neotectonic linear zone.

There is an Apvardai along the meridian break zone between points No 4 and 5. The five above-mentioned main tectonic blocks are defined by seismic, geological and neotectonic methods. Points No 6 and 7 are also important. They are separated by a meridian direction neotectonic zone with a higher concentration of helium gasses in the underground water. It means the contemporary activity of the zone. Point No. 7 is separated from point No. 10 by a NW-SE direction tectonic fault. Both points are separated by the South Drūkšiai hill from the latitude direction fault zone of Visaginas. Points No 8 and 9 will show possible activity of the North Zarasai step and relative movements to the Drûkðiai bend. Point No. 9 is in the east wing of the Skirnas meridian direction of the fault zone.

The outline of the arrangement of points is presented in the Illustration 3. Points are arranged so that it would be possible to control the main blocks of the earth-crust. Two cycles of measurement were performed: in September 1998 and in 1999. The measurement was carried out with eight GPS devices of the Firm ASHTECH Z–Surveyor and Z–12. Contributors of Vilnius Gediminas Technical University and Danish specialists Nellemann & Bjornkjar performed the measurement. The program of measurement consisted of 4 sessions. The duration of one session was 24 hours. All 4 sessions were conducted on the 1, 2, 4, 6, 9 and 10 points of measurement. 2 sessions were conducted on 3, 5, 7 and 8 points. The program of measurement is presented in the Table 3.

Table 3: Program of measurement

Session	Point number									
	1	2	3	4	5	6	7	8	9	10
A	x	x		x	x	x	x		x	x
B	x	x		x	x	x	x		x	x
C	x	x	x	x		x		x	x	x
D	x	x	x	x		x		x	x	x

The processing of the results of the measurement was performed with program packets GPPS and FILLNET.

It turned out that after processing of the results of measurements the average quadratic errors of measurement of the distances between points of the polygon (precision of measured chords) were within the limits of 0.3-2.4 mm. The average quadratic errors of spatial coordinates did not exceed 1.2 mm. Plane rectangular coordinates and their changes ΔX, ΔY in the period from 1998 to 1999 are presented in the table 4.

Table 4: Coordinates of points

Point Number	X98 (m)	Y98 (m)	X99 (m)	Y99 (m)	ΔX (m)	ΔY (m)
1	66373.772	62793.511	66373.772	62793.511	0.000	0.000
2	63911.515	62186.465	63911.528	62186.469	0.013	0.004
3	64276.247	57780.546	64276.255	57780.555	0.008	0.009
4	61073.407	62485.044	61073.417	62485.044	0.010	0.000
5	60745.917	55036.724	60745.930	55036.728	0.013	0.004
6	66983.435	47762.041	66983.429	47762.049	−0.006	0.008
7	71076.298	50898.683	71076.301	50898.685	0.003	0.002
8	75083.175	52536.489	75083.165	52536.487	−0.010	−0.002
9	72475.732	65068.225	72475.736	65068.226	0.004	0.001
10	71040.621	61144.137	71040.626	61144.137	0.005	0.000

Making the vector schemes of geodesic point movement at the difference of coordinates obtained in separate cycles of measurement, is the simplest method of representation of horizontal deformations of the earth-crust. However, such a method of definition and representation of horizontal movements of the earth-crust is not monosemantic, as the results depend on the selection of fixed points. That is why the methods of definition of the parameters of deformations obtained on the basis of tensor analysis were applied to the analysis of horizontal deformations of the earth-crust in the region of Ignalina Nuclear Power Plant [8], [9], [10]. The advantages of these methods is that they are invariant with respect to the systems of coordinates, i. e. it does not matter what point was conventionally taken for stable within the analysis of deformations. The model of dislocation of points in the geodynamic polygon is applied for the definition of components of the tensor of deformations [9], [10]

$$\Delta x_i = x'_i - x_i = \acute{a} + e_{11} x_i + e_{12} y_i, \quad (6)$$

$$\Delta y_i = y'_i - y_i = \hat{a} + e_{21} x_i + e_{22} y_i, \quad (7)$$

where $e_{11}, e_{12}, e_{21}, e_{22}$ are components of the tensor of deformations

$$T = \begin{Vmatrix} \frac{\partial v_x}{\partial x} & \frac{\partial v_x}{\partial y} \\ \frac{\partial v_y}{\partial x} & \frac{\partial v_y}{\partial y} \end{Vmatrix} = \begin{Vmatrix} e_{11} & e_{12} \\ e_{21} & e_{22} \end{Vmatrix}, \quad (8)$$

x'_i, x_i, y'_i, y_i are the coordinates obtained in the first and the second cycles of measurement.

Tensor (8) is invariant to the systems of coordinates, i.e. it does not depend on the selection of the primary conventionally stable point. Components e_{ij} describe the natural deformations of the earth-crust plot. In equations (6) and (7), parameters á and â characterize the dislocation of some plot of the earth-crust as of absolute solid, also including the dislocation of the system of coordinates. Since we are interested in the natural deformations of the earth-crust, it is sufficient to define coefficients e_{ij}.

The estimation of coefficients e_{ij} is carried out by the method of final elements. The synonymy of solution is obtained when non-recovered triangles are taken for final elements.

Maximum and minimum relative lengthening are the main characteristic of deformation of the earth-crust [8], [9]

$$\begin{bmatrix} D_1 \\ D_2 \end{bmatrix} = \frac{1}{2}(e_{11} + e_{22}) \pm \sqrt{(e_{11} - e_{22})^2 + (e_{12} + e_{21})^2}, \quad (9)$$

where: D_1 – maximum lengthening, D_2 – minimum lengthening.

Relative dilatation

$$\Delta = e_{11} + e_{22}. \quad (10)$$

Maximum and minimum relative lengthening is perpendicular between each. The direction of maximum lengthening in the respect of abscises axis

$$\varphi = \frac{1}{2}\operatorname{arctg}\left(-\frac{e_{12}+e_{21}}{e_{11}-e_{22}}\right)+\begin{bmatrix}90°, \text{ when } (e_{11}-e_{22})>0 \\ 0°, \text{ when } (e_{11}-e_{22})<0\end{bmatrix}. \qquad (11)$$

Maximum and minimum relative lengthening, perpendicular between themselves were used for the estimation of characteristics of deformations of the earth-crust (9), dilatation (10) and the direction of maximum lengthening (11). It is expedient that they are directly related with respect to the characteristics of geodynamic tensions of the earth-crust. Lengthening is proportional to tensions, and their directions and signs are related with directions and signs of minimal and maximal tensions.

Parameters of deformations were calculated by the method set forth above (9), (10), (11) for five triangles of the polygon showed in the Illustration 3. Results of the calculation are presented in the Table 5.

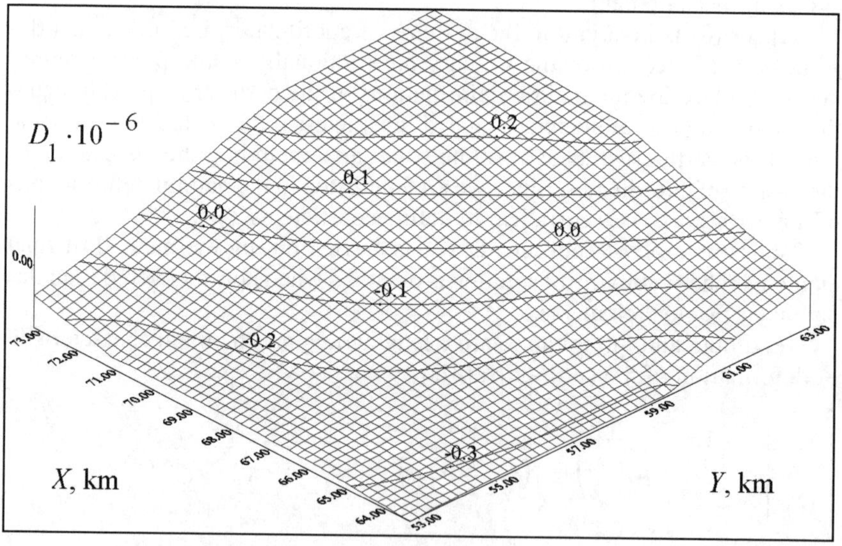

Figure 4: Isogram of the maximum relative lengthening

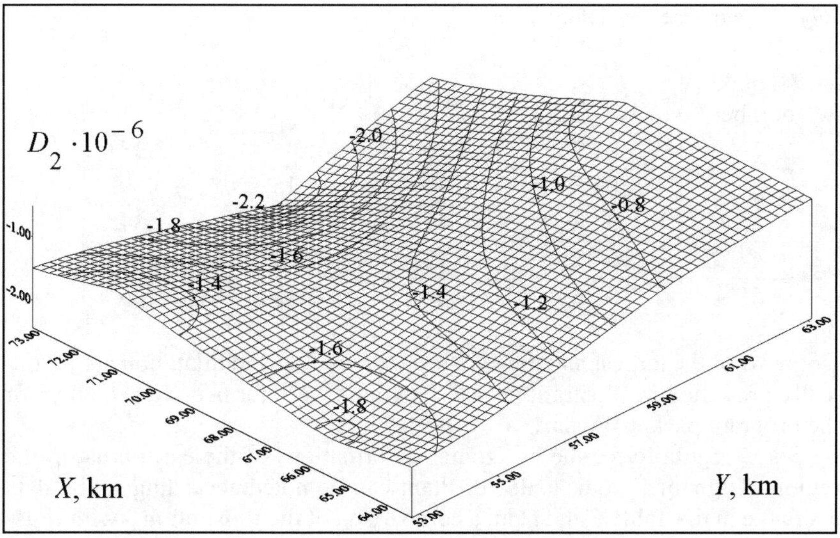

Figure 5: Isogram of the minimum relative lengthening

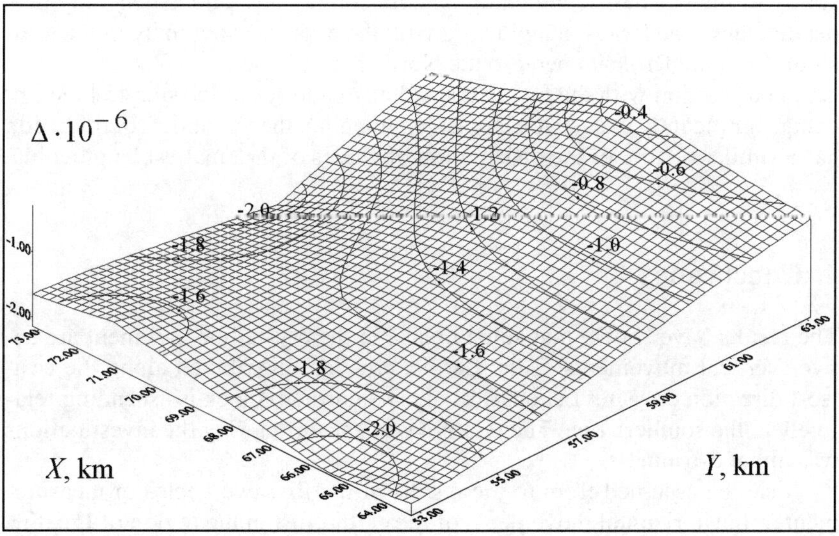

Figure 6: Isogram of relative dilatation

Table 5: Parameters of deformation

Triangle number	$D_1 \cdot 10^{-6}$	$D_2 \cdot 10^{-6}$	$\Delta \cdot 10^{-6}$	φ°
1	–0.27	–1.19	–1.46	3.38
2	0.28	–2.43	–2.15	6.89
3	0.24	–0.51	–0.27	–1.34
4	–0.32	–0.77	–1.09	–43.81
5	–0.28	–1.85	–2.13	26.26

Isograms of the largest and the smallest lengthening and dilatation are graphically presented in Illustrations 4-6. Isograms were formed with the help of the program packet Winsurf.

Some regularity of the horizontal deformations of the earth-crust in the region of Ignalina Nuclear Power Plant can be noted according to the data available in the Table 5 and from a comparison of the Illustration 3 with Illustrations 4-6. The most maximum relative lengthening grows almost rectilinear from the southeast to the northwest, and the directions of isolines of their changes conform generally to the directions of the tectonic breaks. The variability of the smallest lengthening is more complicated and is generally perpendicular to the largest lengthening with the expressed anomaly in the transition from the Drūkšiai bend to the North-Zarasai rise.

In connection with the fact that absolute quantities of the smallest lengthening significantly exceed the largest lengthening, the variability of dilatation has a similar picture of the distribution of values of the smallest lengthening.

4. Conclusions

The results have shown that some breaks in the crystalline basement are active. Vertical movements of the Earth's crust are distributed along the east-west direction crossing Lake Drūkšiai. The northern block is ascending relatively to the southern one. The relative vertical shift during the investigations was about 5-6 mm.

It can be assumed from to the results of the first two cycles of measurement of the horizontal movements of the earth-crust in the region of Ignalina Nuclear Power Plant that horizontal deformations of the earth-crust in this territory are related with the tectonic building of the territory.

The largest relative lengthening of the territory and at the same time the largest tectonic tensions of the earth-crust in the region of the nuclear power plant are perpendicular to the main tectonic breaks. The smallest lengthening and tensions are parallel to the breaks.

Results of the analysis of the first two cycles of measurement of horizontal movements of the earth-crust in the region of Ignalina Nuclear Power Plant do not contradict the conclusions made according to the data of vertical movements of the earth-crust about the contemporary activity of some tectonic breaks of the crystal foundation.

It is necessary to continue the measurement, in order to draw more detailed conclusions about the characteristics of horizontal deformations of the earth-crust in the study area.

References

[1] A. Zakarevičius, V. Marcinkevičius. Ar juda žemės blokai po Ignalinos atomine elektrine // Geologijos akiračiai, 1992, Nr. 3, pp. 17-19.

[2] S. Kazakevičius, J. Stepanovienė. Ignalinos atominės elektrinės geodinaminio poligono niveliacija // VTU mokslo darbai. Geodezijos darbai. T. 19, V.: Technika, 1993, pp. 56-65.

[3] A. Zakarevičius, J. Stepanovienė, R. Ramanauskas. Vertikalių žemės paviršiaus judesių Ignalinos atominės elektrinės teritorijoje analizė // VTU mokslo darbai. Geodezijos darbai. T.19, V.: Technika, 1993, p. 101-108.

[4] A. Zakarevičius. Vertikalių žemės paviršiaus judesių Ignalinos atominės elektrinės geodinaminiame poligone analizė // VTU mokslo darbai. Geodezijos darbai. T. 20, V.: Technika, 1994, pp. 7-19.

[5] A. Zakarevičius, J. Česnauskienė. Vertikalių žemės paviršiaus judesių Ignalinos atominės elektrinės geodinaminio poligono šiaurinėje dalyje // Geodezija ir kartografija, Nr. 1 (21), V.: Technika, 1995, pp. 77-81.

[6] A. Zakarevičius. Adaptyvinio eksponentinio modelio panaudojimas analizuoti Ignalinos AE geodinaminio poligono niveliavimo rezultatus // Geodezija ir kartografija, Nr. 2 (22), V.: Technika, 1995, pp. 81-89.

[7] A. Zakarevičius. Dabartinių vertikalių žemės plutos judesių Lietuvos teritorijoje tyrimas. V.: Technika, 1994, 276 p.

[8] Н. П. Есиков. Тектонофизические аспекты анализа современных движений земной поверхности. Новосибирск, “Наука”, 1979. 182 с.

[9] T. Sagiya, S. Miyazaki, T. Tada. Continuous GPS array and present-day crustal deformation of Japan. Pure and Applied Geophysics, 157. Basel: Birkhäuser Verlag, 2000, pp. 2303-2322.

[10] A. Zakarevičius, A. Stanionis. Horizontaliųjų žemės plutos deformacijų nustatymas afiniojo koordinačių perskaičiavimo būdu // Geodezija ir kartografija, XXIX t., Nr. 1, V.: Technika, 2003, pp. 13-19.

Teil III
Beiträge zu neuesten Entwicklungen in anderen Ländern

HANS MATTSSON

Nature Conservation in Sweden

Introduction

Sweden has a total area of 450,000 km^2 and just over 50 per cent of it is productive forest land. Lakes and wetlands are very much in evidence. Ten per cent of the area of Sweden consists of lakes and watercourses, and another 10 per cent of wetlands. Agricultural acreage is small, about 8 per cent, and urban activities occupy about 2 per cent of the total area. The remaining 15 per cent of land consists mainly of mountainous country (SCB 1998). The national aims for management of these land and water areas can be found in the Environmental Code where it is stated:

> Land and water areas shall be used for that or those purposes for which the areas are most suitable with respect to character and location as well as existing needs. Priority shall be given to such use, which entails a good management from a public perspective.

Another section of the Environmental Code reads as follows:

> Land and water areas of importance in respect of their natural beauty or cultural value, or with respect to outdoor recreation, shall as far as possible be protected against measures that may substantially damage the natural or cultural environment. Particular attention shall be paid to the need for areas for outdoor recreation in the proximity of urban areas.
> Areas of national importance for the conservation of nature or of cultural monuments and finds, or for outdoor recreation, shall be protected against measures referred to in the first paragraph.

Nature conservation is, to a large extent, related to forest land including forest wasteland (impediments) but also to lakes, the sea, wetlands, mountain areas and meadows. This article focuses on the legislation in place for nature conservation, with an emphasis on forests. Some reference is also made to the connection between nature protection and the issue of landowner compensation. The emphasis placed on forests is because these extensive areas are of particular importance in Sweden from the point of view of nature conservation. At the same time, forests are also economically important because one-sixth of the country's export earnings (SCB 2003) is derived from forest products.

This article will therefore begin with a description of the ordinary management of forestry land, and will then continue with a comment on those regulations that contribute to a more specific nature conservation regime. The most central enactments in support of nature conservation are the Environmental Code (miljöbalken) and the Forestry Act (skogsvårdslagen) and the main content of the two acts will thus be described.

Management of Forest Land

Due to the importance of forests, Swedish legislation has long included provisions limiting the individual person's right of disposal over forests. Following the growth of forest industry during the second half of the 19th century, the foremost objective has been to safeguard supplies of timber. The 1923 Forestry Act introduced, for the first time, a more coherent regulatory system for ensuring sustainable production of timber. A new Forestry Act, passed in 1948, included stricter provision with regard to the duty of taking rejuvenation measures and also in respect of the restrictions on felling rights. In 1974 it was laid down that consideration should also be paid to the interests of nature conservation (SOU 1978).

The existing Forestry Act, which came into force in 1979, contains provisions on the conduct of forestry activities and the considerations that management has to take into account, which includes those of nature conservation. The Act was thoroughly revised in 1993, one reason for the revision being the poor record of consideration for environmental aspects. The amendments resulted, among other things, in the environmental objectives being defined more clearly and put on a par with production targets.

An evaluation of the changes in the Forestry Act of 1993 took place in 1997 (SOU 1997). The evaluation concluded that some deficiencies existed that involved the fulfilment of the goals for both production and environment. The problems in relation to production were that planting and thinning had been neglected whilst the criticism concerning environment was mainly about tree cutting on non-productive land. Minor changes were therefore introduced into the Act (Prop 1998) to enable the authorities to better check and regulate the aforementioned activities.

The key principle of the Forestry Act is that forestry should be conducted on both a rational and an ecologically sustainable basis. This requires that consideration be given to various environmental factors so that the Forestry Act harmonises with both the global forest principles and the Convention on Biological Diversity. The environmental objective for forestry is contained in the Government Bill (Prop 1993) and states that:

> The natural productive capacity of forest land shall be preserved. Biological diversity and genetic variety in the forest shall be safeguarded. The forest shall be used in such a way that flora and fauna species naturally belonging there are

> enabled to survive under natural conditions and in viable stocks. Threatened species and environments shall be protected. The cultural environmental values of the forest and its aesthetic and social values shall be safeguarded.

As a result of this aim the opening section of the Forestry Act defines its purpose as follows:

> The forest is a National resource. It shall be managed in such a way as to provide a valuable yield and at the same time preserve biodiversity.
> Forest management shall also take into account other public interests.

In order to achieve its goal, the Act contains general rules of management with which the landowner has to comply, together with powers that allow the authorities to introduce special rules and restrictions on forestry. Included among the provisions of the Act are:

- Establishment of forest on forest land
- Felling of forest on forest land
- Requirement of permits for felling etc.
- Broadleaf forests
- Insect pests
- Consideration for nature conservation, heritage conservation and reindeer husbandry
- Environmental impact assessments

Environmental considerations

The Forestry Act contains rules which, directly or indirectly, can be used to safeguard the interests of nature conservation and heritage conservation. The Act empowers the Government or an authority appointed by the Government, i.e. the National Board of Forestry (Skogsstyrelsen) to issue regulations on the consideration to be shown in the management of forests in respect of the size and distribution of felling sites, the remaining stands, the leaving of trees and groups of trees, manuring, ditching and the routing of lorry roads.

But the Act together with the Forestry Ordinance (skogsvårdsförordningen) also contains other provisions, which can enable the State to influence questions of management:

- After final felling, it is the landowner's duty to ensure that replanting takes place.
- In the case of large forest land holdings (exceeding 1,000 hectares), the National Board of Forestry is entitled to prescribe that not more than a certain proportion of the acreage may be felled annually. This is designed to promote an even-age distribution of forest stocks.

- The owner of the forest land shall inform the County Forestry Board (skogsvårdsstyrelsen) concerning felling which is to take place on his land[1], ditching for which permission is not required under the Environmental Code and measures planned in order to provide for the interests of nature conservation and heritage conservation.
- The stipulation of permits for felling operations in protected forests or forest which is difficult to rejuvenate. Conjointly with the application, an account shall be given of the allowances to be made for the interests of nature and heritage conservation.
- Felling of sub-montane forest may not take place if it is contrary to or incompatible with interests of vital importance for nature or heritage conservation. If permission is given, the National Board of Forestry shall decide which allowances are to be made in the course of felling operations. Compensation for damage resulting from such decisions is payable as provided in the Nature Conservation Act.
- Felling of broadleaf forest (defined as forest containing at least 70 per cent leaf-trees) is subject to permission from the County Forestry Board. When applying for such permission, thc applicant shall indicatc what measures he intends taking to provide for the interests of nature and heritage conservation. In granting permission, the County Forestry Board may resolve on the manner in which felling and replanting are to take place. Other measures, which can affect the mix of trees, are also prohibited. This protection for broadleaf forests has been introduced mainly to prevent the broadleaf forests of southern Sweden being superseded by the faster-yielding spruce.
- The State (the National Board of Forestry) can prescribe that there shall be an environmental impact assessment (EIA) if new management methods or new regeneration materials are to be used. This need presupposes that the activity is on a certain scale and will continue for a considerable period of time. The National Board of Forestry decides those points that are to be covered by the EIA and also who is to pay for it. A large forest company ought normally to be capable of paying the cost of the EIA itself, whereas a smaller landowner is unlikely to have such resources.
- If necessary, the County Forestry Board can issue injunctions or prohibitions to secure compliance with the Forestry Act. These may carry contingent fines.

1 In forestry units exceeding 50 hectares, felling may not take place on such a scale that more than half the unit's productive acreage will consist of clear-felled land or juvenile forest (less than 20 years old).

Compensation questions

The Forestry Act does not contain any rules of compensation, except for its provisions concerning measures in sub-montane forest. Otherwise, no decisions are permitted whereby current land use is substantially impeded for the landowner. Regulations on felling and on the maintenance of non-productive forest land (impediments), on the other hand, never confer entitlement to compensation, the reason being that these lands are such that no economic restrictions are considered necessary.

Remarks

The Forestry Act is the law which is used for regulating both productive and environmental aspects of "everyday" forestry. It provides good opportunities for influencing and supervising questions of preservation. Breaches of regulations are punishable by fines or by up to six months' imprisonment. No liability is imposed, however, for minor offences.

Consequently, the provisions of the Forestry Act give the authorities, at least in theory, adequate opportunity to steer the actions of landowners in such a way that an acceptable solution will also be reached in respect of environmental aspects. Since, however, the economic circumstances of the individual landowner may only be partially affected, the possibility of regulating practical forest management by peremptory decisions are circumscribed. It is mainly through consultation and counselling that the Forestry Act affords a basis for influencing the environmental aspects of forestry. Areas of outstanding interest, however, have to be made part of a nature reserve or set aside as biotope protection areas under the rules of protection of the natural environment in the Environmental Code in order to be adequately protected.

Protection of the Natural Environment

Protection of the natural environment has been regulated by the Environmental Code since 1998 as it is a national asset that must be properly cared for. Although public access to the countryside (allemansrätt / "Everyman's Right") should be unrestricted in accordance with the law, due regard and circumspection should be taken to ensure that no damage occurs. Whilst unnecessary damage should be prevented, if intervention is required, it should aim to alleviate as far as possible any damage inflicted. The purpose of the Code, however, is not only to facilitate outdoor recreation (social nature conservation) but also to protect species, biotopes, ecosystems and landscapes (preservative nature conservation).

In order for the purpose of the Code to be achieved, an authority may order restrictions on landowners' use of their properties. In addition, regulations can be issued concerning the management of protected areas, and in certain cases, a public authority may itself undertake management measures. It should also be mentioned that in the examination of nature protection matters under the Environmental Code, both private and public interests should be taken into account and restrictions should be no more severe than necessary to meet the demands required for the protection provided.

The Environmental Code can be said to provide two essentially different forms of active protection. One of these is based on area protection being imposed on the landscape, and the other on the necessity to obtain permission before certain controlled enterprises may take place. The area protection institutes in the Code are listed below[2]. The order of the list is according to compensation rules where no compensation is given for the four first types of areas:

- Shoreline protection areas
- Fauna and flora protection areas
- Environmental protection areas
- Natural monuments
- Nature reserves
- Cultural reserves
- Biotope protection areas
- Special protected areas
- National parks

Certain controlled enterprises for which special permission has to be obtained and which can affect land are:

- Agriculture
- Water undertakings
- Quarrying of gravel etc.
- Other enterprises which can alter the natural environment

Area protection institutes and controlled enterprises are dealt with below, one by one, and this is followed by a concise review of the principles of compensation for restrictions on disposition.

2 Water protection areas (vattenskyddsområden) are excluded from the list. The reason is that this type of protection is not aimed at protecting the natural environment but at safeguarding water supply for consumption.

Area protection institutes in the Environmental Code

Right of public access and shoreline protection

Right of public access to the countryside (allemansrätt / "Everyman's Right") is of overriding importance to a number of provisions under the Environmental Code and is even inscribed in the constitution (regeringsformen). Thus, it is this right that will be described in the first instance, together with shoreline protection.

Right of public access means the right to roam in the countryside, to bathe, pick berries and mushrooms, camp for a short time and so on. This began as a customary right. The concept was not codified until 1975, when it became part of the opening provision of the Nature Conservation Act[3]. However, the powers deriving from the right of public access are not defined by statute but are determined indirectly in the light of other laws. The Penal Code (brottsbalken), for example, shows that the right has its limitations. That Code lays down that if a person "unlawfully goes across a building lot and plantation or across other land that can thereby be damaged, he can be convicted of taking an unlawful path" . Another restriction is evident from the following provision of the Penal Code whereby "a person who in forest or field unlawfully takes growing trees or grass or from growing trees takes twigs, branches, bark, leaves, bast, acorns, nuts or resin, or a windthrow, stone, gravel, turf or other such things which have not been prepared for use, shall be convicted of trespass" .

The right of otherwise crossing another person's land also applies to skiing, riding and cycling, so long as the natural environment is not harmed. On the other hand, right of public access does not include the right of driving a motor vehicle off-road. Right of public access also applies to an area of water, where it includes the right of travelling by boat. Hunting and fishing are not included in the right of public access, but fishing with hand-held tackle is permitted off all coasts, private coastal waters included.

The Environmental Code not only contains a number of provisions to secure the continuing exercise of this right of public access but also includes provisions for the general protection of the countryside. Public right of access to an area can therefore be restricted, for example, in favour of scientific nature conservation.

To guarantee access to areas of importance for outdoor recreation, provisions also exist whereby landowners can be ordered to provide a gate or other aperture in fencing which would otherwise exclude the general public from an attractive area. If fencing is put up for the sole purpose of excluding the

---

3 The Nature Conservation Act was included in the new Environmental Code 1998.

general public, the landowner can be ordered to remove it completely. Wildlife fencing may not be erected without permission in areas where the general public is allowed to roam freely.

The Environmental Code contains provisions on shoreline protection (strandskydd), aimed at securing access to places for bathing and outdoor recreation through the right to public access to the countryside, and also for securing a good environment for flora and fauna on land and in the water. This is a general area protection, concerned primarily with regulating building development along coasts and shores. Shoreline protection prevails by the sea and by lakes and watercourses. It automatically includes land and water areas up to 100 metres from the shoreline, but the area can be enlarged by up to 300 metres.

There are, however, a number of exemptions from shoreline protection. The County Administrative Board (länsstyrelsen), for example, may waive the ban on building development. Another important exception is that facilities or measures of importance for agriculture, fisheries, forest management or reindeer husbandry do not come under the shoreline protection provisions. In addition, a shoreline protection area that is manifestly lacking in importance for the purposes of shoreline protection can also be exempted from the provisions.

Within a shoreline protection area, no new building may be erected and an existing building may not be essentially altered in such a way as to acquire a new purpose. Nor may facilities be constructed whereby land is used as developed land or public access to the shore is otherwise impaired. Furthermore, measures may not be taken that can impair suitable environments for fauna and flora.

No compensation is payable to landowners on account of shoreline protection.

Fauna and flora protection areas

A plant species can be placed under protection if it is in danger of extinction. The same applies if a plant species is at risk of being plundered. A large number of plant species, including several orchids, have been placed under protection through these provisions. In some parts of Sweden, even more common spring flowers such as hepatica, cowslip and lily-of-the-valley, which would otherwise be picked in large quantities by the general public, have also been placed under protection.

Plant protection applies both to the landowner and to the general public. This means that a landowner is not entitled to pick a protected plant growing on his or her property. If, however, a protected plant species grows in an area where the landowner has obtained building permission for a house or another structure, the plant may be removed.

If there is a danger of an animal species disappearing, a prohibition may be issued against killing, injuring or trapping animals of that species. The provisions of the Environmental Code merely supplement other legislation, because provisions on hunting and fishing are mainly found in other enactments.

Under the Hunting Act (jaktlagen), wildlife in general is protected and species may only be hunted where this is permitted by specific provisions. Protection also applies to eggs and nests. The Hunting Ordinance (jaktvårdsförordningen) contains provisions on the times of the year when hunting of different animal species is permitted. Thus, in the absence of particulars concerning open seasons, hunting is prohibited all the year round. It should be pointed out that hunting and fishing rights belong to the landowner.

If further protection is needed, the County Administrative Board, acting in consultation with the Swedish Environmental Protection Agency (Naturvårdsverket), may issue regulations for wider protection in certain defined fauna and flora protection areas (djur- och växtskyddsområden). This may concern the right to hunt and also the right of the landowner and the general public to roam the area. Otherwise, the landowner has freedom of movement in the area regardless of what may otherwise be prescribed.

Some 1000 wildlife sanctuaries of this kind have been formed. They occupy 1,100 km^2, half of it water. This form of protection is most often used to protect the nesting areas of birds and areas frequented by seals. In addition to this, some 300 plants species and animal species are protected all over the country (SCB 2001)

The question of compensation for the formation of wildlife sanctuaries has not been regulated and so nature reserves should be formed if intervention results in current land use being substantially impeded . The same applies to plant protection.

Environmental protection area

A major land or water area may be declared by the Government to be an environmental protection area (miljöskyddsområde) if special rules are required to improve the current environmental situation in that area. The reason can be that it is exposed to pollution or does not satisfy an environmental quality norm and therefore the Government or the County Administrative Board may issue regulations concerning protective measures, limitations and other precautionary measures for that area. Compensation is not payable and the number of such areas is small.

It should be mentioned that if pollution is considered a potential risk to health or environment, an area can be declared an environmentally hazardous zone (miljöriskområde). Land use can then be restricted and measures decided upon that the landowner will be required to take. Compensation is not paid.

Natural monuments

Specially distinctive natural objects which, by virtue of their importance for the natural environment, their beauty or their otherwise remarkable characteristics, should be protected or cared for, can be placed under protection as natural monuments. This protection refers to the natural object but can also include the area of land necessary for its protection. Natural monuments include ancient oak trees, erratic blocks and potholes. No compensation is payable to the landowner.

Some 1,400 natural objects have been placed under protection as natural monuments (SCB 2001).

Nature reserves

Areas of countryside can also be protected through the formation of nature reserves. The purpose of the protection is to preserve biological diversity, conserve and protect valuable natural environments and satisfy the needs of the area for outdoor activities. An area required to protect, reinstate or create new valuable natural environments for species of special protective value may also be declared as a nature reserve. Given this variety of purposes, nature reserves vary considerably in character and size. They can, for example, comprise large areas of importance for outdoor recreation or small ones of great ecological value.

Ownership conditions are immaterial where the formation of nature reserves is concerned. Thus reserve provisions can be applied to privately owned land, even if the landowner objects. To secure the purpose of the reserve, restrictions can be imposed on the landowner's right of use of the area. Restrictions may consist of a ban on building development or fencing, as well as the prohibition of arable farming, planting, felling of timber, ditching, hunting, fishing etc. The use of pesticides and artificial fertilisers can also be prohibited. Access to the area can be prohibited for the whole or part of the year. A special management plan has to be drawn up for a nature reserve and a warden appointed.

If the purpose of the reserve so requires, measures can also be taken to make it more accessible. This can mean the construction of roads, parking facilities, hiking trails, chalets, camp sites, bathing points, sanitary facilities or suchlike. The landowner can also be required to acquiesce in felling, clearance, grass cutting, planting etc. and to the opening of the area to the general public if it is not already accessible. If, on the other hand, the area requires protection, public access to it can be regulated, limited or abolished completely.

If the landowner's right of disposal over his property is such that current land use is substantially impeded , he is entitled to compensation for the damage suffered. Restrictions on manuring, ditching rights, felling rights etc. can, for example, have the effect of substantially impeding current land use. If land is utilised for facilities, the landowner is also entitled to compensation for the area thus claimed. Should the encroachment be so extensive as to entail exceptional detriment, a request can be made for the purchase of the entire property. If, on the other hand, the authorities want to take over ownership, they have the right to expropriate the land.

Where a landowner is unable, due to the formation of a nature reserve, to put the land to a new use, no compensation is payable if the alternative use had been subject to the grant of a permit before the formation of the nature reserve. This is because current land use is not normally deemed to include the right to a change of land use. This can, for example, apply to a desire to construct a summer cottage in the area.

It is worth noting that felling of timber within an area is deemed to form part of the right of current land use and is therefore not subject to the grant of a permit (only to a duty of notification, as has already been described). If, however, the County Administrative Board receives notice, for example, of the final felling of timber, it can prevent felling by forming a nature reserve or by forming a temporary nature reserve for up to five years. In the meantime, the County Administrative Board can investigate whether or not the area is of such value for conservation that a nature reserve should be formed. Annual compensation can be paid during the investigation period if permission is not granted for a measure that would otherwise be permissible.

About 2,300 nature reserves have been formed and they occupy a combined area of 40,000 km^2, equalling nearly 10 per cent of Sweden's total area. Most of the area is in the mountains in northern Sweden. Only one per cent of productive forest land is protected. About 6,000 km^2 consists of water areas (SCB 2001).

Cultural reserves

An area can be declared a cultural reserve if the purpose is to protect a valuable, culturally characteristic landscape. The same rules apply to this type of area as to natural reserves. There are only a few reserves of this kind.

Biotope protection areas

In smaller areas of land and water (biotopes) which constitute the habitats of threatened fauna and flora species or are otherwise particularly deserving of protection, enterprises such as excavation, earth moving, timber-felling and clearance may not be undertaken if they have the potential for damaging the natural environment. This is an absolute prohibition, effective immediately there is a risk of damage.

The Government or an authority appointed by the Government decide which biotopes are to be included in protected biotope areas. The decision-making body depends on the type of biotope. The principle is that if the biotopes are easily identifiable, they can be mentioned directly in a schedule to the Environmental Code (Protected Areas) Ordinance (*förordning om områdesskydd enligt miljöbalken*), in which case they are automatically protected.[4] Biotopes which are hard to delimit are designated and delimited. This is done by the County Forestry Board in the case of forest land and the County Administrative Board in the case of arable and grazing land. Transitional zones between different types of land are normally handled by the County Administrative Board.[5] The method of protection is prompted by considerations of legal security: that is, easily defined biotopes are listed in general terms in a national Ordinance, while those which are less easily definable are delimited through a specific official decision.

Measures capable of damaging the biotope may not be taken, and whoever carries on a working operation within a protective biotope is liable for the damage caused. There are no restrictions on the taking of measures which benefit the biotope. To make sure that nothing is done which is not permissible, the landowner can consult the responsible authority before work is carried out. The County Administrative Board and the County Forestry Board can make such arrangements as needed to protect the biotope areas.

If the taking of a measure within a biotope-protected area is prohibited and this means that current land use is substantially impeded , the owner is entitled to compensation. If, on the other hand, a regulation has been issued whereby permission has to be sought in order to take a certain measure which would otherwise be permissible, compensation can only be claimed if permission is not obtained or if permission is made subject to conditions. If exceptional detriment occurs to the property owner, he can request that the property be purchased.

A survey of key biotopes in Sweden indicates that they are 30,000 in number and that their average size is upwards of one hectare. The number of protected biotopes, however, is just over 2,000 (SCB 2001).

---

4 The areas listed in the Environmental Code (Protected Areas) Ordinance mainly concern wetlands in arable land and special groups of trees and thus do not concern continuous areas of forest.

5 The County Administrative Boards have nature conservation and agriculture units, whereas expertise on forest questions is to be found at the separate County Forestry Boards.

Special protected areas

The Government may declare an area to be a special protection area under the EC Bird Protection Directive and the EC Species and Habitat Directive (Natura 2000). If, in such an area, a measure is to be taken that significantly influences the environment, permission is required. Should permission be refused and current land use is substantially impeded , the landowner is entitled to compensation. Permission can be conditional, in which case it is the landowner that has to bear the cost of any required measures. If the encroachment is of exceptional detriment, the landowner can request that the property be purchased.

About 3.500 such areas have been established. According to plans available from the Swedish Environmental Protection Agency, about 10 per cent of the total area of Sweden is covered by Natura 2000 areas. However, most of the areas are or will be covered by other forms of protection such as nature reserves and nature parks (SNV 2003). One advantage of using these other kinds of protection areas is that special management plans can be adopted and wardens appointed.

National park

State-owned land can be designated a national park for the purpose of preserving a large, continuous area of a certain type of landscape in its natural state or essentially unaltered. For every national park, regulations are issued concerning the right of crossing or frequenting it. If a national park is to be formed on land not owned by the State, then the State has to purchase or expropriate the land in order to obtain the title to it.

Sweden has 28 national parks totalling 6,500 km^2 in area. Eighty per cent of this area is accounted for by three large national parks in the Swedish mountains (SCB 2001).

Permissions for certain enterprises

In accordance with the Environmental Code, permission is also required for certain types of enterprises.

Agriculture

If agricultural land and grazing land are to be taken out of production (set aside), notice to this effect must be given to the County Administrative Board at least eight months in advance. No notification is needed, however, if the measure is only of slight importance for the agricultural enterprise in question. The notification enables the County Administrative Board, through the authority of the Environmental Code, to impose area protection on acreage which would otherwise be spoiled or afforested. The legislation also makes it possible to issue regulations concerning allowances to be made in agricultural operations for the protection of the agrarian and cultural landscape and of flora and fauna. But these regulations may not be so extensive that current land use is considerably impeded.

Water undertakings, in particular land drainage

As water management is important for the environment, it is covered by extensive regulations in the Environmental Code. However, only the regulations relating to land drainage will be discussed in this article as wetland areas are important for the preservation of fauna. In order to safeguard wetland areas, permission has to be obtained for land drainage (ditching, reclamation of lakes etc.). This provision does not apply to normal drainage. Protective drainage in a forest, to prevent the water table from rising after clear-felling, does not require permission either, but has to be notified to the County Forestry Board. This Board then has to decide whether it is a case of normal protective ditching, whether the measure comes under the land drainage provisions or will otherwise alter the natural environment and can thus be regulated. The Government can prohibit all land drainage in cases where the preservation of wetlands is of particular urgency.

Land drainage enterprises do not affect very great areas. In 1992, permits were granted for 30 km^2. The notified area affected by protective ditching that year was 75 km^2, and four years later it was only 6 km^2.

Quarrying of gravel etc.

Under the Environmental Code, extraction of stone, gravel, sand, clay, soil, turf or other types of soil may not be undertaken without permission. An exception is made for extraction for the domestic needs of the tenant of the land. This provision means that permission is not necessary if a farmer needs to extract sand for casting operations relating to farm buildings and structures. But if he extracts the same sand for sale, permission is needed.

The County Administrative Board can stipulate that an extraction permit application be accompanied by an "extraction plan". This can show when, where and how extraction is to take place, where the extracted material is to be sold, when it is estimated that the extraction will cease, what the extraction area will look like after extraction has been concluded and so on. If an extraction permit is granted, special regulations can be issued concerning those measures which must be taken to counteract the negative impact of the extraction on the natural environment. The authority may also stipulate that economic security is to be provided for measures of this kind.

An extraction permit can be refused on several grounds. For example, certain gravel ridges must be preserved for scientific reasons while other gravel ridges may need to be preserved because they are important groundwater reservoirs. Extraction of stone, gravel or other material can also be prevented if it leads to the destruction of an attractive area of countryside.

A landowner is deemed not to have a guaranteed right of availing himself of such natural resources as sand, gravel, stone etc. on his land. This means that refusal of an extraction permit does not confer a right of compensation.

Other enterprises which can alter the natural environment

Timber-felling, afforestation, introduction of alien tree species, cultivation of natural meadow land or wooded pasture, filling in of wetlands, ditching and ditch clearance, extraction of gravel for domestic use, construction of service lines above or in the ground, road construction and the handling of chemical products are examples of enterprises which can change the natural environment. Over and above the rules concerning control of land drainage and quarrying activities, the Environmental Code contains a general rule of consultation with respect to enterprises which can substantially alter the natural environment. Before any such enterprise takes place, the entrepreneur shall consult the County Administrative Board. Responsibility for the occurrence of consultation devolves on the party undertaking the enterprise. If damage occurs, he is liable.

In addition, the Government may prescribe that consultation notice shall always be given for specified enterprises. Similarly, the County Administrative Board may prescribe a general duty of consultation throughout the county or in a certain part of it where certain types of enterprise are concerned. In the course of this consultation, the County Administrative Board may order that the party undertaking an enterprise be first required to carry out any measures necessary to limit or counteract damage to the natural environment. In addition, the County Administrative Board may prohibit the enterprise if limited measures are judged insufficient to safeguard the interests of nature conservation.

If an enterprise is entirely forbidden or limitations are being imposed on the enterprise, the entrepreneur may be entitled to compensation if the current land use is substantially impeded .

The felling of timber, like other measures of forest maintenance, is dealt with elsewhere in this article. In cases of this kind, the County Forestry Board is notified, and, this done, the landowner is deemed to have given consultation notice under the Environmental Code. If the County Administrative Board has designated an area as valuable from a conservation viewpoint, the County Forestry Board in turn notifies the County Administrative Board of the enterprise.

Penalties

Offences against the provisions of the Environmental Code are punishable by fines or imprisonment. It should be added that the dumping of litter in the natural environment is prohibited, and that this prohibition applies to both landowners and usufructuaries as well as to the general public.

Compensation rules in the Environmental Code

The procedure for protection orders

The procedure for ordering protective measures and for handling the compensation questions thus involved is as follows. Before a protection order is made under the Environmental Code, the responsible authority shall investigate the possibilities of issuing protective provisions. Landowners shall be notified of the order contemplated and shall be given the opportunity of raising objections. Before an authority makes a decision which can occasion compensation claims, it may also order the landowners to give notice of compensation claims within a certain time, failing which their claims will lapse. The same rules apply to usufructuaries that are adversely affected economically.

The authority has a duty of negotiation. Whether or not compensation is to be payable, like the amount of compensation to be paid, is thus primarily a matter for negotiation between the decision-making authority and landowners claiming compensation or demanding the purchase of their property. In a matter of nature conservation, it is the County Administrative Board which must try to reach an agreement with the landowner. A similar task devolves upon the municipality if it is the municipality and not the State which is responsible for a decision.

If the parties are unable to agree, the matter can be adjudicated by an environmental court (miljödomstol) on the basis of an application made by the person who is claiming compensation or demanding purchase. Application must be made not more than one year after a protection order has been in force for an area. The same applies to a request for compensation in connection with a permit procedure. The County Administrative Board or the municipality can, in some case, also apply to the court to get the amount of compensation settled.

Compensation can be paid for a loss of market value and for other damage suffered by the landowner but a percentage will be deducted to offset the costs that the landowner is obliged to bear. How this figure is calculated, however, is not stipulated by law and the basis of the calculation has instead been left to the environmental courts to decide through court cases (Sjödin et. al. 2002 and Bengtsson et. al. 2000).

Compensation is normally paid by the State. If, however, the decision-making power is vested in the municipality and the latter has made a decision for which compensation is payable, the compensation is paid instead by the municipality.

Summary of the rules on compensation

As mentioned earlier, the general public is entitled to access private land on condition that plant life is undisturbed and domestic peace (hemfrid) is not broken. No compensation is paid to landowners because this is regarded as an ancient right.

Again, if permission is required for a change in the use made of land and an application is rejected, no compensation is payable to the landowner. In cases of this kind, the right of alteration of land use is not deemed part of the right of ownership, since the measure is always subject to permission. This means that a landowner receives no compensation if he is refused an extraction permit. The same applies if he is refused permission to drain land over and above ordinary drainage and if permission is refused for the putting up of a sign, a hoarding or suchlike, or a wildlife enclosure. Refusal of a building permit application within a shoreline protection area also does not qualify for compensation.

Where the Environmental Code allows other public intervention, it is the form of the protection and the extent of the intervention that determines whether or not compensation is payable.

In order to understand the principles of compensation, we can begin with the concept of current land use . A landowner is normally entitled to continue using his land as previously. If, however, he is no longer permitted to take a measure that is already included in such current land use, he can obtain compensation, but only if certain conditions are satisfied. The basic principle is that minor interference with the right of disposal must be tolerated. Establishment of fauna and flora protection areas and natural monuments count as minor intervention. Only if current land use is substantially impeded is the landowner entitled to compensation, and if exceptional detriment occurs, he can request that the property be purchased. If current land use is substantially impeded, three situations can be distinguished which can occasion compensation or purchase (Prop 1997 and Prop 2001):

- Actual land use is substantially impeded as a direct consequence of a protection order.
- Through protection orders, measures are made subject to permission in spite of normally being included in the right of current land use.
- Enterprises are refused or conditioned after consultation with an authority, in spite of normally being included in the right of current land use.

In the first instance, compensation or purchase can be requested directly in connection with a protection order being made for the property or part thereof. This applies when a nature reserve, cultural reserve or biotope protection area is established.

The second instance refers to cases where a regulation has been made for a nature reserve, cultural reserve, protected biotope or special protected area whereby a certain measure may not be taken without permission, even though it would normally be permitted. In this case, the question of compensation cannot arise until permission to carry out a measure has been applied for and the application then refused or a conditional permit granted.

The third instance concerns enterprises which are not subject to permission and are included in what may normally be undertaken within the framework of current land use but are queried by the authority in the course of obligatory consultation because they significantly alter the natural environment. If the enterprise is prohibited or conditioned, compensation or purchase may come into question.

One further compensation case concerns land utilised by an authority for the construction of facilities such as roads, parking places and fences or measures taken such as the felling of trees, cleaning of bushes, and grass cutting. The landowner is always entitled to compensation for encroachment and damage in these cases. This question can arise within a nature reserve and cultural reserve.

References

Bengtsson, B., Bjällås, U., Rubenson, S. And Strömberg, R. 2000. Miljöbalken – En kommentar. Del 1 och II med supplement t.o.m. 2003. Nordstedts blå bibliotek. Nordstedts Juridik AB. Stockholm.

Prop 1993. En ny skogspolitik. Regeringens proposition 1992/93:226. Stockholm.

Prop 1997. Miljöbalk. Regeringens proposition 1997/98:45. Stockholm.

Prop 1998. Uppföljning av skogspolitiken. Regeringens proposition 1997/98:158. Stockholm.

Prop 2001. Skyddet för vissa djur- och växtarter och deras livsmiljöer . Regeringens proposition 2000/01:111. Stockholm.

SCB 1998. Markanvändningen i Sverige. Third edition. Statistiska Centralbyrån/Statistics Sweden.

SCB 2001. Skyddad natur 31 dec 2001. Statistiska meddelanden MI 41 SM 0201. Statistiska Centralbyrån / Statistics Sweden.
SCB 2003. Statistisk Årsbok för Sverige 2003/Statistical Yearbook of Sweden 2003 Statistiska Centralbyrån / Statistics Sweden.
Sjödin, E., Ekbäck, P., Kalbro, T. And Norell, L. 2002. Markåtkomst och ersättning. Nordstedts Juridik AB. Stockholm.
SNV 2003. Natura 2000. Värdefull natur i EU . Naturvårdsverket. Stockholm.
SOU 1978. Skog för framtiden. Betänkande av 1973 års skogsutredning. Statens Offentliga Utredningar 1978: 6. Stockholm.
SOU 1997. Skydd av skogsmark. Behov och kostnader. Huvudbetänkande av miljövårdsberedningen. Statens Offentliga Utredningar 1997: 97. Stockholm.

Referred statutes

The numbers within the brackets refers to the Swedish official numbering system for codes and acts in Svensk författningssamling, SFS. The translation of the code and act names is unofficial.

Constitution, part (regeringsformen SFS 1974: 152)
Environmental Code (miljöbalk SFS 1998: 808)
Environmental Code (Protected Areas) Ordinance (förordning om områdesskydd enligt miljöbalken m.m. SFS 1998: 1252)
Forestry Act (skogsvårdslag SFS 1979: 429)
Forestry Ordinance (skogsvårdsförordning SFS 1993: 1096)
Hunting Act (jaktlag SFS 1987: 259)
Hunting Ordinance (jaktvårdsförordning 1987: 905)
Penal Code (brottsbalk 1962: 700)

AINA PALABINSKA

Die Nutzung von Landwirtschaftsflächen in Lettland und ihre Zukunftsperspektiven

Einführung

Nach der Wiedererlangung der Unabhängigkeit Lettlands (04.05.1990) wurde der Bedarf der Bodenreform durch die neu entstandene politische und ökonomische Situation bestimmt. Sie forderte die Schaffung von Bedingungen eines allmählichen Übergangs zur Marktwirtschaft. Als Beginn der Bodenreform kann der 13. Juni 1990 betrachtet werden. An diesem Tag fasste der Oberste Rat den Beschluss über die „Agrarreform in der Republik Lettland". Diesem Beschluß nach sollten auch unter anderen wichtigen Aufgaben der Volkswirtschaft die Verhältnisse des Bodens verändert und die Bodenreform umgesetzt werden.

Dem gefassten Beshluss folgte eine Reihe von Gesetzen, die die Grundbedingungen und Ordnung für die Durchführung der Bodenreform beinhalteten: „Über die Bodenreform in ländlichen Regionen der Republik Lettland" (21.11.1990); „Über die Bodenkomissionen" (10.07.1990); „Über die Bodennutzung und Bodenordnung" (21.06.1990); „Über die Privatisierung des Bodens in ländlichen Räumen" (09.07.1992); „Über die Beendigung der Bodenreform in ländlichen Regionen" (13.11.1997) [2].

Die Verteilung des Bodens nach der Bodenreform

Das Ziel der Bodenreform in ländlichen Räumen war Bodenprivatisierung und Wiederbelebung der traditionellen Lebensweise auf dem Lande Lettlands. Gemäß dem Gesetz „Über Bodennutzung und Bodenordnung" wird der Bodenfonds der Republik Lettland in zwei Kategorien geteilt [4]:

- Böden der ländlichen Regionen;
- Städtische Böden.

Die Bodenreform umfasste alle Böden der ländlichen Räume der Republik und ihre Nutzer. Sie wurde in zwei Etappen durchgeführt:

- die Hauptaufgabe der ersten Etappe der Bodenreform (1990-1996) war die Bodenübergabe zur Nutzung;
- die wichtigste Aufgabe der zweiten Etappe, beginnend ab 1. Januar 1993, war Bodenprivatisierung.

Eine bedeutende Wende im Laufe der ersten Etappe der Bodenreform erfolgte durch den gefassten Beschluss „Über die Bodenprivatisierung in ländlichen Regionen" (09.07.1992). Diesem Gesetz nach erlangten die ehemaligen Eigentümer, und gemäß dem wieder erneuerten Zivilgesetz von 1937, die große Zahl ihrer Erben das Recht, spätestens bis 01.09.1997 ihre Eigentums- und Nutzungsrechte auf Boden wieder zu erneuern, in der ersten Etappe der Reform in die Nutzung juristischer Personen und staatlicher Wirtschaftsbetriebe übergeben worden war.

Ungeachtet der genannten und anderen Veränderungen wurde die erste Etappe der Bodenreform im großen und ganzen termingemäß (01.11.1996) beendet. Das Tempo der Bodenprivatisierung nahm zu – durch Wiedererneuerung der Eigentumsrechte von ehemaligen Besitzern und ihrer Erben, sowie durch Bodenverkauf (Zahlungsmittel – Privatisierungszertifikate).

Aufgrund des Gesetzes „Über die Beendigung der Bodenreform in ländlichen Räumen" (13.11.1997) und der nachfolgenden Veränderungen (15.04.1999) wurde die Ordnung für die Beendigung der Bodenreform in ländlichen Räumen bestimmt. Die Bodenkommissionen ordneten im Grunde genommen die Bodennutzungen. Nicht nur natürliche, sondern auch juristische Personen, sowie Selbstverwaltungen und staatliche Institutionen ordneten formgemäss nach dem Recht der Bodennutzung für die ihnen gehörenden Bodenflächen [5].

Heutzutage (01.01.2003) sind freie oder unzugeteilte Staatsbodenflächen präzisiert worden, die am 01.01.2003 1,9% der Bodenfläche der ländlichen Regionen betrugen (Abbildung 1).

Die Hauptaufgabe der Bodenreform war die Neuordnung der rechtlichen und wirtschaftlichen Verhältnisse sowohl in städtischen als auch in ländlichen Räumen. Durch Bodenübergabe zur Nutzung wurden im Laufe der Bodenreform gute Bedingungen für den allmählichen Übergang zum Eigentum an Boden geschaffen. Damit waren zwei politisch bedeutende Ziele erreicht worden – durch Wiedererneuerung der Eigentumsrechte für ehemalige Eigentümer oder ihre Erben wurde ein großer Beitrag für die Erneuerung der Unabhängigkeit Lettlands geleistet. Die Bedeutung von Liegenschaften in der Gesellschaft wuchs heran.

Abbildung 1: Flächenverteilung im ländlichen Raum

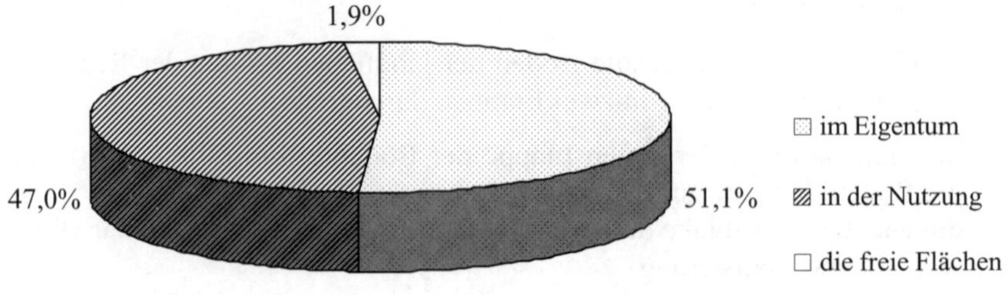

Es sind schon 10 Jahre her, dass der Oberste Rat der Republik Lettland den Beschluss über die „Agrarreform in der Republik Lettland" (13.06.1990) gefasst hat; demgemäss wurde beschlossen, die Bodenreform in der Republik Lettland umzusetzen. Daraus darf geschlossen werden, dass die 10 Jahre der Bodenreform als die bedeutende Phase für ihre Umsetzung angesehen werden kann [1].

Seit Anfang der Reform (1990) ist genügend Zeit vergangen, damit die Bodennutzer ihre Eigentumsverhältnisse ordnen und die Privatisierung realisieren konnten. Die Verteilung der Bodenflächen der ländlichen Räume zum 1.Januar 2003 zeigt, dass von der Gesamtfläche der Böden in ländlichen Räumen 51,1 % privatisiert und als Eigentum fixiert sind, 47 % der Nutzung übergeben, die freien oder die unzugeteilten staatlichen Böden betragen 1,9 % (Abbildung 1, 2).

Abbildung 2: Verteilung der Fläche der Republik Lettland

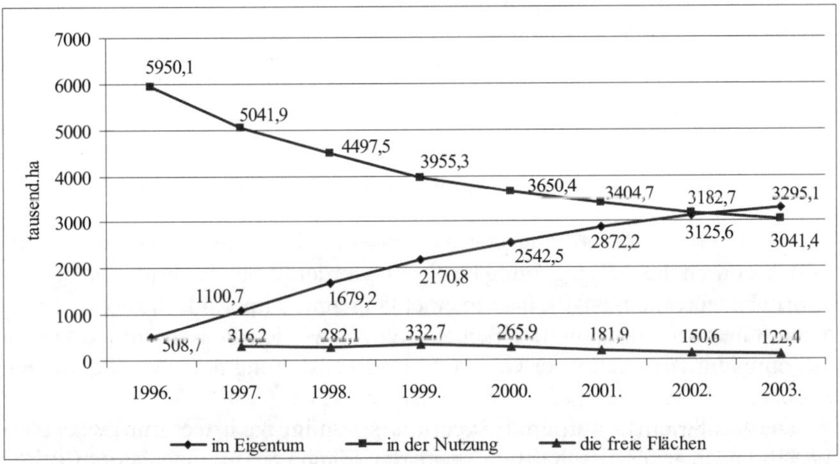

Auf 01.01.2003 betrugen vom Bodenfonds der ländlichen Regionen Lettlands (6,4 Mio ha) die Wälder 44,9 %, d. h. 2,9 Mio ha, landwirschaftlich genutzter Boden – 38,7 %, d. h. 2,5 Mio ha. Von den gesamten landwirtschaftlich genutzten Flächen sind 1,7 Mio ha, d. h., 69,6 % privatisiert, in Nutzung – 680,8 Tausend ha, d. h. 47,6 %.

Im Laufe der Reform hat sich die Bodenverteilung zwischen juristischen und physischen Personen kraß verändert. Bis Anfang der Reform (01.01.1990) waren laut Gesetz „Über die Bauernhöfe in Republik Lettland" (1989) schon 2,9 Tausend Bauernhöfe entstanden. Im Laufe der Reform wuchs die Zahl der Bauernhöfe von 2,9 Tausend (01.01.1990) auf 101,8 Tausend (am 01.01.2003). Mit der Vergrößerung der Zahl von Bauernhöfen vergrößerte sich auch die Bodenfläche, die in Nutzung natürlicher Personen übergeben worden war – von 4,7% (am 01.01.1990) auf 92,1% (am 01.01.2003), d.h. aus der Gesamtfläche landwirtschaftlicher Nutzung 4,0 Mio ha.

In derselben Zeit hatten sich die Bodenflächen, die an juristische Personen zur Nutzung übergeben wurden, verringert von 95,3% am 01.01.1990 auf 7,9%, am 01.01.2003 – von der Gesamtfläche der landwirtschaftlich genutzten Böden 4,0 Mio ha (Abbildung 3).

Abbildung 3: Verteilung der Landwirtschaftsflächen

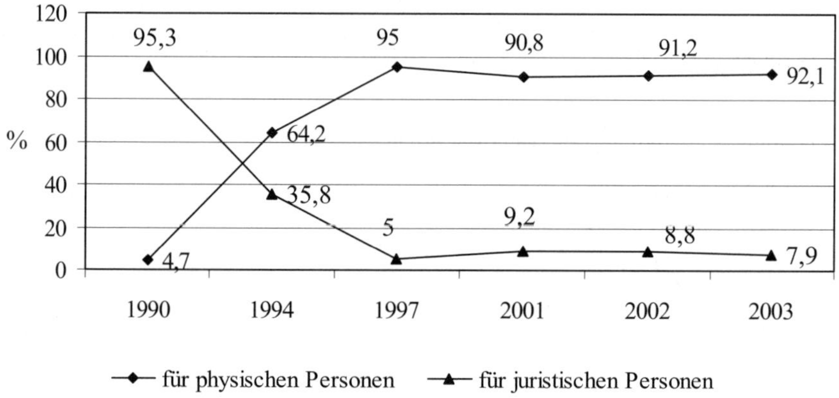

Die Ordnung für Beendigung der Bodenreform in ländlichen Räumen Lettlands ist durch die Gesetzgebung bestimmt worden. Aber das Ende der Bodenreform bedeutet keinesfalls, dass in einer längeren Zeitperiode die herausgebildeten Bauernhöfe unverändert bleiben. Künftig werden die Bauernhöfe sowohl zahlenmäßig, als auch ihrer Orts- und Raumgestaltung nach Veränderungen erleben.

Die Größe eines mittleren Bauernhofs beträgt nach Gesamtfläche etwa 14,5 ha, nach landwirtschaftlich genutzter Fläche 8,5 ha, nach Forstflächen 4,6 ha.

Ungenutzte landwirtschaftliche Nutzflächen (LNF)

Laut Erlaß des staatlichen Bodendienstes der Republik Lettland vom 7. Juli 2000 „Über Monitoring der landwirtschaftlich genutzter Flächen in Gemeinden“ und „Über Methodik des Bodenmonitorings landwirschaftlich genutzter Flächen in Gemeinden“ wird seit 1995 in Lettland die Überwachung der landwirtschaftlich genutzten Flächen durchgeführt, mit dem Ziel, die für Produktionszweche ungenutzten landwirtschaftlichen Bodenflächen zu erfassen.

Von der Gesamtfläche der landwirtschaftlich genutzten Böden Lettlands (2,47 Mio ha) wurden 2001 2,38 Mio ha, d. h.96,1 % erfasst.

Bei der jährlichen Erkundung der Bodennutzer von landwirtschaftlich genutzten Flächen in Gemeinden wurden folgende Flächen festgestellt:

- die im laufenden Jahr nicht für landwirschaftliche Produktion genutzt wurden;
- Saatflächen, durch Unkraut bewachsen;
- gehölzbewachsene landwirtschaftliche Nutzflächen;
- überfeuchte LNF;
- transformierte (umgesiedelte) in landwirtschaftlich nicht genutzte Flächen.

Abbildung 4: Verteilung der freien Staatsflächen

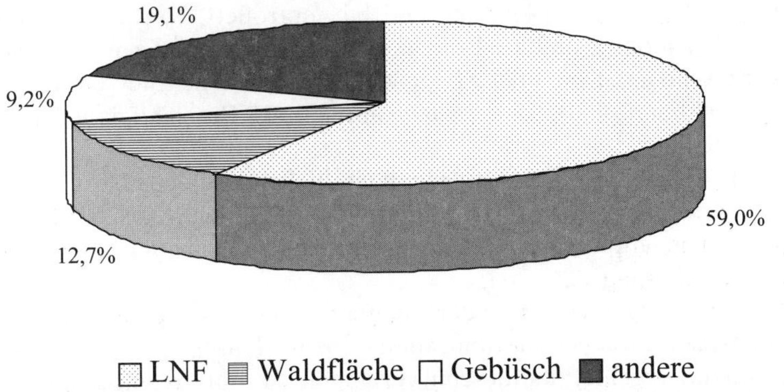

Bei der Zusammenfassung der Ergebnisse des Monitorings der landwirtschaftlich genutzten Böden von Gemeinden wurde festgestellt, daß von den landwirtschaftlich genutzten Böden (2,77 Mio ha) im Jahre 2001 [6]:

- 537,2 Tausend ha (21,7 %) landwirtschaftlich nicht genutzt waren, 255,7 Tausend ha (16,4 %) davon meliorierte Böden;
- etwa 44,6 Tausend ha (1,8 %) LNF der Verwachsung durch Gehölz unterworfen sind, darunter 14,0 Tausend ha meliorierte Böden;
- 138 Tausend ha (5,6 %) überfeuchte Flächen, etwa 120,1 Tausend ha (7,7 %) Meliorationsflächen;
- etwa 2,0 Tausend ha LNF zu nicht landwirtschaftlichen Flächen transformiert.

Beim Vergleich mit Ergebnissen des Bodenmonitorings in Gemeinden vom 1996-2001 wird ersichtlich, dass sich der Gesamtwert der für landwirtschaftliche Produktion nicht genutzten Flächen in Lettland um 147,2 Tausend ha vergrößert hat.

Die gehölzbewachsenen Flächen haben sich um 24,6 Tausend ha vergrößert.

Im Ergebnis der Bodenreform ist der Bodenfonds Lettlands unter juristischen und natürlichen Personen, Selbstverwaltungen und Staat aufgeteilt worden. Nicht zugeteilte, also freie Böden, bilden 1,9% von der Gesamtfläche Lettlands.

Die Ergebnisse des Bodenmonitorings zeigen, dass eine einheitliche Bodenpolitik in Lettland erarbeitet werden soll, damit die notwendige Aufmerksamkeit der Bodennutzung und dem Bodenschutz gewidmet wird, der eine einzigartige Ressource darstellt. Boden hat nicht nur ökonomische, sondern auch politische, soziale und ökologische Bedeutung. Die einheitliche Bodenpolitik ist deshalb wichtig, weil sie fest mit der Politik der Wirschaft, der regionalen Entwicklung, dem Umweltschutz verbunden ist. Ebenso ist ein wichtiger Bestandteil der Bodenpolitik die Planung der Entwicklung von Bodennutzung, die durch die künftige Europas noch mehr an Bedeutung erlangt.

Im Vergleich mit dem Jahre 1996 haben sich die landwirtschaftlich nicht genutzten Flächen um 147,2 tausend ha vergrößert. Die nicht genutzten LNF werden moorig und sind mit minderwertigen Gehölzen überwachsen (144,6 Tausend ha). Eine solche Lage verursacht nicht nur die Verarmung von Umwelt und Landschaft, sondern auch allmähliche Beschädigung der Meliorationssysteme.

Das niedrige Einkommensniveau in ländlichen Regionen und der Mangel an Motivierung zur Nutzung der Liegenschaften für landwirtschaftliche Produktion und on rationeller Bodennutzung beeinflussen die wirtschaftliche und soziale Orientierung.

Der Verlauf der Gestaltung der Bodenreform und des Eigentums hat sich in den letzten Jahren verlangsamt. Immer mehr Menschen verzichten auf die Bodennutzungsrechte, stattdessen verlangen sie eine Kompensation. Der Grund dafür sind nicht klar formulierte Grundprinzipien der Bodenpolitik, sowie eine niedrige Motivation der Bodennutzer zur Registrierung der Eigentumsrechte.

Mit der jährlichen Vergrößerung der ungenutzten LNF ist die Frage über die Verantwortung der Liegenschaftseigentümer und Nutzniesser der Bodennutzung aktuell geworden. Im Gesetz „Über die Bodennutzung und Bodenordnung" wird die Verantwortung nur für Bodennutzer vorgesehenin dem Falle, dass der Boden nicht genutzt wird.

Es besteht eine Reihe von Rechtsakten, die die Planung der Bodennutzung regeln; der Umsetzungsmechanismus ist jedoch nicht genügend erarbeitet. Praktisch gilt das für die Realisierung der Detailplanungen und die Bestimmung von Schutzstreifen.

Die Unklarheit für die künftige Nutzung der städtischen und ländlichen Böden, für die Umsetzung der territorialen (regionalen) Reform, Beendigung der Bodenreform und Mangel an notwendigen Ressourcen sind Gründe dafür, dass die Flächen der örtlichen Selbstverwaltungen laut der bestehenden Gesetzgebung keine vollständig erarbeiteten und festgesetzten Raumplanungen (Regionalplanungen) haben.

Schlussfolgerungen

Die Erarbeitung einer einheitlichen Bodenpolitik ermöglichte:

1. Erhöhung des Liegenschaftswertes:
   Verminderung des Parzellierungsgrades des ländlichen Bodeneigentums;
   Förderung der Vereinigung der LNF;
2. Erhaltung des Bodens und anderer Naturressourcen:
   Erhaltung und Förderung von Bodenfruchtbarkeit;
   Förderung der Erhaltung der staatlich besonders wichtigen LNF.
3. Dauerhafte Bodennutzungsfähigkeit und Erhöhung ihrer Fruchtbarkeit:
   Sicherung der Eigentumsnutzung laut den Raumplanungen, in denen wirtschaftlich begründete Ziele für Eigentumsnutzung bestimmt sind.
   Sicherung einer rationellen Nutzung der für landwirtschaftliche Nahrungsproduktion nicht genutzter Böden.

Literatur

1. Zeme: mana, tava, mūsu. VZD. Rīga, 2002.
2. Zemes reformai Latvijā 10 gadi. VZD. Rīga, 2000.
3. Палабинска А. О закреплении результатов земельной реформы. Материалы международной конференции "Земельная реформа и землеустройства 98". К: Академия. 1998.
4. Par zemes privatizāciju lauku apvidos / LR likums, 07.09.1992.
5. Par zemes reformas pabeigašnu lauku apvidos / LR likums, 27.11.1997.
6. Zemes pārraudzība, VZD. 2001.

TOMASZ BAJEROWSKI, RYSZARD CYMERMEN,
MAGDALENA NOWAK-RZASA, CESARY KOWALCZYK

Modern Planning Elaborations as an Instrument of Environmental Protection in Poland

1. Introduction

The natural environment in which we live and whose natural resources we process is renewable, but not unlimited or inexhaustible, so the way in which it is exploited depends on both our awareness and legal regulations, aimed at restricting certain actions and promoting others, aimed at environmental protection and renewal. It should be also noted that the natural environment free from man's regulatory activities would be subjected to constant changes connected with the struggle for existence carried on by particular animal and plant species. This could result in the formation of areas characterized by a totally different ecological value.

A modern approach to environmental protection must include elements of active environmental planning – one of its forms may be the establishment of nature reserves under strict protection. All environmental protection procedures require plans for their implementation. Each spatial system, represented by the opposition of two fundamental values – ecological and economic – devoid of conscious and planned actions, would enter one of "natural" relationships of the predator-prey or competition-parasitization type. Such relationships are well-known in ecology. However, none of them is advantageous from the perspective of the natural environment or ecological value. The only beneficial relationship seems to be symbiosis, but – in a modern world – it requires undertaking planned actions. Symbiosis is, to a high extent, a synonym of the concept of sustainable development, aimed at providing benefits to all planning space users.

2. Sustainable development planning

2.1. Model approach

The concept of sustainable development that can be found in the Constitution of the Republic of Poland includes two main elements allowing to create an appropriate mathematical model. The term "development" means that a widely understood value of the phenomenon modeled must increase. Maintaining the existing state cannot be perceived as development. The term "sustainable" means that the factors affecting development must remain in stable, balanced relationships (functional dependences). The factors crucial to spatial planning can be divided into natural (ecological) and anthropogenic (mainly economic). The former create the ecological value, whereas the latter – a generally understood economic value of planning space. So far they have been in opposition to each other – an increase in the ecological value of space has been accompanied by a decrease in its economic value, and – even more frequently – vice versa. However, it should be stressed that both these values are anthropogenic concepts, and apply to the relations between the natural environment and man (human activities) only. Nature as such knows nothing about its ecological value. Unplanned or improperly planned relationships between the natural environment and man are usually of the above mentioned predator-prey type, where human beings and their present civilizational position perform the role of predator. In classical ecology the predator-prey system is of a cyclical character – the number of predators and preys oscillates around a certain mean value, depending on the fact which of the two groups dominates in a given territory at a given time. Generally, their number can be interpreted as some admitted, conventional value (both are quantitative categories). In this case the correlation between the ecological and economic value of a given area is also such a system, with the economic value as predator and the ecological value as prey (Figures 1, 2).

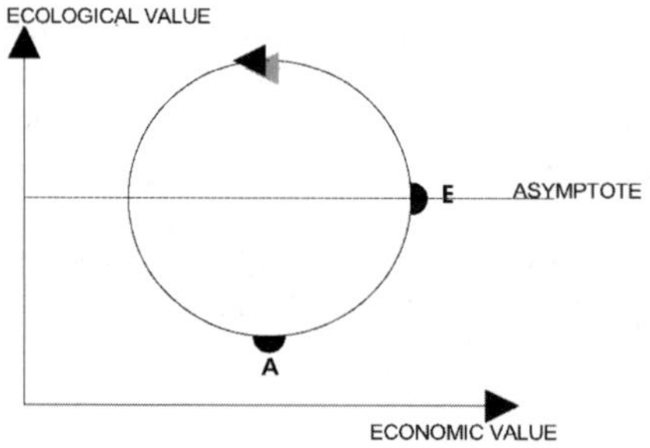

Figure 1.

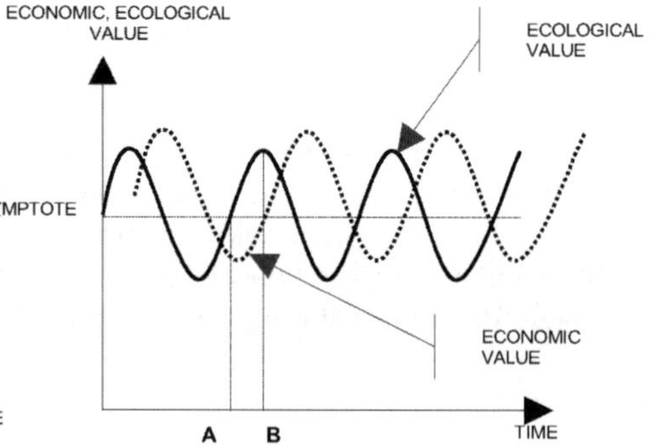

Figure 2.

In this system the ecological and economic value of a given area increases only in the short segment A and B; the maximum increase in the ecological value is followed by an increase in the economic value. However, this is not the sustainable development aimed at.

The state of equilibrium may be achieved by eliminating the fluctuations (cycles) of both values (Figures 3, 4).

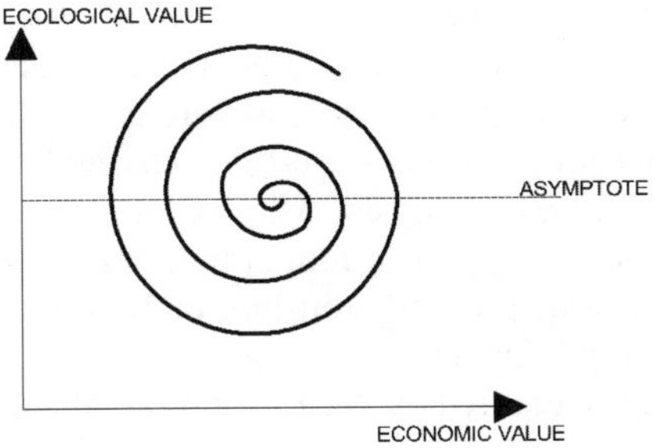

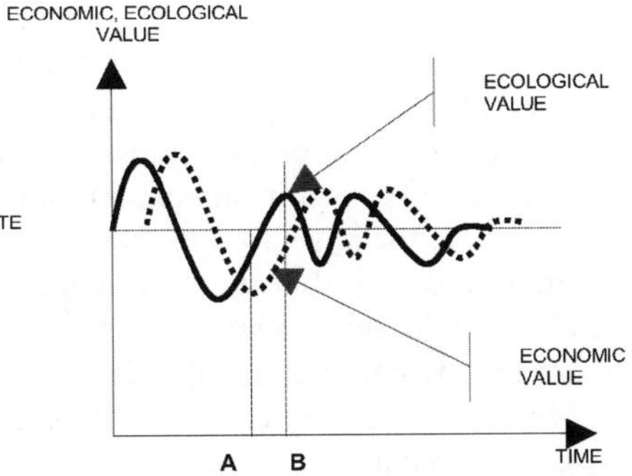

Figure 3.

Figure 4.

To achieve development, the asymptote r must be given a positive direction ratio (Figures 5, 6).

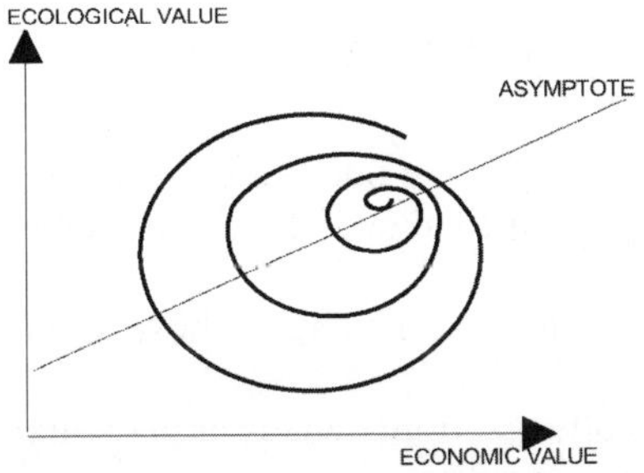

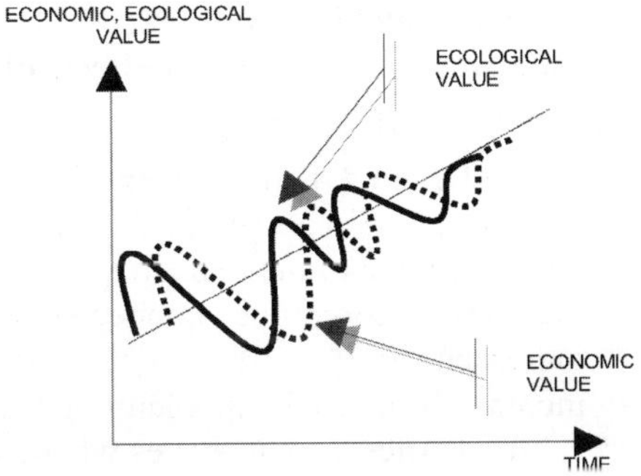

Figure 5.

Figure 6.

This is a theoretical (model) description of sustainable development. However, such models are simplified representations of complex behavior of actual spatial systems, including those existing in planning space, i. e. the whole space subject to planned transformations.

2.2. sustainable development within the framework of spatial planning – a proposal

In all developed countries space planning is regulated by numerous legal acts. In Poland the first regulation concerning environmental protection – prohibiting yew-tree cutting – dates back to the year 1423, and the first Environmental Protection Act was adopted by the Polish Parliament on March 10, 1934.

However, the idea of sustainable development as a basis for all actions taken in space is included only in the recent Act of March 23, 2003 on spatial planning and management. According to this Act, spatial management plans must be accompanied by environmental impact prediction and estimation of the financial effects of their implementation (Art. 17 items 4 and 5). Such regulations provide the basis for making a cost-benefit analysis of the undertakings planned. However, this analysis cannot be understood in economic categories only. It should be an analysis of the predator-prey relationship, allowing to determine the parameters affecting this relationship.

Moreover, the predictions and forecasts required by the above Act make it easier to compare the ecological and economic costs and benefits resulting from other, similar undertakings completed in the past, to draw conclusions and to answer the question whether the changes in space use postulated in a given plan are consistent with the sustainable development model.

Local spatial management plans, as acts of local law, have to be accompanied by two documents relating to environmental protection and affecting the final form of these plans, i. e.:

- environmental impact prediction,
- estimation of the financial effects of the local plan implementation.

The financial effects include:

- costs of claims paid or, in special cases, of buying out real estates from their owners and perpetual usufructuaries, if the value of these real estates decreased due to the plan adoption;
- incomes of communes connected with selling real estates whose value increased;
- incomes connected with planning fees collected from owners or perpetual usufructuaries of real estates whose value increased;

The estimation of the financial effects of the local plan adoption allows to determine the costs to be incurred by communes and their expected incomes, as well as an increment in the value of real estates (denoted in Figure 7 with the symbol ΔV ECONOMIC) in a minimal area equal to the commune area.

As regards the financial effects of the plan implementation, they can be easily expressed as a monetary value. In order to predict the environmental impact of the plan adoption, i. e. changes in the ecological value of a given area, this ecological value must be determined before the local spatial management plan is adopted (W1), and then when its provisions had been taken into consideration (W2). At the final stage of making this prediction the ecological value increment (denoted in Figure 7 with the symbol ΔW ECOLOGICAL) has to be determined, by subtracting the ecological value before the plan implementation from the ecological value after the adoption of its provisions, or adding these two values. An important aspect of ΔW ECOLOGICAL determination is the necessity to consider only the value increment resulting from the plan provisions, and not its increment in the time interval T1-T2.

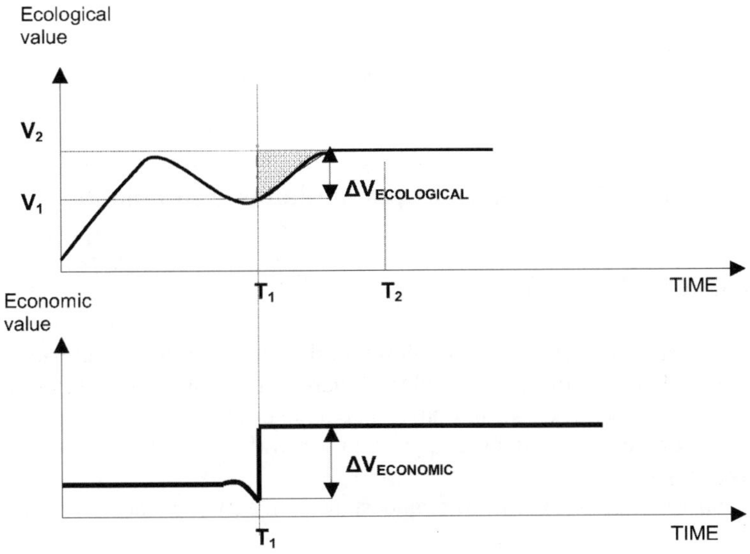

Figure 7. Adoption of a local spatial management plan (T1) versus increments in the economic and ecological values.

The determination of the economic value increment on the basis of the estimation of thc financial effects of the local plan adoption and the ecological value increment estimated in the environmental impact prediction allows to draw up a spatial management plan oriented at sustainable development, whose model is presented in Figure 5, where both the economic and ecological values increase in time. It should be emphasized that a simultaneous increment in both these values is not indispensable. It is important that it takes place for each of them in a time interval longer than the cycle duration – in Figure 5 the distance of measurement of these values is over 360°.

The total ecological value of the natural environment may be measured as a sum of the ecological values of its particular components. The higher the ecological value the more limited the use of a given area.

The economic value of environmental components is described with absolute numerical quantities (in natural or value units), or as relative comparisons determining the hierarchy of significance of particular components.

The higher the ecological value of particular environmental components, the higher the total ecological value of the natural environment. The evaluation of environmental components may be based on:

- estimation of the labor expenditures incurred for using environmental resources for economic purposes,
- estimation of the labor expenditures incurred for environmental resources reproduction,

- results of using environmental resources in the production process,
- estimation of the future labor expenditures that will be incurred by the society to eliminate the losses caused by environmental pollution.

3. Environmental protection in spatial planning in Poland

3.1. Planning System in Poland

The legal basis of spatial planning in Poland is the Spatial Planning and Management Act. It establishes the principles of pursuing spatial policy by local government units and state administrative agencies in the areas they govern, as well as the procedures of land designation for various purposes and real estate management.

In Poland local government units and state administrative agencies are responsible for spatial planning, i. e.:

- the tasks to be performed by communes are: to carry out spatial policy in the areas they govern, to adopt local spatial management plans and resolutions concerning the conditions and directions of spatial management;
- the tasks to be performed by district governments are: to make analyses and conduct studies concerning spatial management and development of the district;
- the tasks to be performed by provincial governments are: to carry out spatial policy in the province and to adopt provincial spatial management plans;
- the tasks to be performed by the Council of Ministers are: to carry out spatial policy in the territory of Poland and to formulate concepts of spatial management of the country.

Table 1 presents planning policy instruments used at different levels of administrative division of Poland.

Table 1

Planning policy instruments

Specification	Area
• study of conditions and directions of spatial planning in a commune • local spatial management plan	• commune • commune or a smaller area
• provincial spatial management plan	Province
• concept of spatial management of the country	Country

3.2. Environmental protection versus particular instruments of planning policy

According to the Act of March 27, 2003 on Spatial Planning and Management, the following factors must be considered while taking any actions in the field of spatial planning and management. (art. 1 item 2):

- environmental requirements, including water resources management, and protection of farmland and forest land;
- preservation of cultural goods and protection of historical monuments;
- architectural values and landscape amenities.

These are three regulations from among nine enumerated in art. 1 item 2 of the above Act. They are directly related to the problem of environmental protection. Apart from general regulations concerning spatial planning, environmental protection is also to be taken into account while working out particular plans, formulating concepts or making planning studies.

3.2.1. Environmental protection versus planning policy at the national level

The Government Strategic Studies Center is responsible for formulating concepts of spatial management of the country, taking into consideration the objectives included in strategic documents, as well as the principles of sustainable development based on environmental, cultural, social and economic conditions.

Concepts of spatial management of the country determine the conditions, goals and directions of sustainable development, as well as the actions that must be taken to achieve it, paying special attention to the protection of the natural environment and historical monuments, and to protected areas.

3.2.2. Environmental protection versus planning policy at the Provincial level

Provincial spatial management plans are based on the strategy of province development, and particular attention is paid to:

- protected areas, including environmental protection areas, nature conservation areas and cultural landscape conservation areas, health-resort protection, protection of historical monuments and preservation of cultural - goods;
- flood hazard areas.

Spatial management plans worked out at the provincial level are also based on the concept of spatial management of the country.

3.2.3. Environmental protection versus planning policy at the communal level

As already said, communes are responsible for making a study of conditions and directions of spatial management, and drawing up local spatial management plans. These two types of documents differ in character and scope:

- the study concerns the territory of the whole commune, whereas the local plan only its part,
- it is compulsory to make the study, whereas the local plan is treated as compulsory only in some cases,
- the study does not constitute a part of local law, whereas the local plan is an act of local law.

In both cases the problems of environmental protection are taken into account, i. e.:

a) study of conditions and directions of spatial management.

The following conditions are taken into consideration in the study:

1. land designation and management, provision of technical infrastructure;
2. state of spatial order and its protection;
3. state of the natural environment, including the state of agricultural and forest production space, amount and quality of water resources, environmental protection, nature and cultural landscape conservation;
4. state of cultural heritage, cultural goods and historical monuments;
5. presence of objects and areas protected on the strength of separate provisions;
6. presence of natural geological hazard areas;
7. presence of documented mineral deposits and underground water resources;
8. presence of mining areas defined on the strength of separate provisions.

Special attention is paid to:

1. directions of changes in the spatial structure of the commune and land designation;
2. directions and indices of land management and use, including open spaces not designed for building;
3. principles of protection of the natural environment and its resources, nature and cultural landscape conservation, and health-resort protection;
4. principles of protection of cultural heritage, cultural goods and historical monuments;
5. areas for which local spatial management plans are to be made, including areas which require changes in the designation of agricultural and forest land and its release for non-agricultural and non-forest purposes;
6. directions and principles of management of agricultural and forest production space;
7. flood- and earth sliding-hazard areas;
8. objects or areas with safety pillars in mineral deposits.

b) local spatial management plan,

The following elements have to included in a local spatial management plan:
1. land designation and lines separating lands designed for different purposes or with different management principles;
2. principles of protecting and introducing spatial order;
3. principles of environmental protection, nature and cultural landscape conservation;
4. principles of protection of cultural heritage, cultural goods and historical monuments;
5. limits and methods of management of areas or objects protected on the strength of separate provisions, including mining areas, flood- and earth sliding-hazard areas;
6. special conditions of land management and constraints on its use, including building prohibition.

Depending on the needs, the following elements may be included in a local spatial management plan:
1. boundaries of areas of rehabilitation of the existing buildings and technical infrastructure;
2. boundaries of areas requiring transformations or reclamation.

An integral part of a local spatial management plan is its environmental impact prediction. According to the Regulation by the Minister of Environmental Protection of November 14, 2002, such a prediction is to be made in a cartographic and descriptive form.

It includes the following elements:

1) as for effects:
1. effects of the plan implementation on the natural environment, connected with land designation, gas and dust discharge into the air, waste production, sewage discharge into water or land, use of environmental resources, soil or land contamination, adverse relief transformation, noise emission, electromagnetic field emission and a major risk of potential accidents,
2. effects of the plan implementation on the air, land surface, soil, mineral deposits, surface and underground waters, climate, animals and plants, and their interactions, ecosystems and landscape;

2) as for assessment:
1. state and functioning of the natural environment, its resources, resistance to degradation and regeneration capacity, resulting from the conditions determined in ecophysiographic documentation, and a tendency towards changes in the case the local spatial management plan is not implemented;
2. functional and spatial solutions, as well as other provisions of a local spatial management plan, especially those concerning:
   - conformity of the planned land use and management with the conditions defined in ecophysiographic documentation,

  - conformity with environmental regulations, especially those included in legal acts on creation of protected objects and areas, and environmental protection plans,
  - efficiency of biological diversity protection,
  - proper proportions between areas characterized by different forms of use, and other areas.
3. land management conditions defined in a local spatial management plan, resulting from the needs of environmental protection, rational natural resources management and protection of farmland and forest land,
4. threats to the environment, including dangers to public health, that may occur in the area specified in a local spatial management plan and in the areas within the impact range resulting from its implementation,
5. effects for the existing forms of environmental protection and other protected and conservation areas,
6. landscape changes.

3) as for the possibility of applying solutions eliminating or reducing the adverse environmental impact, including the negative effects on landscape, which may be caused by the implementation of a local spatial management plan, as well as – depending on the needs – as for the possibility of adopting proposals other than those included in the plan, conducive to environmental protection.

The prediction concerns the area specified in a local spatial management plan and the areas within the impact range resulting from its implementation; the prediction should be made along with a draft of a local spatial management plan.

4. Summary

Spatial planning, whose aim is to localize in space various objects or forms of man's activities, has a significant effect on the natural environment, preserving or destroying its valuable components.

This effect may assume the form of:

- making provisions on environmental protection superior to economic activities,
- taking the state of the natural environment into account while establishing certain planning procedures,
- introducing environment protecting restrictions to design solutions.

Certainly, spatial planning cannot be perceived as the only effective instrument of environmental protection, but wise planning policy promotes environmental protection actions, in this way preventing its degradation.

Particular instruments of environmental protection differ in impact and character. They are connected with certain technical, organizational, economic and legal actions. Also spatial planning is to play an important role in environmental protection. New legal provisions in this respect, combined with traditional planning, may become effective instruments of rational use of the environment. An example may be here the new Spatial Planning and Management Act.

Andrzej Nowak, Wojciech Żebrowski

Polish Model for Afforestation and Tree Planting in Rural Areas

Abstract

The paper deals with the problem of farmland forestation. It discusses the national forestation project drawn up by the Forestry Research Institute, and legal regulations concerning a forestation. Particular attention is paid to afforestation limitations stemming from the Act of Farmland Designation for afforestation. The advantages and disadvantages of natural afforestation in the process of land structure changes are also presented.

1. Introduction

The economic and social living conditions of the agricultural population in Poland have been undergoing radical changes since 1989. Until that time the government pursued preferential policy towards agriculture, which manifested itself, among other, in:

- state-guaranteed purchase of all agricultural production
- small fluctuations of purchase prices of agricultural produce (resulting from the fact that they were subsidized by the government)
- preferential credits for farmers (including amortization of unpaid installments)
- public subsidies for the state agricultural sector, i.e. state-owned farms (among which only some were bringing profit)
- very high volume of agricultural production, clearly exceeding the needs of the domestic market.

This shows that there are really big differences between a planned and free-market economy. The year 1989 brought transformation of the economic system in Poland, which was connected with serious, sometimes long-term, changes in agriculture. The liquidation of state-owned farms seemed to be easy to conduct, on condition that certain legal regulations were introduced. In 1991 the property of former state-owned farms was taken over by the State Treasury Agency for Agricultural Property. After 12 years the Agency managed to dispose of a part of this property only. The volume of agricultural production decreased, approaching the boundary value, i.e. the requirements of the domestic market. This made the prices of agricultural produce approach the actual market prices.

The process of afforestation of low-productivity farmland was carried out with different intensity in the post-war period. Its results are presented in Table 1.

Table 1. Farmland afforestation process in particular periods in the years 1947-1999 (ths ha)

Period	State forests	Non-state forests	Total
1947-1949	59.30	8.60	67.90
1950-1959	197.70	177.70	375.40
1960-1969	272.60	212.70	485.30
1970-1979	98.10	74.70	172.80
1980-1991	75.70	36.41	82.11
1992-1999	74.18	38.76	112.94
Total	747.58	548.87	1296.45

Source: Puchniarski, 2000

At present woodiness in Poland is above 28 %. Forests owned by the State Treasury constitute approx. 80 % (including 76.7 % owned by the State Forest Enterprise), private forests – approx. 17 %, and municipal (urban) forests – less than 1 %.

2. Aims of afforestation and tree planting in Poland

At first the aim of afforestation was to repair the extensive damage caused by the Second World War. Later on it turned out that both afforestation and tree planting can play other important roles, connected with:

- management of poor-quality land no longer used for agricultural purposes
- prevention of soil degradation caused by excessive deforestation
- introduction of the so called protection zones
- landscape planning.

Following this new approach to the problem of afforestation and tree planting, the Forestry Research Institute proposed at the end of the 1990s a national afforestation project, taking into account both ecological aspects and legal requirements.

The problem of afforestation may be viewed from different perspectives: "wood consumers" focus on the production and economic aspects, whereas representatives of other social groups perceive forests as an element of the natural environment, affecting the quality of their life and supposed to satisfy their needs in the sphere of recreation, culture and science. On the other hand, even environmentalists and ecologists express various opinions on the problem of afforestation.

The forest area should be increased taking into account the beneficial effects of forests on the whole environment. The following elements seem to be of primary importance (Puchniarski, 2000):

- increasing the hydrological role of forests in improving the water balance in Poland, and controlling extreme stages of surface, ground and deep waters,
- prevention of soil erosion and stepping,
- increasing the role of forests in air, water and soil cleaning of chemical substances, reducing the rate of eutrophication of lakes and other open waters,
- using forests as a soil-forming and soil degradation preventing factor,
- using forests as a factor improving local hydrological conditions in adjacent areas,
- satisfaction of the growing social needs as regards rest and recreation, improving the living conditions of urban populations.

For the above reasons the location of areas designed for afforestation cannot be accidental. They should be located on watersheds, especially in the case of diversified relief and increased surface runoff. In the spatial system forests are supposed to protect land used for agricultural purposes, and form the so called ecological corridors.

Moreover, forests contribute to prevention of pollutant movement in the air, water and soil, and fixation of harmful chemical elements and compounds, including carbon dioxide. That is why their role is especially important in industrialized areas, e. g. around large urban agglomerations.

Drawing up the afforestation project, the Forestry Research Institute proposed a model for spatial solutions in communes, taking into account the following elements:

- rational land use structure, especially as regards poor-quality soil management,
- functions performed by forests in improving the water balance, preventing soil erosion and stepping,
- presence of protected landscape areas, including protected catchment-basins and recreation areas,
- controlling the negative effects of population migrations and unemployment, improving the living conditions in industrialized, urbanized and contaminated areas.

25 criteria were considered altogether, five of them restricting afforestation (the presence of: urban agglomerations, intensive farming, raw material extraction, high woodiness, swampy grounds). All communes in Poland were evaluated on the basis of these criteria.

The above criteria were divided into five groups:

1) optimization of space structure use, taking into account:
- percentage of soils of classes VI and VIz,
- agricultural suitability of soils according to the production space evaluation, proposed by the Institute of Soil Cultivation and Fertilization in Puławy,
- forest location in areas unsuitable for farming due to specific relief and surface features, according to the evaluation system proposed by the Institute of Soil Cultivation and Fertilization in Puławy,
- forest location on major watersheds,
- forest location within the range of water supply to major groundwater reservoirs.

2) enhancing the environmental function of forests in the prevention of environmental degradation by their location in the following areas:
- threatened by surface water erosion,
- threatened by landscape stepping, according to the Sielaninov hydrothermal index,
- characterized by low woodiness (below 16% on lowlands and 30% in the mountains),
- located within the range of water supply to major streams and watercourses of the second and higher contamination, and the Baltic Coastland,
- of water deficit for agriculture and urban agglomerations.

3) implementation of government ecological policy, taking into account:
- areas pf high natural values, according to the concept of spatial management of the country,
- protected areas: national parks, landscape parks and protected landscape areas,
- forest spatial system, according to the concept formulated by Łonkiewicz (1993),
- protected catchment areas,
- areas of high recreation values.

4) prevention of adverse social and demographic processes, taking into account areas characterized by:
- domination of lands over forests (according to survey data),
- domination of intensive farming and large commercial farms,
- ecological and flood hazards,
- long-term population migrations,
- a high rate of unemployment.

5) restrictions to land designation for afforestation, taking into account the presence of:
- urban agglomerations,
- areas of mineral extraction,

- intensive farming areas, characterized by high values of agricultural production space,
- areas of high woodiness (above 36 % on lowlands and 50 % in the mountains),
- swampy grounds and wetlands.

Taking into consideration all of the above criteria, the national a forestation project was adopted for implementation. The area of high preference in this project are communes located in central Poland, characterized by relatively low woodiness and water deficits, exposed to water and wind erosion, and stepping. However, due to a high farming level and a low share of poor-quality grounds the model afforestation program cannot be introduced there. Another area of high preference is the north-western region, classified as a watershed and frontal moraine exposed to erosion, a part of the Baltic Coastland characterized by relatively low environmental degradation. There are large areas designed for afforestation there, most of them state-owned. Also the north-eastern belt (referred to as the so called "Green Lungs of Poland") seems to be designed for afforestation, due to its high ecological values and a high share of poor-quality soils. The tasks defined in the above project can be performed there much earlier than stated in its provisions.

The mountain area is also important, since it cannot be used for agricultural purposes due to its relief, surface features and soil structure. Forests can play protective functions there, in relation to both soils and waters. There are large areas that can be planted with trees.

A high supply of grounds for afforestation is usually observed in regions of secondary importance from the perspective of environmental needs. This concerns first of all the northern zone of Poland. The areas of primary importance can be found in Central Poland, where poor-quality land is scarce. Therefore, in this region afforestation projects will include small, scattered areas only, and a major role in landscape planning will be played by tree plantings, being a substitute for forests.

According to the national afforestation project, a total area of approx. 1.5 mln ha will be planted with trees by the year 2050. This will increase woodiness in Poland to 33 %. For planning purposes, Poland was divided – within the framework of the national forestation project – into macro regions characterized by different regional policy (Łonkiewicz, 1995).

The north-western macro region includes the Baltic Coastland, Pomeranian Lake District, and District of Ziemia Lubuska. The main aim of afforestation in this region should be to strengthen the existing ecological-physiographic structures, and especially to simplify the boundaries of large forest complexes, to unite small and medium-sized afforested areas and to form ecological corridors between them. The creation of new, small areas designed for afforestation seems to be less justified.

The north-eastern macro region includes the Masurian Lake District, East-Baltic Lake District, Podlasie-Belorus Plateau and North-Mazovian Lowland. This area is suitable for the development of spatial structures and creation of ecological relationships uniting the water system, swampy grounds and forests into a coherent whole. Due to diversified relief and strong winds the areas designed for planting with trees should be large, to reduce ground overdrying. Most of this area is referred to as the so called "Green Lungs of Poland", which means that ecological objectives should be considered the most important there. Afforestation is also necessary due to vulnerability to erosion, the presence of protected catchments and a high contribution of poor-quality soils.

The central macro region includes Central Polish Lowlands and the eastern part of the Wielkopolska Lakeland. The area has been highly developed and transformed. It is characterized by a high level of farming and numerous restrictions as regards farmland designation for afforestation, despite the presence of poor-quality soils. In this region the main aims of afforestation should be to strengthen the provincial systems of landscape protection, to enrich the agrocenotic space with forest landscape components (i. e. to form a belt system based mainly on watersheds), and to create small forest complexes and tree planting areas. The fundamental function performed by afforestation should be to improve hydrological condition and prevent environmental degradation.

The mid-southern macro region includes the Silesian Lowland, Silesian-Cracovian Upland, Mid-Małopolska Upland, East-Małopolska Upland, Polesie and Wołyń-Podole Upland. In this region it is important to develop and use in a rational way the existing forest complexes, especially national parks, their protection zones and areas performing special functions, and to create ecological relationships, afforest watersheds and erosion–threatened areas. Tree planting is also necessary.

The southern macro region includes the Sudeten Mountains, Western Carpathians with Eastern Beskid, Nysa Kłodzka River Valley, Głubczyce Plateau and Niemodlin Plains. A forestation is expected to enhance the physiotactic, soil- and water-protective functions of forests, and to develop the existing macrospatail structure of active biotic resources. It is also important to simplify the field-forest boundaries, taking into account the specific relief and the need to preserve meadows and pastures as ecologically significant landscape components. A high level of tree planting is desirable in areas characterized by large-scale air pollution.

3. Natural succession and its role in tree planting and afforestation planning

The difference between reforestation and afforestation is that in the case of afforestation a new forest generation of trees is introduced into a non-forest area, whereas in that of reforestation – a new forest generation replaces an old one. Reforestation may be artificial (planting, seeding) or natural (self-seeding, regrowth). Some elements of this process may be used in afforestation.

Nowadays reforestation by regrowth is of marginal importance, whereas reforestation by self-seeding has many advantages, which contributes to its growing popularity and practical application. The most important ecological and economic arguments for this method are: the use of native tree species, well adapted to local environmental conditions, natural habitat diversity, higher resistance of new forest communities to environmental stresses, saving on the costs of planting, a shorter production period. The first variant of using self-seeding concerns the situation when it has already occurred. Then the soil on the remaining area should be prepared and the gaps should be filled with seedlings. This provides the possibility of adding new species to the existing composition. The second variant concerns the situation when self-seeding has not occurred yet, but can be expected soon. Then afforestation should be carried out by the biogroup method, planting only small plots of land (several areas each) with seedlings and leaving open space between them, to be filled by self-seeding.

Spontaneous, uncontrolled processes taking place in the natural environment are not always consistent with economic objectives and not always lead to the most beneficial effects. It would be too risky to base forest development plans only on natural succession and communities of potential natural vegetation, i. e. plants that would appear under given habitat conditions without man's interference. Such interference is necessary, for both environmental and economic reasons. The starting point for taking rational actions in this field must be a clearly formulated goal of afforestation, followed by appropriate selection of measures and methods for tree breeding, protection or use. As regards economic use of former farmland, two methods may be employed. According to the first one, certain succession changes may be initiated and properly directed, and their rate can be accelerated, especially as concerns the first stage of the process, aimed at the formation of forest ecosystems; the other method is based on the creation of artificial forest communities.

In order to create relatively stable and dynamically balanced forest communities, natural succession should be adopted as the standard while implementing a forestation projects. Natural succession, as a progressive process, ensures unlimited vegetation development in space and time, as well as optimal – for given environmental conditions – species and spatial structures of forest ecosystems. Such forest communities are characterized by inertia, i. e. the ability to resist changes and maintain relative equilibrium, and high flexibility, i.e. the ability to restore equilibrium, within a relatively short time, after being affected by stress factors. These characteristics may also be molded by planned forest economy, but this is connected with certain investment costs. Forest communities developed as a result of natural succession require less expenditure, but also bring less profit to their owner.

Secondary succession is a special case of natural succession, observed at transformed habitats where forest vegetation had been earlier destroyed, i.e. fields under cultivation, pastures, gardens, etc. An example of this process is the return of forest to former farmland. The term "secondary succession" is also used when the process starts from semi-natural communities or idle land, and lasts even as long as 350 years.

The rate of forest restitution by ecological succession depends on the presence, in the immediate vicinity, of forest tree species for seeding, and the time that had passed since forest removal (habitat condition may be so poor that even after a long period of time the results of afforestation may seem unsatisfactory, from both the ecological and economic perspective) (Gorzelak, 1999).

4. Legal aspects

Legal regulations concerning forest management require due observance especially in the case of afforestation, because the first generation of trees is unstable and susceptible to various threats. As regards afforestation, two acts are of primary importance, i.e. The Act on Forests and the Act on Farmland Designation for afforestation.

4.1. Act on Forests

This Act concerns all land owners, and afforestation of non-forest grounds is treated uniformly, irrespective of the fact whether their owner is the State Treasury, or some other subject. Article 14 of the Act says that:

- forest resources can be increased, among other, by afforestation,
- the following types of land may be designed for afforestation: wasteland, farmland no longer used for agricultural purposes, areas located near heads of a river or stream, on watersheds, along river sides and lake shores, around water bodies, quick sands and dunes, steep slopes, valley-sides, bluffs, depressions, waste-tips and exploited areas of peat-, sand-gravel- and clay-digging,
- the size and distribution of the areas that are to be planted with trees, as well as the method of the process realization, are defined in the "National Afforestation Project",

- land designed for afforestation is specified in a local spatial management plan or in a decisions on conditions of land development and management,
- the duty to afforest land owned by the State Treasury weights on forest inspectors, whereas the duty to afforest the other land – on its owners or perpetual usufructaries,
- land owners or perpetual usufruct Aries may be granted public subsidies to cover the costs of afforestation.

The weakest link in the process of increasing forest resources is the statement that land for afforestation is to be specified in a local spatial management plan. It follows that chief officials of rural districts cannot convey the land whose afforestation is not provided for in the local spatial management plan to the State Forest Enterprise, whereas the land designed for afforestation in the local plan cannot be released for other purposes.

4.2. Act on Farmland Designation for Afforestation

This Act, implemented on January 1, 2002, is expected to stimulate farm owners to afforest land considered unsuitable for agricultural production. According to this Act, farmland designed for afforestation has to fulfill at least one of the following conditions (article 3 item l):

- it has to belong to class V, VI, VIz, or IV – provided that its area does not exceed 10 % of the total area of land designed for afforestation,
- it has to be located on a slope whose inclination exceeds 15 %,
- it has to be considered degraded according to the Act of February 3, 1995 on protection of farmland and forest land (Journal of Laws of 1997 no 16, then 78, with later amendments),
- afforestation has to be provided for in a local spatial management plants or in a decision on conditions of land development and management.

The total area of such land must be 0.4 to 30 ha, and the minimum division area – 0.1 ha. All persons interested should apply to the chief official of a given rural district for land designation for afforestation. The application submitted must contain extracts from documentation of land registration, and information about the expected dates of afforestation commencement and completion. The chief official of a rural district has to issue the decision within 60 days. The applications are considered by order of submission. Upon giving a favorable opinion, the chief official is obliged to ask the Commune Council for their approval for change of land use – from farmland to forest land (article 3 item 7). If the Council does not issue a decision on this matter within 45 days, it is tantamount to its favorable opinion. The land is reclassified ex officio on the first positive evaluation of the afforestation results (two years from the moment of its completion) (article 9).

The annual afforestation limit is to be fixed until March 31 by the chief official of a rural district, after seeking the opinion of forest inspectors, commune administrators / mayors, the Provincial Fund for Environmental Protection and Water Economy, and the Agency for Agriculture Restructuring and Modernization (article 2). If any applications cannot be opinioned positively due to the above limit, the chief official is obliged to inform land owners who submitted them about the expected date of their consideration and the planned date of beginning afforestation in successive years (article 3 item 8).

Afforestation is carried out on the basis of an afforestation plan (article 4 item l), drawn up by the forest inspector within 30 days from the receipt of the chief official's decision. Tree planting is conducted by the land owner according to a simplified forest management plan, or the chief official's decision specifying tasks to be performed in the field of forest husbandry. The forest inspector is also obliged to provide the land owner with seedlings – the costs are to be covered by the Provincial Fund for Environmental Protection and Water Economy. Trees can be planted three meters from the adjacent parcel; if the adjacent parcel is forest land or land that can be classified for afforestation – then the distance can be 1.5 m. The afforestation results can be considered satisfactory if at least 70 % of seedlings take root, afforestation is carried out on the basis of an afforestation plan, tree planting is conducted according to a simplified forest management plan or the chief official's decision, and the distances mentioned above are kept (article 4 item 3).

The land owner is obliged to notify the chief official of a rural district that the afforestation process had been completed. The chief official is obliged (within three months from that moment if afforestation had been completed in the first half of the year, or within eight months if it had been completed in the second half of the year) to make an inspection (article 6 item l) and issue an administrative decision permitting the land owner to grow forest crops, or – in case the land owner had not fulfilled the necessary conditions – refuse to issue such a decision.

Upon receipt of the above decision, the land owner acquires the right to receive a monthly cash equivalent in lieu of change of land use – from farmland to forest land (article 7 item l); the equivalent is subject to annual valorization:

- for afforestation area of 0.4 to 10 hectares – PLN 150 per month for each hectare,
- for afforestation area of 10 to 20 hectares – additional PLN 50 for each hectare,
- for afforestation area to 30 hectares – additional PLN 25 for each hectare.

The equivalent value may be additionally increased by 50 % if land designation for afforestation is connected with the liquidation of a farm that has been managed by the land owner, his testator or donor for at least five years, and if he does not own other farmland or a homestead covering an area of more than

0.8 ha. The application for increasing the equivalent can be submitted to the chief official of a rural district after the first evaluation of afforestation results, made by the chief official or within the framework of a task assigned by the forest inspector after two years from the completion of the afforestation process (article 8 item 1). The right to receive the equivalent may be granted for not more than 20 years, or until the acquisition of the right to draw an old age pension or a disability pension.

The Act discussed is applied to practice to a limited extent only, due to lack of budgetary funds.

5. Conclusions

1. The problem of farmland afforestation still requires particular attention, especially that is has not been sufficiently regulated by law.
2. The criteria enumerated in the Act of Farmland Designation for Afforestation, providing the basis for application submission, concern only land type, disregarding other circumstances that may affect the objectives and preferences of the afforestation process. The applications are considered by order of submission only, ignoring environmental, spatial and social conditions.
3. The criteria of urgency of farmland afforestation may be as follows: provisions of a local spatial management plan and a study of the directions and conditions of land development and management, soil quality, structure and area of grounds and their links with the existing forest complexes.
4. At present social factors are disregarded in the procedure of submitting applications for afforestation. They should be taken into account while determining the order of application consideration and granting cash equivalents to applicants.

6. References

Czygier P.: Kryteria pilności zalesień gruntów rolnych. Praca magisterska UW-M w Olsztynie, Wydział Geodezji i Gospodarki Przestrzennej, 2003.

Gorzelak A.: Zalesianie terenów porolnych. Warszawa, IBL 1999.

Łonkiewicz B.: Krajowy program zwiększania lesistości. Warszawa, IBL 1995.

Puchniarski T.H.: Krajowy program zwiększania lesistości. Poradnik od a do z. Warszawa, PWRiL 2000.

Act of September 28, 1991 on Forests (Journal of Laws no 101, item 444 with later amendments)

Act of June 8, 2001 on Farmland Designation for Afforestation (Journal of Laws no 73, item 764, Journal of Laws of 2003 no 46, item 392).

Erich Weiss

Die ausgleichspflichtige Inhalts- und Schrankenbestimmung des Grundeigentums beim Umweltschutz

1. Zur inhaltlichen Bedeutung einiger Begriffe für die Bundesrepublik Deutschland

Bei der vorgegebenen Thematik erscheinen m. E. zunächst folgende umweltpolitische Begriffe noch näher erklärungsbedürftig: Natürliche Ressourcen, Nachhaltiger Schutz, Gewährleistung des Grundeigentums.

1.1 Natürliche Ressourcen

Die wesentlichen Komponenten der natürlichen Ressourcen des Menschen sind die Lithosphäre, die Hydrosphäre, die Atmosphäre und ein ständig gleichbleibender Energiestrom. Insbesondere Flora und Fauna eines Ökosystems stehen unter komplizierten physikalischen, biochemischen und energetischen Mitwirkungsprozessen von im wesentlichen Boden, Wasser, Luft und Sonnenenergie über Nahrungsketten bzw. -netze miteinander in Verbindung. Auf vergleichbarer Stufe mit der Flora und Fauna steht der Mensch hinsichtlich seiner vitalen Daseinsbedürfnisse wie Atmung, Wasser- und Nahrungsaufnahme etc. Das bedeutet, seine Lebens- bzw. Überlebenschancen hängen von seiner stabilisierenden Teilnahme am System natürlicher Stoffkreisläufe bei gleichbleibendem Energiestrom ab.

Das Zurückdrängen der natürlichen Umwelt durch den Menschen, das Zerstören ihrer Zusammenhänge, ihrer Wechselwirkungen und Gleichgewichte, die Überlastung ihrer Puffer- und Regenerationsfähigkeit lassen sich energetisch als Folge eines durch den Menschen künstlich erhöhten Energiestromes in etwa den letzten 200 Jahren darstellen. Die entsprechende Energie stammt aus dem Verbrauch der fossilen Brennstoffe, also der in Kohlenstoffverbindungen gespeicherten Sonnenenergie. Auf diesen erhöhten Energiestrom ist die Biosphäre nicht angelegt; entsprechendes gilt m. E. auch für nachwachsende Rohstoffe als Energiequelle.

Beispiele natürlicher und naturnaher Ökosysteme, die in unserem Lande zu erhalten und zu schützen sind, werden regelmäßig in mehr oder weniger aggregierter Form aus dem Zusammenhang der Biosphäre ausgegliedert und nach verschiedenartigsten Gesichtspunkten, zum Beispiel als Gewässer beim

Gewässerschutz oder als Boden beim Bodenschutz, gegenenenfalls auch zusammen beim Natur- und Landschaftsschutz angesprochen.

1.2 Nachhaltiger Schutz

Der Umweltschutz in der Bundesrepublik Deutschland bezeichnet die Gesamtheit aller Maßnahmen zur Vermeidung und Verminderung der Umweltgefährdung. Er verfolgt dabei (nach dem Umweltgutachten des Rates von Sachverständigen für Umweltfragen in der Bundesrepublik Deutschland von 1987 [Randziffer 29]) folgende drei bedeutsame Ziele:

- Beseitigung bereits eingetretener Umweltschäden,
- Ausschaltung oder Minderung aktueller Umweltgefährdungen,
- Vermeidung zukünftiger Umweltgefährdungen durch Vorsorgemaßnahmen.

Umweltschutz ist insoweit (nach Breuer, R. [1988][1], nach Erbguth, W. [1984][2] oder nach Hoppe, W. [1980][3]) eine problembezogene Querschnittsaufgabe mit den wichtigen Handlungsebenen der Umweltpolitik, der Umweltökonomie und des Umweltrechts.

Die Umweltpolitik in der Bundesrepublik Deutschland umfaßt dabei bereits nach dem Umweltprogramm der Bundesregierung aus dem Jahre 1971 die Gesamtheit aller notwendigen Maßnahmen, um

- den Menschen eine Umwelt zu sichern, wie er sie für seine Gesundheit und für ein menschenwürdiges Dasein braucht,
- Boden, Wasser und Luft, Flora und Fauna vor nachteiligen Wirkungen menschlicher Eingriffe zu schützen und
- Schäden oder Nachteile aus menschlichen Eingriffen zu beseitigen.

Fachlich erstreckt sich die Umweltpolitik insbesondere auf die Bereiche des Bodenschutzes, des Gewässerschutzes, des Natur- und Landschaftsschutzes sowie der Luftreinhaltung. Räumlich erstreckt sich die Umweltpolitik auf alle politischen Handlungsebenen lokalen bis globalen Umfanges.

Oberstes Ziel der Umweltpolitik in der Bundesrepublik Deutschland ist es dabei (nach dem Umweltgutachten des Rates von Sachverständigen für Umweltfragen von 1974 [S. 5]), das Leben und die Gesundheit aller Menschen in Gegenwart und Zukunft vor Schäden zu bewahren. Um dieses Ziel zu erreichen, bedarf es (nach den Leitlinien zur Umweltvorsorge der Bundesregierung vom 19. September 1986)[4] folgender ergänzender Ziele,

---

1 Breuer, R. (1988): Umweltschutzrecht; in: v. Münch, J. (Hrsg.): Besonderes Verwaltungsrecht, 8. Aufl., Berlin, New York, S. 601 ff.

2 Erbgut, W. (1984): DÖV, S. 699 ff.

3 Hoppe, W. (1980): Staatsaufgabe Umweltschutz; in: VVDStdRL 38, S. 211 ff.

4 Bundestagsdrucksache 10/6028, S. 6.

- dem Schutz der Ökosysteme in ihrer Gesamtheit,
- dem Schutz der natürlichen Ressourcen Boden, Wasser, Luft,
- dem Schutz und der Erhaltung von Flora und Fauna,
- der Vermeidung nachteiliger Klimaveränderungen,
- dem Schutz des Eigentums, der Senkung volkswirtschaftlicher Kosten und der Erhaltung des kulturellen Erbes.

Hinzu treten sicher gleichrangig weitere Ziele des sozialen, ökonomischen und kulturell-wissenschaftlichen Bereichs. Die Handlungsprinzipien der Umweltpolitik in der Bundesrepublik Deutschland werden (nach Hoppe, W. [1980][5] oder nach Kloepfer, M. und Meßerschmidt, K. [1986][6]): vornehmlich geprägt vom Vorsorgeprinzip, vom Verursacherprinzip sowie vom Kooperationsprinzip. Nach den Leitlinien zur Umweltvorsorge der Bundesregierung vom 19. September 1986[7] umfaßt dabei die Umweltvorsorge alle Handlungen,

- die der Abwehr konkreter Umweltgefahren,
- die im Vorfeld der Gefahrenabwehr der Vermeidung oder Verminderung von Risiken für die Umwelt und
- die vorausschauend der Gestaltung unserer zukünftigen Umwelt, insbesondere dem Schutz und der Entwicklung der natürlichen Lebensgrundlagen dienen.

Versteht man nun (nach Kloepfer, M. und Meßerschmidt, K. [1989][8]) das Vorsorgeprinzip im Sinne einer Ressourcenvorsorge, so handelt es sich beim Grundsatz der Nachhaltigkeit um die Bekräftigung der ressourcenökonomischen Intention dieses Prinzips. Unter dem Verzicht auf den vorstehend genannten Aspekt der Gefahrenabwehr werden in der Fachwelt nun im allgemeinen folgende zwei Systemvarianten des Vorsorgeprinzips unterschieden, eine risikovorsorgende und eine planerische Variante.

Allein hieraus ergeben sich vielfältige Anknüpfungspunke unter dem Trinitätsaspekt von Bodeneigentum, Bodenordnung und Bodenwirtschaft. Gleichwohl sei hier auch der Begriff der nachhaltigen Entwicklung (aus dem bekannten Bericht der Brundtland-Kommission) eingeführt: „Nachhaltige Entwicklung ist eine Entwicklung, welche die Bedürfnisse der Gegenwart befriedigt, ohne zu riskieren, daß die künftigen Generationen ihre Bedürfnisse nicht befriedigen können.“

---

5 Hoppe, W. (1980): Staatsaufgabe Umweltschutz; in: VVDStdRL 38, S. 211 ff.

6 Kloepfer, M. u. Meßerschmidt, K. (1986): Innere Harmonisierung des Umweltrechts, Bundesumweltamt Berlin – Bericht 6/86, S. 67.

7 Bundestagsdrucksache 10/6028, S. 7.

8 Kloepfer, M. u. Meßerschmidt, K. (1986): Innere Harmonisierung des Umweltrechts, Bundesumweltamt Berlin – Bericht 6/86, S. 81.

Der englische Begriff „sustainable development" bedeutet „Aufrechterhaltbarkeit"; andere übertragen diesen Begriff mit „Zukunftsbeständigkeit".

Der Zukunftsbeständigkeit wird dabei eine soziale, eine ökonomische und eine ökologische Komponente mit gewissen hierarchischen Strukturen zugeordnet:

- Zukunftsbeständigkeit des Gemeinwesens:
  Konsens über Grundwerte, gesunde Lebensbedingungen und Verteilungsgerechtigkeit zwischen den derzeit lebenden Menschen und zwischen den Generationen. Voraussetzung hierfür ist die
- Zukunftsbeständigkeit des Wirtschaftssystems:
  Stützung auf menschliche Arbeit und erneuerbarer Ressourcen statt auf Verbrauch nicht erneuerbarer Ressourcen; Ökonomie mit niedriger Entropie. Voraussetzung hierfür ist die
- Ökologische Zukunftsbeständigkeit:
  Erhaltung der Artenvielfalt, der menschlichen Gesundheit sowie der Sicherung von Luft-, Wasser- und Bodenqualitäten, die ausreichen, um das Leben und das Wohlergehen der Menschen sowie das Tier- und Pflanzenleben für alle Zukunft zu sichern. Die Verbrauchsrate von erneuerbaren Rohstoff-, Wasser- und Energieressourcen darf nicht höher sein, als die Neubildungsrate, nicht erneuerbare Ressourcen dürfen nicht schneller verbraucht werden, als sie durch dauerhafte, erneuerbare Ressourcen ersetzt werden; die Emission von Schadstoffen darf nicht größer sein, als die Fähigkeit von Luft, Wasser und Boden, diese Schadstoffe zu binden und abzubauen.

1.3 Gewährleistung des Grundeigentums

Der Boden umfaßt in diesem Sachzusammenhang im Wesentlichen die feste Erdoberfläche. Er ist begriffsinhaltlich bedeutsam als originärer Produktionsfaktor, als standortgebundene Rohstoffquelle, als naturgegebene Energiequelle u. v. a. m. Als besonderer Produktionsfaktor ist er zumindest im Grundsatz unvermehrbar, unbeweglich, aber auch nicht abnutzbar. Aus seinem wirtschaftlichen Einsatz resultiert eine Bodenrente, wie z. B. aus der Vermietung der Mietzins[9].

Kennzeichnend für Eigentum ist, daß man eine Sache haben kann, sie gebrauchen und über sie verfügen kann. Als geometrisch abgegrenzten Teil der Erdoberfläche bezeichnet man den Eigentumsgegenstand als Grundstück. Bodeneigentum ist also die rechtliche Zuordnung von Grundstücken zu Per-

---

9 Meyers Enzyklopädisches Lexikon (1972); Bibliographisches Institut Mannheim, Bd. 4, S. 407.

sonen, die durch die Merkmale der Privatnützigkeit und der ausschliessenden Verfügungsbefugnis bestimmt werden. Inhaber solcher Eigentumsrechte können alle natürlichen Personen sein, juristische Personen jedoch nur, soweit sie dazu fähig sind.

Die besonderen Eigenschaften des Bodeneigentums erklären seine herausragende Bedeutung in den verschiedenen gesellschaftspolitischen Entwicklungen der Geschichte: Das Bodeneigentum ist ein Menschenrecht, das in einem inneren Zusammenhang mit der Garantie der persönlichen Freiheit des Menschen steht. Ihm kommt im Gesamtgefüge der Menschenrechte die Aufgabe zu, dem Träger des Menschenrechts einen Freiheitsraum im vermögensrechtlichen Bereich sicherzustellen und ihm damit eine eigenverantwortliche Gestaltung des Lebens zu ermöglichen. Die Garantie des Bodeneigentums als Rechtseinrichtung dient der Sicherung eines Menschenrechts. Das Menschenrecht des einzelnen setzt das Rechtsinstitut Bodeneigentum voraus; es wäre nicht wirksam gewährleistet, wenn ein Gesetzgeber an die Stelle des Bodeneigentums Privater etwas setzen könnte, was den Namen Bodeneigentum nicht verdiente[11].

Gewährleistet wird das Grundeigentum in der Bundesrepublik Deutschland einschließlich Erbrecht nach Artikel 14 Grundgesetz (GG). Diese Vorschrift ergänzt insoweit auch das allgemeine Persönlichkeitsrecht nach Artikel 2 GG. Zusammen mit den Vorschriften zur Vereinigungsfreiheit nach Artikel 9 GG und zur Freiheit der Berufswahl nach Artikel 12 GG ergeben sich daraus die wesentlichen Grundlagen für eine freie und private wirtschaftliche Betätigung sowie als Folge die freie und soziale Marktwirtschaft in der Bundesrepublik Deutschland. Die Gesetzgebung, die vollziehende Gewalt und die Rechtsprechung sind nach Artikel 1 Abs. 3 GG an diese unmittelbar geltenden Rechtsvorschriften gebunden. Nach Artikel 19 Abs. 2 GG darf ein solches Grundrecht in keinem Fall in seinem Wesensgehalt angetastet werden. Nach Artikel 79 Abs. 3 GG ist eine Änderung der in den Artikeln 1 bis 20 des Grundgesetzes niedergelegten Grundsätze gar unzulässig. Auch nach Artikel 222 des EG-Vertrages gilt diese Eigentumsordnung bisher als unantastbar[12]. Gleichwohl wird diese Eigentumsordnung als wesentlicher Bestandteil einer freien und sozialen Marktwirtschaft heute sicher bereits von den vielfältigen und vielgestaltigen der weit überwiegend EU-gesteuerten Marktmechanismen mitbestimmt.

10 BVerfGE 95, 267 (300) – m. w. N. – „Altschuldenregelung“.

11 BVerfGE 24, 367, 289.

12 Gather, H.-H. (2000): Der Schutz des Eigentums im Grundgesetz; in: DWW 2000, S. 78.

2. Neue Probleme bei der Erfüllung der Schutzaufgabe in der Bundesrepublik Deutschland

2.1 Zur Ausgangslage

Der Gesetzgeber muß bei der Inhalts- und Schrankenbestimmung des Grundeigentums nach Artikel 14 Abs. 1 Satz 2 GG, und dazu ist wohl der Umweltschutz im weitesten Sinne regelmäßig zu rechnen, einerseits das Grundeigentum Privater nach Artikel 14 Abs. 1 Satz 1 GG und andererseits die Sozialpflichtigkeit nach Artikel 14 Abs. 2 GG beachten (BVerfGE 25, 112-Nds. Deichgesetz). Es wird jedoch seit einiger Zeit Kritik daran geäußert, daß sich das natürliche Spannungsfeld zwischen dem Grundeigentum Privater und dem Interesse bzw. dem Wohl der Allgemeinheit immer deutlicher zugunsten der Allgemeinheit, also zu Lasten des Grundeigentums Privater entwickelt habe. Trotz des steigenden Gemeinwohlbedürfnisses am Schutz von Natur und Landschaft dürfe aus dem Eigentum nicht immer mehr herausgenommen werden, bis man an die äußerste Grenze der Privatnützigkeit gestoßen sei. Entsprechendes gilt sicher auch für weitere diesbezügliche Schutzgüter.

2.2 Enteignung oder Inhalts- und Schrankenbestimmung des Grundeigentums

Die heute vom Verfassungsgericht der Bundesrepublik Deutschland vertretene Konzeption geht davon aus, daß Enteignung sowie Inhalts- und Schrankenbestimmungen in einem aliud-Verhältnis zueinander stehen. Danach sind Legal- und Administrativenteignung sowie Inhalts- und Schrankenbestimmung jeweils eigenständige Rechtsinstitute, die das Grundgesetz deutlich voneinander unterscheidet. Dabei wird dieser Unterschied allein anhand formaler Kriterien getroffen, so daß es keinen Überschneidungsbereich geben kann und die jeweilige Intensität des Eingriffs ohne Belang für die Beurteilung bleibt.

Zur Erinnerung:

Der verfassungsrechtliche Enteignungsbegriff in Deutschland geht auf die preußische Verfassungsurkunde aus dem Jahre 1850 zurück. Sie definierte die Enteignung als den Entzug von Grundeigentum durch die Verwaltung zur Übertragung des entzogenen Gutes auf ein dem öffentlichen Wohl dienendes Unternehmen. Sie wurde in den nachfolgenden Jahrzehnten vom deutschen Reichsgericht nach und nach ausgestaltet: Dabei wurde die Vorstellung aufgegeben, Enteignungsobjekt könne nur das Grundeigentum sein; es wurde die Enteignung direkt durch Gesetz ermöglicht; schließlich wurde auch jede sonstige Beeinträchtigung des Eigentümerrechts, mit der Sache nach Belieben zu verfahren (§ 903 Bürgerliches Gesetzbuch [BGB]), unter die Enteignung subsumiert.

In der Weimarer Verfassung vom Jahre 1919 fehlte noch die Möglichkeit, sich unmittelbar gegen den Entzug des Eigentums zu wehren („Dulde und liquidiere!" nach O. Mayer). Deshalb dehnte das deutsche Reichsgericht den Enteignungsbegriff noch weiter aus, um wenigstens den daraus resultierenden wirtschaftlichen Verlust kompensieren zu können.

Der Bundesgerichtshof, also das höchste Zivilgericht der Bundesrepublik Deutschland (und mit gewissen Unterschieden im Detail auch das Bundesverwaltungsgericht) knüpfte an diese Rechtsentwicklungen an, ohne die zwischenzeitlich veränderte Verfassungslage durch das Grundgesetz vom Jahre 1949 (mit seinem Primärrechtsschutz des Eigentums, dem gesetzgeberischen Ausgestaltungsauftrag von Inhalt und Schranken des Eigentums sowie dem diesbezüglichen Verwerfungsmonopol des Bundesverfassungsgerichts) deutlich wahrzunehmen. Also wurde der Eigentumsschutz nach wie vor über die Entschädigungspflichten verwirklicht: Da eine Entschädigung nur bei einer Enteignung zugebilligt werden konnte, wurde die Abgrenzung von Sozialbindung des Eigentums sowie seiner Enteignung, also die Schwere des jeweiligen Eingriffs als Entscheidungskriterium angesehen. Wesentliche Folge dieser Sichtweise war die sogenannte „Umschlagtheorie". Die Sozialbindung des Eigentums, also seine Inhalts- und Schrankenbestimmung, sowie dessen Enteignung waren danach keine grundsätzlich voneinander zu unterscheidenden Sachverhalte, sondern sie unterschieden sich nur durch die jeweilige Intensität der Belastung des Eigentums. Bewegte sich eine Maßnahme nicht mehr in den Grenzen der Sozialpflichtigkeit, weil sie für den Eigentümer zu belastend wirkte, schlug sie in eine Enteignung um; es wurde entschädigt.

Damit lag die Entscheidung, was eine Enteignung war und eine Entschädigung zur Folge hatte, nicht mehr allein beim Gesetzgeber, sondern vermehrt bei den Gerichten. Dieses war jedoch mit den Grundzügen des Verfassungsrechts im Grundgesetz der Bundesrepublik Deutschland nicht vereinbar und wurde deshalb mit der Naßauskiesungs-Entscheidung des Bundesverfassungsgerichts im Jahre 1981 entscheidend geändert.

Entscheidendes Merkmal einer Enteignung ist danach ein finaler Zugriff auf das (Grund-)Eigentum zur Durchbrechung der bisher bestehenden Eigentumsordnung. Sie setzt einen normativ-abstrakten oder exekutiv-einzelfallbezogenen Regelungsakt der öffentlichen Gewalt voraus, also eine einseitige, verbindliche, rechtsbegründende Ordnung von Lebenssachverhalten. Maßgebend ist insoweit die Entzugswirkung; daß damit ein Güterbeschaffungsvorgang zugunsten der öffentlichen Hand verbunden ist, ist dabei kein Wesensmerkmal der Enteignung[14].

---

13 Roller, G. (2001): Enteignung, ausgleichspflichtige Inhaltsbestimmung und salvatorische Klauseln; in: NJW 2001, S. 1003 ff.

14 BVerfGE 83, 201 (221).

Demgegenüber ist die Inhalts- und Schrankenbestimmung dadurch gekennzeichnet, daß Inhalt und Schranken des (Grund-)Eigentums generell und abstrakt für die Zukunft bestimmt werden.

Durch diese Einengung des Enteignungsbegriffes und die Erweiterung der Inhalts- und Schrankenbestimmung, verbunden mit einer Ausgleichspflichtigkeit, ist das oben beschriebene Grundproblem im Spannungsfeld von Grundeigentum Privater und Allgemeinwohlinteressen jedoch grundsätzlich nicht gelöst, sondern nur verlagert worden – nämlich, wann darf nun eine Inhalts- und Schrankenbestimmung ausgleichslos oder ausgleichspflichtig ausgestaltet werden?

2.3 Ausgleichspflichtige Inhalts- und Schrankenbestimmung des Grundeigentums

Als ausgleichspflichtige Inhalts- und Schrankenbestimmung des (Grund-) Eigentums wird ein Institut bezeichnet, bei dem der Gesetzgeber zwar eine generelle Bestimmung von Inhalt und Schranken des (Grund-)Eigentums vornimmt, diese Bestimmung aber derart belastend für einzelne Eigentümer wirkt, daß die Regelung an sich gegen den Verhältnismäßigkeitssatz verstoßen würde. Nur durch die Einräumung eines Ausgleichs, der nicht notwendigerweise ein finanzieller sein muß, wird die Belastung derart gemildert, daß die Regelung insgesamt als verhältnismäßig angesehen werden kann. Wegen des inneren Zusammenhanges mit dem Verhältnismäßigkeitsgrundsatz wird diese Ausgleichsform teilweise auch als Verhältnismäßigkeitsausgleich bezeichnet[15].

Das Bundesverfassungsgericht hat diese Entwicklung bereits im Pflichtexemplarbeschluß des Jahres 1981 vorgezeichnet[16] (man vgl. dazu auch den Naßauskiesungsbeschluß von 1981[17]. Der Bundesgerichtshof[18] und das Bundesverwaltungsgericht[19] kennen seit einigen Jahren diese ausgleichspflichtige Inhalts- und Schrankenbestimmung ebenfalls an; in den Bereichen von Natur- und Landschaftsschutz sowie Denkmalschutz haben sie bereits größere Bedeutung erlangt.

Gleichwohl stellt das Bundesverfassungsgericht klar, daß eine Inhalts- und Schrankenbestimmung des (Grund-)Eigentums durch den Gesetzgeber im Grundsatz durchaus ausgleichslos zulässig ist, die Ausgleichspflicht in diesem Zusammenhang jedoch ein mögliches Mittel sein kann, eine ansonsten

---

15 Roller, G. (2001): Enteignung, ausgleichspflichtige Inhaltsbestimmung und salvatorische Klauseln; in: NJW 2001, S. 1003 ff.

16 BVerfGE 58, 137 = NJW 1982, S. 633.

17 BVerfGE 58, 300 (331) = NJW 1982, S. 745.

18 BGHZ 102, 350 (359) = NJW 1988, S. 478.

19 BVerfGE 84, 361 (367) = NJW 1990, S. 2572.

ausnahmsweise drohende Unverhältnismäßigkeit des Gesetzes zu vermeiden. Damit wird vom Bundesverfassungsgericht aber ausdrücklich und deutlich die Bestandsgarantie des (Grund-)Eigentums vor eine Vermögenswertgarantie eingestuft[20].

2.4 Anforderungen an gesetzliche Ausgleichsvorschriften

Im Zusammenhang mit der Entscheidung des Bundesverfassungsgerichts zum Denkmalschutzgesetz des Landes Rheinland-Pfalz – Direktorenvilla[21] – sind zugleich einige Maßgaben zur Gesetzgebung ausgleichspflichtiger Inhalts- und Schrankenbestimmungen des (Grund)-Eigentums deutlich geworden[22].

Danach muß ein diesbezügliches Gesetz Vorkehrungen vorsehen, die eine unverhältnismäßige Belastung des (Grund-)Eigentümers real vermeiden und die Privatnützigkeit des (Grund-) Eigentums soweit als möglich erhalten (Übergangsregelungen, Ausnahmen, Befreiungen usw.).

Die Verwaltung muß bei der Aktualisierung der (Grund-)Eigentumsbeschränkung zugleich über den gegebenenfalls erforderlichen Ausgleich zumindest dem Grunde nach entscheiden.

Außerdem ist der (Grund-)Eigentümer, der einen ihn in seinem Grundrecht aus Artikel 14 GG beeinträchtigenden Verwaltungsakt für unverhältnismäßig hält, gehalten, diesen im Verwaltungsrechtsweg anzufechten.

Als Folge dieser Entscheidung sind in der Bundesrepublik Deutschland sicher noch einige Landesgesetze auf den Gebieten des Bodenschutzes, des Wasserschutzes, des Natur- und Landschaftsschutzes sowie des Denkmalschutzes den neuen verfassungsrechtlichen Vorgaben anzupassen. Ähnliche Maßnahmen sind auch für die noch vorhandenen sogenannten salvatorischen Klauseln notwendig.

3. Ausblick

Grundlage vielgestaltiger und vielfältiger fachplanerischer Bodenordnungsmaßnahmen in den ländlichen Gebieten und am Rande städtischer Verflechtungsgebiete ist das Flurbereinigungsgesetz der Bundesrepublik Deutschland mit seinen rein privatnützig ausgestalteten Instrumenten des Freiwilligen Landtausches (nach §§ 103a ff. FlurbG), der Beschleunigten Zusammenlegung (nach

20 BVerfG; in: NJW 1999, S. 2877 (2879).

21 BVerfGE 100, 26 = NJW 1999, S. 2877.

22 Rinne, E. (2000): Bestätigung und Präzisierung der BGH-Rechtsprechung zur ausgleichspflichtigen Inhaltsbestimmung des Eigentums durch das BVerfG; in NVwZ, Heft 4, S. 410.

§§ 91 ff. FlurbG), der Vereinfachten Flurbereinigung (nach § 86 FlurbG) sowie der Regelflurbereinigung (nach den §§ 1; 4 u. 37 Abs. 1 FlurbG), also einer klaren Inhalts- und Schrankenbestimmung des Grundeigentums auf der einen Seite und der fremdnützig ausgestalteten Unternehmensflurbereinigung (nach den §§ 87 ff. FlurbG), also einer klaren Enteignungs- und Entschädigungsvorschrift auf der anderen Seite. Welche Konsequenzen ergeben sich aus den aufgezeigten Problemverlagerungen für Maßnahmen nach dem Flurbereinigungsgesetz der Bundesrepublik Deutschland zum Zwecke des Bodenschutzes, des Gewässerschutzes, des Natur- und Landschaftsschutzes oder des Denkmalschutzes unter den Aspekten der Gewährleistung des Grundeigentums? Hier erscheint noch eine sorgfältige Problemschau und Problembewertung wünschenswert, aber wohl auch notwendig. Eine Änderung des Flurbereinigungsgesetzes erscheint jedoch aus diesen Entwicklungen heraus nicht stringent herleitbar.

Schlußendlich ist auch für die anfangs eingeführte natürliche Ressource Energie eine umfassende Bilanz bezüglich der Wirkung von zum Beispiel Bodenordnungs- bzw. Dorferneuerungsmaßnahmen nach dem Flurbereinigungsgesetz der Bundesrepublik Deutschland und seiner bisherigen Finanzierungsgrundlage, dem Gesetz über die Gemeinschaftsaufgabe zur Verbesserung der Agrarstruktur und des Küstenschutzes der Bundesrepublik Deutschland möglich und gegebenenfalls auch nötig. Für die Verringerung des „Parzellensuchverkehrs" in der freien Landschaft durch eine weitere eigentumsrechtlich mögliche und betriebswirtschaftlich sinnvolle Zusammenlegung des ländlichen Grundbesitzes sowie für die Verringerung des sogenannten „Hof-Feld-Verkehrs" durch notwendige allseits akzeptable Erschließung der Wald-, Feld- und Grünlandflächen mittels Anlage und Ausbau eines hinreichenden ländlichen Wegenetzes liegen seit etwa 30-35 Jahren erste und sehr ermutigende Energiebilanzen vor. Ihre aggregierte Betrachtungsweise erscheint jedoch aus heutiger Sicht nicht mehr ganz befriedigend; dieses gilt aber für gewisse Ökobilanzen in gleicher Weise. Hier könnten noch wesentlich mehr Transparenz und mehr Glaubwürdigkeit und damit mehr nachhaltige Akzeptanz geschaffen werden.

Hans Neuhofer

Bodenschutz und Altlastensanierung in Österreich

Es bedarf keines Nachweises, dass weltweit die Gesunderhaltung und Sanierung der Böden eine existentielle Voraussetzung für die Menschen, Tiere und Pflanzen ist. Sowohl in der Europäischen Union als auch in Österreich ist die Gesunderhaltung und Sanierung der Böden eingehend rechtlich geregelt.

1 Bodenschutz in der Europäischen Union

Ein Ziel der Umweltpolitik der Europäischen Gemeinschaft ist „die umsichtige und rationelle Verwendung natürlicher Ressourcen" (Art 174 Abs 1 EGV[1]); hiezu zählt auch der Boden. Die Umweltpolitik der Europäischen Gemeinschaft „beruht auf den Grundsätzen der Vorsorge und Vorbeugung, auf dem Grundsatz, Umweltbeeinträchtigungen mit Vorrang in deren Ursprung zu bekämpfen, sowie auf dem Verursacherprinzip".

Diese Ziele und Grundsätze der Umweltpolitik der Europäischen Gemeinschaft finden in zahlreichen Verordnungen und Richtlinien der zuständigen EG-Organe ihren Niederschlag[2]. Nach Art 249 EGV (= Vertrag über die Europäische Gemeinschaft) gilt die Verordnung „unmittelbar in jedem Mitgliedsstaat". Die Richtlinie ist „für jeden Mitgliedsstaat an den sie gerichtet wird, hinsichtlich des zu erreichenden Zieles verbindlich, überlasst jedoch den innerstaatlichen Stellen die Wahl der Form und Mittel". Nach der Rechtsprechung des Europäischen Gerichtshofes (EuGH) hat das Gemeinschaftsrecht der Europäischen Gemeinschaft gegenüber dem nationalen Recht Vorrang, auch gegenüber nationalem Verfassungsrecht[3]; nationales Recht, das dem Gemeinschaftsrecht widerspricht, ist nicht anzuwenden.

1 Vertrag zur Gründung der Europäischen Gemeinschaft in der konsolidierten Fassung im Abl der Europäischen Union 1997 C 340, 1 und geändert durch Abl der Europäischen Union 2001 C 80, 1.

2 Vgl. hiezu die RL 86/278/EWG des Rates vom 12.6.1986 über die Verwendung von Klärschlamm in der Landwirtschaft, Abl L 181 v 4.7.1986, 6 (Klärschlammrichtlinie); RL 91/414/EWG v 15.7.1991 über das Inverkehrbringen von Pflanzenschutzmitteln, Abl L 230 v 19.8.1991, 1 (Pflanzenschutzmittelrichtlinie).

3 Vgl. hiezu im Näheren u.a. Öhlinger / Potacs, Gemeinschaftsrecht und staatliches Recht (1998, 85).

Soweit der Bodenschutz gemeinschaftsrechtlich geregelt ist, geht das Gemeinschaftsrecht dem nationalen Recht vor. Das nationale Recht hat auch hinsichtlich des Bodenschutz geltendes Gemeinschaftsrecht als Vorgaben zu beachten und es darf der nationale Gesetzgeber keine Regelungen treffen, die mit dem Gemeinschaftsrecht in Widerspruch stehen.

Das sechste EU-Umweltaktionsprogramm 2001-2010 der Europäischen Kommission (2001) sieht einen erhöhten Bodenschutz durch ein „systematisches Bodenschutzkonzept" vor[4].

2 Bodenschutz in Österreich

Dieser Beitrag beschäftigt sich – nach einer Einführung über die Vorgaben des Gemeinschaftsrechts der Europäischen Gemeinschaft – mit der rechtlichen Regelung des Bodenschutzes in Österreich.

2.1 Bodenschutz als Teil des umfassenden Umweltschutzes

Das Bundesverfassungsgesetz über den umfassenden Umweltschutz, Bundesgesetzblatt (BGBl) Nr 491/1984, verpflichtet den Bund, die Länder und Gemeinden zur Beachtung eines „umfassenden Umweltschutzes", insbesondere zu Maßnahmen „zur Reinhaltung der Luft, des Wassers und des Bodens sowie zur Vermeidung von Störungen durch Lärm". Der Bodenschutz ist verfassungsgesetzlich geboten[5]; die Gesetzgeber des Bundes und der Länder sind verpflichtet, im Rahmen ihres Kompetenzbereiches gesetzliche Regelungen für den Schutz des Bodens zu treffen.

2.2 Bundes- und Länderkompetenzen für den Bodenschutz

Die Kompetenzverteilung zur Gesetzgebung und Vollziehung für den Schutz des Bodens bestimmt sich nach dem Bundes-Verfassungsgesetz (B-VG)[6]. Der Bodenschutz kann unter verschiedenen Aspekten betrachtet werden und fällt

---

4 Vgl. im Näheren Seite 39 und im weiteren den 4. Jahresbericht (der Europäischen Kommission) über die Durchführung und Durchsetzung des Umweltrechts der Gemeinschaft (2002), insbesondere zu den Kapiteln Wasser, Natur, chemische Stoffe und Abfälle.

5 Vgl. eingehend zum Bundesverfassungsgesetz über den umfassenden Umweltschutz, Kerschner (Hrsg), Staatsziel Umweltschutz (1996).

6 Zur Kompetenzverteilung zwischen Bund und Ländern vgl eingehend u. a. Walter / Mayer, Bundesverfassungsrecht, 9. A (2000), S. 131 ff.

demnach entweder in die Kompetenz des Bundes (Art 10 B-VG) oder der Länder (Art 15 Abs 2 B-VG).

2.3 Bodenschutz durch den Bund

Im Rahmen der Bundeskompetenz finden sich zahlreiche Bodenschutzbestimmungen, insbesondere in den folgenden Gesetzen:

- Forstgesetz 1975, BGBl 440/1975 idgF, insbesondere über die Erhaltung des Waldes, über das Verbot der Waldverwüstung;
- Wasserrechtsgesetz 1959, BGBl 215/1959 idgF, insbesondere über die Reinhaltung von Gewässern;
- Mineralrohstoffgesetz, BGBl I 38/1999 idgF, insbesondere über den Abbau von Mineralien;
- Abfallwirtschaftsgesetz 2002, BGBl I 102/2002 idgF, insbesondere über das Verbot der Ablagerung von Abfällen sowie über die Errichtung und den Betrieb von Deponien und Abfallbehandlungsanlagen;
- Altlastensanierungsgesetz, BGBl 299/1989 idgF, insbesondere zur Sicherung und Sanierung von verseuchten Böden und zur Sicherstellung der Finanzierung
- usw.

2.4 Bodenschutz durch die Länder

Im Rahmen der Landeskompetenz finden sich Bodenschutzbestimmungen, insbesondere in folgenden Gesetzen:

- Bodenschutzgesetze, zur Erhaltung von gesunden Böden und zur Bodenverbesserung;
- Raumordnungsgesetze, insbesondere für die Widmung von Grundstücken für bestimmte Zwecke;
- Naturschutzgesetze, insbesondere zur Erhaltung bestimmter Gebiete z. B. von Mooren, Trockenrasen, Feuchtgebieten usw.;
- Almschutzgesetze, insbesondere zur Erhaltung der Almböden;
- usw.

Die umfassendste Regelung des Bodenschutzes finden sich in den Bodenschutzgesetzen der Länder[7]; damit möchte ich mich am Beispiel des Oö. Boden-

7 Bodenschutz(Klärschlamm-)gesetze der österreichischen Bundesländer (LGBl): Burgenland: 87/1990, 40/1992, 75/2000, 32/2001; Niederösterreich: 6160; Oberösterreich: 115/1991, 19/1997, 63/1997, 104/1997, 37/1998, 13/1999, 34/1999, 2/2000, 83/2001, 84/2002; Salzburg: 80/2001, 85/2002; Steiermark: 66/1987, 58/2000; Vorarlberg: 41/1985, 58/2001; Wien: 8/2000.

schutzgesetzes 1991, LGBl 115/1991 iVm LGBl 63/1997 idgF (Oö. BoSG) beschäftigen.

Die Bodenschutzgesetze der neun Bundesländer Österreichs haben inhaltlich weitgehend ähnliche Regelungen wie das hier behandelte Oö. Bodenschutzgesetz 1991.

2.5 Ziele und Grundsätze des Bodenschutzes

Ziele des Bodenschutzes sind

- Erhaltung des Bodens und
- Schutz der Bodengesundheit vor schädlichen Einflüssen, insbesondere durch Erosion, Bodenverdichtungen oder Schadstoffeinträgen sowie
- Verbesserung und Wiederherstellung der Bodengesundheit (§ 1 Abs. 1 Oö. BoSG)

Jedermann hat sich so zu verhalten, dass Beeinträchtigungen der Bodengesundheit möglichst vermieden werden (§ 1 Abs. 1 Oö. BoSG).

3 Klärschlamm, Müll- und Klärschlammkompost

Die Ausbringung von Klärschlamm auf bestimmten Gebieten oder zu bestimmten Zeiten ist absolut verboten, ansonsten bei Einhaltung der gesetzlichen Vorsichtsmaßnahmen in einem bestimmten Umfang erlaubt. Wer gegen die Verbote der Klärschlammausbringung verstößt, begeht eine Verwaltungsübertretung und wird nach Maßgabe des § 49 Oö. BoSG von der zuständigen Verwaltungsbehörde bestraft.

Grundsätzlich dürfen Klärschlamm, Müll- und Klärschlammkompost nur aus Anlagen ausgebracht werden, die im Gebiet des Landes Oberösterreich liegen (§ 3 Oö. BoSG), soweit in Staatsverträgen und in Vereinbarungen nach Art. 15a B-VG nichts anderes vereinbart ist (§ 11 Oö. BoSG). Darüber hinaus kann unter den Voraussetzungen des § 11 Oö. BoSG im Einzelfall mit Bescheid die Ausbringung von Klärschlämmen, Müll- und Klärschlammkompost auch aus Anlagen bewilligt werden, die nicht im Gebiet des Landes Oberösterreich gelegen sind.

3.1 Klärschlamm-Ausbringungsverbot

Nach den näheren Bestimmungen des § 6 Oö. BoSG ist die Ausbringung von Klärschlamm – ausgenommen aus Senkgruben und Kleinkläranlagen –, von Müll- oder Klärschlammkompost auf bestimmten Böden und mit bestimmten Pflanzen verboten.

Verboten ist die Ausbringung von Klärschlamm sowie von Müll- oder Klärschlammkompost auf verkarsteten Böden, Wiesen, Weiden, Bergmähdern, Almböden und Feldfutterkulturen sowie auf wassergesättigten oder durchgefrorenen Böden sowie auf Böden mit geschlossener Schneedecke und auf Böden, die für Gemüse-, Beerenobst- oder Heilkräuterkulturen verwendet werden.

3.2 Zulässige Ausbringung von Senkgrubeninhalten und Klärschlämmen aus Kleinkläranlagen

Die Ausbringung von Senkgrubeninhalten und von Klärschlämmen aus Kleinkläranlagen (bis 50 Einwohnerwerte) ist ausnahmsweise auf bewirtschafteten landwirtschaftlichen Kulturflächen erlaubt, wenn es sich um häusliche Abwässer und Klärschlämme mit biologischer Abwasserreinigung aus ausschließlich häuslichen Abwässern handelt (§ 7 Oö. BoSG)[8]. Die Ausbringungsmenge ist pro Hektar und Jahr beschränkt, bei Senkgrubeninhalten mit 50 m³.

Das Ausbringungsverbot auf verkarstete Böden, Almböden, auf wassergesättigten oder durchgefrorenen Böden sowie auf Böden mit geschlossener Schneedecke und weiters auf Gemüse-, Beerenobst- und Heilkräuterkulturen ist auch für Senkgrubeninhalte verboten. Unter sehr erschwerten Voraussetzungen kann auf Antrag die Ausbringung von Senkgrubeninhalten und Klärschlämmen aus Kleinkläranlagen auf Almböden und/oder verkarsteten Böden bewilligt werden.

Die Bestimmungen des § 7 Oö. BoSG enthalten weitere sehr eingehende Bestimmungen über die Ausbringung von Senkgrubeninhalten und Klärschlämmen aus Kleinkläranlagen. Unter sehr einschränkenden Bestimmungen können nach der eingehenden Regelung in § 8 Oö. BoSG auch erhöhte Ausbringungsmengen von Senkgrubeninhalten und Klärschlämmen aus Kleinkläranlagen (bis 100 m³ pro Hektar und Jahr) bewilligt werden.

3.3 Zulässige Ausbringung von Klärschlamm, Müll- oder Klärschlammkompost

Soweit kein Ausbringungsverbot für Klärschlämme, Müll- oder Klärschlammkompost besteht, setzt die Ausbringung auf Böden eine Eignungsbescheinigung für den Klärschlamm, Müll- oder Klärschlammkompost sowie ein Bodenuntersuchungszeugnis voraus.

---

8 Vgl. hiezu Neuhofer, Neuregelung der Abwasserentsorgung in Oberösterreich durch das Oö. Abwasserentsorgungsgesetz 2001, bl 2002/4, 129 ff.

3.3.1 Klärschlamm-Eignungsbescheinigung

Die Eignung des Klärschlamms sowie des Müll- und Klärschlammkomposts ist durch eine Eignungsbescheinigung der Landesregierung nachzuweisen (§ 3 Oö. BoSG). Die Eignungsbescheinigung ist auf Antrag des Betreibers von Abwasserreinigungsanlagen (und Kompostierungsanlagen) auszustellen, wenn die Eignungsvoraussetzungen des § 3 (2) Oö. BoSG erfüllt sind, insbesondere wenn der Klärschlamm stabilisiert ist, die Schadstoffe (Schwermetalle u. a.) im Klärschlamm die festgesetzten Grenzwerte nicht überschreiten und die Bodengesundheit durch den Klärschlamm nicht beeinträchtigt wird. Die Eignungsbescheinigung darf nur befristet erteilt werden. Die Klärschlammuntersuchungen müssen in bestimmten Intervallen (halbjährlich bzw. jährlich) wiederholt werden. Die Klärschlammuntersuchungen dürfen nur von den von der Landesregierung anerkannten Untersuchungsstellen durchgeführt werden.

Von 1998-2000 wurden in Oberösterreich 583 Klärschlammuntersuchungen durchgeführt. Aufgrund der Analysenergebnisse konnten 537 positive Eignungsbescheinigungen ausgestellt werden. Im weiteren weist der Bodeninformationsbericht 2001 der OÖ Landesregierung[9] aus, dass in Oberösterreich derzeit 89,2 % der anfallenden Klärschlämme zur landwirtschaftlichen Verwertung geeignet sind.

3.3.2 Bodenuntersuchungszeugnis

Die Eignung des Bodens zur Ausbringung von Klärschlamm ist durch ein Bodenuntersuchungszeugnis der Landesregierung nachzuweisen (§ 4 Oö. BoSG).

Vor der ersten Ausbringung von Klärschlamm bzw. Müll- oder Klärschlammkompost hat der Nutzungsberechtigte der betreffenden Grundstücke eine repräsentative Bodenuntersuchung durchführen zu lassen. Die Bodenuntersuchung ist vor einer Klärschlammausbringung zu wiederholen, wenn die letzte Bodenuntersuchung über 10 Jahre zurück liegt oder wenn seit der letzten Bodenuntersuchung 15 t Klärschlamm pro Hektar ausgebracht wurden. Nach den weiteren Regelungen in § 4 Oö. BoSG sind für die Durchführung der Bodenuntersuchungen umfangreiche Analysen der Bodenproben vorgeschrieben.

Die Entnahme von Bodenproben hat durch sachverständige Organe der Behörde zu erfolgen (§ 42 Abs. 5 Oö. BoSG); die Bodenuntersuchung darf nur durch die von der Landesregierung anerkannten Untersuchungsstellen erfolgen (§ 46 Oö. BoSG).

---

9 Bodeninformationsbericht 2001 der OÖ Landesregierung, Seite 19.

Die landwirtschaftliche Verwertung des Klärschlamms zur Düngung zeigt in Oberösterreich eine steigende Tendenz: im Jahr 1999 wurden in Oberösterreich 35 % der angefallenen Klärschlämme auf 6.129 ha landwirtschaftliche Nutzflächen ausgebracht; dies entspricht einer durchschnittlichen Klärschlammmenge von 2,18 t Trockenmasse pro Hektar und Jahr[10].

3.3.3 Kostentragung der Klärschlamm- und Bodenuntersuchung

Die Kosten für die Untersuchung des Klärschlamms und des Bodens trägt der Betreiber der Abwasserreinigungsanlage bzw. der Kompostierungsanlage nach den näheren Bestimmungen des § 12 Oö. BoSG; im Streitfall entscheidet auf Antrag des Nutzungsberechtigten der Ausbringungsfläche die Landesregierung.

4 Überdüngungsverbot

Bei der Düngung der Böden ist nach den näheren Bestimmungen des § 14 Oö. BoSG der Versorgungszustand des Bodens zu beachten; Überdüngungen sind zu vermeiden. Bei der Düngung des Bodens ist auf den Schutz von Oberflächen- und Grundwässern Bedacht zu nehmen. Die Ausbringung von Gülle (Jauche) ist nach den näheren Bestimmungen des § 15 Oö. BoSG mengenmäßig und zeitlich beschränkt. Die Ausbringung von Gülle (Jauche) ist auf wassergesättigtem und durchgefrorenem Boden, bei geschlossener Schneedecke, auf Böden mit Gemüse-, Beerenobst- und Heilkräuterkulturen, auf Almböden und auf verkarsteten Böden verboten.

Es dürfen nur Düngemittel verwendet werden, die nach dem Düngemittelgesetz, BGBl 23/2001, zugelassen sind.

5 Sachkundige Anwendung von Pflanzenschutzmitteln

Es dürfen nur die nach dem Pflanzenschutzmittelgesetz 1997, BGBl I 60/1977, zulässigen Pflanzenschutzmittel verwendet werden und es dürfen nur funktionstüchtige Pflanzenschutzgeräte verwendet werden (§§ 18 und 19). Pflanzenschutzmittel dürfen nur von sachkundigen Personen angewendet werden; der Sachkundenachweis ist zu erbringen durch eine gewerbebehördliche Gewerbeberechtigung, durch Zeugnis einer Landwirtschaftsschule oder eines Ausbildungskurses der Landwirtschaftskammer (§ 17 Oö. BoSG).

---

10 Bodeninformationsbericht 2001 der OÖ Landesregierung, Seite 21.

6 Flächendeckende Bodenuntersuchung des gesamten Landesgebietes – Bodenkataster

Die Landesregierung hat nach den näheren Bestimmungen des § 22 Oö. BoSG für das gesamte Landesgebiet mit einem flächendeckenden Raster von Prüfstandorten Bodenuntersuchungen durchzuführen und die Untersuchungsergebnisse im Oö Bodenkataster[11] festzuhalten; jedermann kann in diesen Oö Bodenkataster Einsicht nehmen.

Die Landesregierung hat durch Verordnung die zu untersuchenden Stoffe im Boden sowie deren Grenzwerte in einer „Bodengrenzwerteverordnung" festzulegen. Die flächendeckenden Bodenuntersuchungen in Oberösterreich haben ergeben, dass sich die landwirtschaftlich genutzten Böden des Landes Oberösterreich großteils in einem zufriedenstellenden bis guten Zustand befinden[12]. Punktuelle Nährstoffungleichheiten sollen nach einer Empfehlung der Bundesanstalt für Agrarbiologie (Linz) durch geeignete Maßnahmen behoben werden; empfohlen werden auch Bodendauerbeobachtungsflächen.

Wird bei einer Bodenuntersuchung festgestellt, dass Bodengrenzwerte überschritten sind oder sonst die Bodengesundheit (z.B. durch Erosion oder Bodenverdichtung) beeinträchtigt ist, hat die Landesregierung zusätzliche Bodenuntersuchungen durchzuführen (§ 25 Oö. BoSG) und Bodenverbesserungen anzuordnen oder vorzunehmen.

7 Bodenverbesserungsmaßnahmen

Wird bei Bodenuntersuchungen eine Beeinträchtigung der Bodengesundheit (z.B. durch Überschreitung der Grenzwerte, durch flächenhafte Erosionen, flächenhafte Bodenverdichtung) festgestellt und ist die Verbesserung der Bodengesundheit im öffentlichen Interesse notwendig, hat die Bezirksverwaltungsbehörde dem Nutzungsberechtigten des Bodens die Vorlage eines Bodenverbesserungsplanes nach Maßgabe des § 27 Oö. BoSG aufzutragen.

Legt der Nutzungsberechtigte einen Bodenverbesserungsplan nicht oder nicht zeitgerecht vor oder ist der Bodenverbesserungsplan nicht ausreichend, hat die Bezirksverwaltungsbehörde die erforderlichen Bodenverbesserungsmaßnahmen mit Bescheid vorzuschreiben; bodenverbessernde Maßnahmen sind nicht vorzuschreiben, wenn sie unverhältnismäßig sind.

Ist die Wiederherstellung der Bodengesundheit in einem größeren Gebiet notwendig, sind die bodenverbessernden Maßnahmen durch eine Verordnung der Landesregierung anzuordnen.

---

11 S. Oö Bodenkataster 1993 mit einer eingehenden Darstellung der Bodenzustandsinventur; derzeit wird der Oö Bodenkataster 1993 überarbeitet.

8 Nutzungsbeschränkungen – Entschädigung

Ist die Bodengesundheit in einem solchen Ausmaß beeinträchtigt, dass der Boden für die Produktion von Ernährungs- und Futterpflanzen ungeeignet (gesundheitsgefährdend) ist, hat die Bezirksverwaltungsbehörde nach Maßgabe des § 28 Oö. BoSG die Bodennutzung für die Produktion von Nahrungsmittel und Futterpflanzen zu verbieten (Nutzungsverbot) oder zu beschränken (Nutzungsbeschränkungen).

Der Nutzungsberechtigte hat nach Maßgabe des § 29 Oö. BoSG gegenüber dem Land einen Anspruch auf Entschädigung. Die Entschädigung „ermäßigt sich insoweit, als der Nutzungsberechtigte oder sein Rechtsvorgänger die Beeinträchtigung" ... verursacht haben; die Beweislast obliegt dem Land. Im Streitfall obliegt die Entscheidung dem zuständigen Gericht.

9 Bodenentwicklungsprogramm

Alle drei Jahre hat die Landesregierung dem Landtag einen „Bodeninformationsbericht" und auf dieser Grundlage ein „Bodenentwicklungsprogramm" vorzulegen.

Im Bodenentwicklungsprogramm 2001[12] wird zunächst die Fortsetzung der bisherigen Bodenverbesserungsmaßnahmen empfohlen; im weiteren werden als vordringliche Bodenverbesserungen genannt: weitere Bodenzustandserhebungen, Bodendauerbeobachtungsflächen, Pilot- und Versuchsprojekte, Verbesserung der Stoffbilanzierung in der landwirtschaftlichen Produktion, ein Grundwasserschutzprogramm usw.

10 Bodenschutzberatung und Bodenschutzförderung

Für die Erhaltung und Verbesserung der Bodengesundheit sind die Bodenschutzberatung (§ 35 Oö. BoSG) sowie die Bodenschutzförderung (§ 36 Oö. BoSG) besonders wichtig.

Die Landwirtschaftskammer für Oberösterreich wird gesetzlich verpflichtet, einen Beratungsdienst für den Bodenschutz einzurichten; die Kosten hiefür trägt das Land Oberösterreich.

Das Land Oberösterreich fördert Maßnahmen der Bodenverbesserung, insbesondere Pflanzenschutzmaßnahmen, biologischen Landbau, bodenschonende Bewirtschaftungsgeräte, Kompostierung von Abfällen, Verbesserung der

---

12 Vgl. im Näheren Bodeninformationsbericht 2001, S. 52 ff.

Klärschlammqualität, Gewinnung und Verwendung von Müll- und Klärschlammkompost usw. Auf die Förderung des Landes besteht kein Rechtsanspruch. Die Förderung der Bodenverbesserung erfolgt im Rahmen der Privatwirtschaftsverwaltung des Landes. Gegen die Ablehnung eines Antrages auf Förderung gibt es kein Rechtsmittel. Das Land Oberösterreich hat von 1998 bis 2000 Bodenschutzmaßnahmen mit dem Betrag von 1,5 Mio Euro gefördert[13].

11 Altlastensanierung

Durch Altlasten, nicht ausreichend gesicherte Abfalldeponien sowie durch die nicht ausreichend gesicherten Betriebsanlagen gibt es in verschiedenen Gebieten Österreichs eine größere Anzahl von verseuchten Böden. Das Altlastensanierungsgesetz – AlSAG, BGBl 299/1989 idgF[14], soll die Sicherung und Sanierung von verseuchten Böden sowie deren Finanzierung sicherstellen[15].

11.1 Altlasten

Altlasten sind nach der Begriffsbestimmung des § 2 (1) AlSAG Altablagerungen (von Abfällen) und Altstandorte (von Anlagen, in denen mit umweltgefährdenden Stoffen umgegangen wurde) sowie durch diese kontaminierte Böden und Grundwasserkörper, von denen erhebliche Gefahren für die Gesundheit von Menschen oder die Umwelt ausgehen; Kontaminierungen durch Luftschadstoffe sind aus dem Geltungsbereich des AlSAG ausgenommen.

11.2 Erfassung der Altlasten im „Verdachtsflächenkataster" und im „Altlastenkataster"

Der Landeshauptmann hat die Altlasten-Verdachtsflächen dem Bundesminister für Land- und Forstwirtschaft, Umwelt und Wasserwirtschaft (BMLFUW) bekannt zu geben; dieser hat die bekannt gegebenen Altlasten bundesweit zu erfassen, abzuschätzen und zu bewerten. Aufgrund der Überprüfungen durch

---

13 Vgl. hiezu im Näheren den Oö Bodenschutzbericht 2001, S. 27.

14 Altlastensanierungsgesetz, BGBl 299/1989 idF 325/1990, 760/1992, 185/1993, 818/1993, 201/1996, I 96/1997, I 151/1998, I 26/2000, I 142/2000, I 27/2001, I 63/2001, I 48/2002, I 155/2002 und I 71 Art. 67/2003.

15 Vgl. zu den neueren Entwicklungen im Altlastensanierungsgesetz, Zahrer, Aktuelle abfallwirtschaftliche Entwicklungen, ÖGZ 2001/1, 8 f.

den genannten Bundesminister hat das Umweltbundesamt ein Verdachtsflächenregister zu führen (§ 13 Abs. 1 AlSAG).

Aufgrund der Gefährdungsabschätzung der Altlasten durch den genannten Bundesminister hat das Umweltbundesamt einen Altlastenkataster zu führen (§ 13 Abs. 2 AlSAG). Der Landeshauptmann hat die Grundstückseigentümer von der Eintragung ihrer Grundstücke als Altlast im Altlastenkataster zu verständigen.

Der Bundesminister für Land- und Forstwirtschaft, Umwelt und Wasserwirtschaft hat jedermann auf Antrag Auskunft darüber zu geben, ob eine Liegenschaft im Verdachtsflächenkataster eingetragen ist und welcher Art die Verdachtsfläche ist (§ 13 Abs. 4 AlSAG). Der Altlastenkataster liegt beim Bundesminister für Land- und Forstwirtschaft, Umwelt und Wasserwirtschaft sowie bei den Ämtern der Landesregierungen zur öffentlichen Einsicht auf. Nach einem Bericht des Bundesministers für Land- und Forstwirtschaft, Umwelt und Wasserwirtschaft wurden zum 1.1.2003 ausgewiesen:

- 2.441 Flächen im Verdachtsflächenkataster;
- 166 Verdachtsflächen als Altlasten im Altlastenkataster;
- 57 Flächen, die unter Beobachtung gestellt wurden;
- 56 Altlasten werden derzeit gesichert oder saniert und weitere
- 42 Altlasten sind als gesichert oder saniert im Altlastenkataster ausgewiesen;
- 28 Verdachtsflächen sind als saniert oder gesichert ausgewiesen.

11.3 Prioritätenklassifizierung

Der Bundesminister für Land- und Forstwirtschaft, Umwelt und Wasserwirtschaft hat die gemeldeten Altlasten nach einer eingehenden Prüfung nach ihrem Gefährlichkeitsgrad und dem sich daraus ergebenden Umfang sowie der Dringlichkeit und Finanzierung erforderlichen Sicherungs- und Sanierungsmaßnahmen eine Prioritätenklassifizierung zu erstellen und fortzuschreiben (§ 14 AlSAG). Die Prioritätenklassifizierung ist im Bundesministerium für Land- und Forstwirtschaft, Umwelt und Wasserwirtschaft sowie in den Ämtern der Landesregierungen zur öffentlichen Einsicht aufzulegen.

11.4 Sicherung und Sanierung der Altlasten

Sofern nicht einem Verpflichteten (nach dem Wasserrechtsgesetz, nach der Gewerbeordnung oder nach dem Abfallwirtschaftsgesetz 2000) die Sicherung und Sanierung von Altlasten aufgetragen werden kann, führt der Bund die

---

16 Bericht des Bundesministeriums für Land- und Forstwirtschaft, Umwelt und Wasserwirtschaft, S. 140.

erforderlichen Sicherungs- und Sanierungsmaßnahmen nach Maßgabe der Prioritätenklassifizierung durch (§ 18 Abs. 1 AlSAG). Dem Bund dürfen hiebei über den Ertrag der Altlastenbeiträge hinaus Belastungen nicht entstehen.

Wer rechtswidrig und schuldhaft die Altlasten verursacht hat oder als Eigentümer der Liegenschaft der altlastverursachenden Ablagerungen zugestimmt oder geduldet hat, hat dem Bund die Kosten für die Sicherung und Sanierung der Altlasten entsprechend zu ersetzen (§ 18 Abs. 2 AlSAG). Besteht das Verschulden des Ersatzpflichtigen nur in einem Versehen, kann das Gericht den Ersatz mäßigen oder auch ganz erlassen.

Der Liegenschaftseigentümer der Altlast sowie die dinglich oder obligatorisch Berechtigten haben die Sicherungs- und Sanierungsarbeiten nach Maßgabe der §§ 16 und 17 AlSAG zu dulden.

11.5 Finanzierung der Sicherungs- und Sanierungsmaßnahmen durch Altlastenbeiträge

Das Altlastensanierungsgesetz soll durch die Einhebung von Altlastenbeiträgen die Finanzierung der Sicherungs- und Sanierungsmaßnahmen gewährleisten (§ 1).

Nach den Bestimmungen des § 4 iVm § 3 AlSAG haben folgende Personen Altlastenbeiträge zu entrichten:

- Personen, die Abfälle ablagern (z.B. auf einer Deponie oder auf einem Zwischenlager oder Geländeunebenheiten mit Abfällen auffüllen);
- Personen, die Abfälle in einer Verbrennungsanlage verbrennen;
- Personen, die aus Abfällen Brennstoffprodukte herstellen;
- Personen, die Abfälle zur Ablagerung oder Behandlung in das Ausland befördern und
- Personen, die „beitragspflichtige Tätigkeiten" der Ablagerung, der Verbrennung oder des Verfüllens von Geländeunebenheiten veranlasst haben.

Die Höhe der Abfallbeiträge bestimmt sich nach der Menge der Abfälle und nach dem technischen Zustand der Deponie und nach dem Zeitpunkt der beitragspflichtigen Tätigkeiten. Für ältere Deponien, die nicht dem letzten Stand der Technik entsprechen, müssen Zuschläge bezahlt werden. Die Abfallbeiträge sowie die Zuschläge sind relativ hoch. Die nachfolgenden Tabellen des Bundesministeriums für Land- und Forstwirtschaft, Umwelt und Wasserwirtschaft geben einen Überblick über die Beiträge und Zuschläge:

Beitragssätze gemäß Altlastensanierungsgesetz für die Ablagerung auf Deponien, die noch nicht dem Stand der Technik entsprechen:

in ATS, je angef. Tonne	Baurestmassen	Erdaushub [1]	kontaminierter Erdaushub [2]	Übrige Abfälle [3,4]
ab 1. Jänner 2001	100 (€ 7,20)	100 (€ 7,20)	200 (€ 14,50)	600 (€ 43,60)
ab 1. Jänner 2004			300 (€ 21,80)	900 (€ 65,00)
ab 1. Jänner 2006				1200 (€ 87,00)
Zuschlag (ATS/t). [1]30 (€ 2,10), [2]200 (€ 14,50), [3]400 (€ 29), wenn kein ausreichendes Dichtungssystem oder vertikale Umschließung vorhanden ist.				
[4]Zusätzlich 400,– ATS/t (€ 29) für die Ablagerung auf einer Deponie mit der Bewilligung zur Ablagerung von Hausmüll oder hausmüllähnlichen Gewerbeabfällen, die über keine, dem Stand der Technik entsprechende Deponiegaserfassung und -behandlung verfügt.				

Beitragssätze gemäß Altlastensanierungsgesetz für Deponien, die dem Stand der Technik entsprechen:

in ATS, je angef. Tonne	Baurestmassendeponie	Reststoffdeponie	Massenabfalldeponie
ab 1. Jänner 2001	80 (€ 5,80)	150 (€ 10,90)	200 (€ 14,50)
ab 1. Jänner 2004	100 (€ 7,20)	200 (€ 14,50)	300 (€ 21,80)

Der Beitragsschuldner hat die Altlastenbeiträge vierteljährlich zu berechnen sowie zeitgerecht dem Hauptzollamt zu melden und zu entrichten (§ 9 AlSAG).

Die Einnahmen aus den Altlastenbeiträgen betrugen für die Jahre 1990 – 2000 4,4 Milliarden ATS (= ca. 319 Millionen Euro)[16]. Durch die Erhöhung der Altlastenbeiträge ab 1.1.2004 und in weiterer Folge ab 1.1.2006 ist mit einer starken Steigerung der Einnahmen aus den Altlastenbeiträgen zu rechnen.

Maruška Šubic Kovač

Legal Basis of Spatial Planning and Environmental Protection in Slovenia

1. Introduction

Environmental protection and spatial development are inseparably linked. The scope, degree and distribution over time of the degrading environmental processes depend on a number of factors, inter alia , on spatial development characteristics, settlement development pattern in particular, spatial arrangement of economic activities and infrastructure, and land use pattern. Nevertheless, just the opposite applies as well. Changes in environmental trends and conditions in the environment as well as its composite parts exert the inverse impacts on the characteristics and spatial development guidelines.

In the past, environmental protection objectives in Slovenia in the field of spatial planning used to be in a subordinated position. As the degree and scope of environmental pollution have in the numerous zones put at risk even the realisation of objectives of spatial planning and development policy, the demand for an equal approach at inclusion of environmental objectives in the developmental spatial planning has become more and more important and necessary.

In the beginning of the 1990'ies, changes in the relationship between environmental protection and spatial development and/or spatial planning in Slovenia were brought about also by the essential changes in the society, the process of approximation to the environmentally more aware European Union, and by the gradual changes in the global developmental scenarios and/or visions (Agenda 21, Agenda Habitat, Kyoto Protocol, and similar). Slovenia has ratified numerous international conventions, international agreements and other instruments, undertaking thereby the commitment to implement on an equal basis the objectives of environmental protection at spatial development planning in the country. The Constitution and legislation in the field of environmental protection and spatial planning needed to be adequately adapted.

This article will analyse:

- some constitutional provisions and parts thereof (Constitution of the Republic of Slovenia, UL RS 33/1991-I), which represent the basis of environmental protection and spatial planning in the Republic of Slovenia,
- fundamental principles of environmental protection in Slovenia, operationalised in the Environment Protection Act (UL RS 32/1993),
- some other important laws, strategies and national programmes in the direct connection with the topic of this article,

- method of adaptation of developmental needs to environmental protection as laid down in the framework law governing spatial planning, i.e. the Spatial Planning Act (UL RS 88/2002).

2. Constitution as the basis for spatial planning and environmental protection

Present understanding of the quality of life in the economically more developed countries includes not only the indexes of material welfare but also social safety, the possibility of equal access to knowledge, freedom, human rights, and a possibility of living in a healthy environment of the present and future human generations and other species.

The basic aims of environmental protection in Slovenia are defined in the Constitution of the Republic of Slovenia (UL RS 33/1991). In the light of environmental protection and spatial planning the following parts of the Constitution are of major importance:

1. In the General Provisions, the Constitution of the Republic of Slovenia indicates the positive obligations of the state, which inter alia include the obligations to preserve natural resources and cultural heritage:

> Article 5: The state … shall provide for the preservation of the natural wealth and cultural heritage and create opportunities for the harmonious development of society and culture in Slovenia.

Positive obligation ensues from the judicature of the European Court of Human Rights and obligates the state to a minimum of active handling.

2. As regards property, the Constitution lays the foundation that property must in addition to the individualistic function, i.e. the realisation of a person's financial freedoms, fulfil also its function for the entire social community. Social community as a whole has a right to impose on the proprietor certain limitations concerning property freedoms:

> Article 67: The manner in which property is acquired and enjoyed shall be established by law so as to ensure its economic, social and environmental function.

The legislator's task is to preserve the balance between the individualistic function of property (realisation of a person's financial freedoms) and the public interests in the field of economy, social relations and environmental protection. Any limitation of property freedoms shall therefore be permissible, provided that it is targeting an aim ensuing from Article 67 (economic, social and environmental function of property). If we paraphrase it, any limitation of property freedoms shall be permissible only, when the limitation of property freedom is adequate, indispensable and in proportion to the benefit that it conveys. The principle of proportionality under the constitutional law shall therefore apply.

Constitutional provisions under Articles 70 to 73 and in part under Articles 67, 69 and 74 impose on the legislator to regulate by law:

- the method of property acquisition and usufrucht so as to maintain its environmental function,
- conditions for the exploitation of natural resources,
- conditions for the acquisition of special rights to use national assets, which include national assets as well as built national assets ,
- conditions of land use,
- method of realisation of the right to a healthy living environment,
- requirements for and scope of compensation that a person, who has damaged the living environment is obliged to provide, and
- method of safeguarding the natural monuments and curiosities.

These constitutional provisions, and Articles 70 to 73 in particular, may be defined as provisions on environmental protection. Obligation of the state is to provide for the maintenance of natural resources and to create the possibilities for a sustainable development in the civilisation as well as culture. Important for environmental protection are those provisions in the Chapter on Economic and Social Relations that refer to the obligation of providing for the environmental function in property acquisition and usufrucht, to the prohibition of carrying out an economic activity in contrast to public benefit, to a healthy living environment, national assets and natural resources, and to the protection of land as well as natural heritage.

3. The Constitution lays down in Article 70:

   Article 70: Special rights to use national assets may be acquired, subject to conditions established by law.
   The conditions under which natural resources may be exploited shall be established by law.
   The law may provide that natural resources may also be exploited by foreign persons and shall establish the conditions for such exploitation.

Subject of protection under this constitutional provision are the national assets and natural resources, whilst the purpose of protection is mainly in the provision of a healthy living environment. National assets (natural in particular) and natural resources are namely part of the ecosystems, i. e. natural communities of living organisms, including humans.

In defining the term of natural resources, the Environment Protection Act (UL RS 32/1993, hereinafter referred to as: EPA) takes as the basis the term of natural goods. Natural goods are constituents of nature, important in order to satisfy man's material and spiritual needs and interests, and may be exclusively or simultaneously natural assets, natural sources or natural worth. Rare, precious or valuable natural goods are natural resources.

National assets may appear as natural good or as natural resource. Naturally, this applies to the natural national assets, which are defined as the surface in public property which is, as the unbuilt part of the ground, underground, freshwater, sea and air national assets, accessible to anybody and under equal conditions.

4. Article 71 of the Constitution lays down land protection:

> Article 71: The law shall establish special conditions for land utilisation in order to ensure its proper use.
> Special protection of agricultural land shall be provided by law.
> The state shall promote the economic, cultural and social advancement of the population living in mountain and hill areas.

The Constitution obligates the legislator to lay down the conditions applicable to the use of land designated for the purposeful utilisation, and to regulate the specific protection of agricultural land. Furthermore, the legislator shall impose on the state the promotion of economy in the mountainous and hilly areas.

This Article applies to all types of land: agricultural land, building land, land under water, and other. Protection of agricultural land is warranted under the Agricultural Land Act (UL RS 59/1996) that defines and classifies agricultural land and regulates its use, safeguards in case of land utilisation change, and trade in land. It regulates also the agrarian operations intended to improve agricultural land or to improve the conditions of its cultivation. Spatial Planning Act (UL RS 88/2002) does not define the term building land as it has been defined in the preceding legislation in this field, but mentions only a building plot, i.e. land consisting of one or more plots of land or pertaining parts, where a facility has been constructed or in envisaged for construction, and which includes developed surface to serve such a facility, or where land development is envisaged in order to serve such a facility. The Act regulates spatial planning and the enforcement of implementation measures in order to implement the planned spatial arrangements, ensures building land development, and the management of the spatial planning database system. Under the Forest Act (UL RS 30/1993), forests are defined as woodland alone or with undergrowth, which provides any function of a forest: environmental, social or productive. Forests are defined as natural resource, and as regards their preservation or introduction and strengthening, the Forest Act regulates their protection, cultivation, utilisation, use and availability for handling.

5. As regards the healthy living environment, Article 72 lays down:

> Article 72: Everyone has the right in accordance with the law to a healthy living environment.
> The state shall promote a healthy living environment. To this end, the conditions and manner in which economic and other activities are pursued shall be established by law.
> The law shall establish under which conditions and to what extent a person who has damaged the living environment is obliged to provide compensation.

In Article 72, the Constitution first ensures the right of any person to a healthy living environment. Under EPA (Article 5), living environment means any parts of nature, which are or could be within the reach of human activity. In this respect, natural environment is part of the primary nature, transformed by man, and living environment is part of the environment under the direct influence of man. Under EPA, built and other technical environment is part of environment as the environment change factor only. It is up to the legislator to define precisely what should be considered a healthy living environment.

The Constitution obligates the state to provide for the healthy living environment. It stipulates that the conditions and manner in which economic and other activities are pursued should be established by law. In this respect, it must take into account the method of property acquisition and usufrucht so as to ensure the environmental function of property as well as its economic and social function. Article 72 makes reference also to Article 74, paragraph 2, prohibiting the carrying out of an economic activity that is contrary to public benefit. Healthy living environment should belong to such public benefit as well.

Although the right to a healthy living environment belongs to the Economic and Social Relations Chapter, it may be defined as the constitutional right of a positive status: it pertains to anyone and the state is obligated to provide it. Environmental protection is a prerequisite in order to realise the right to a healthy living environment.

Obligations of the state ensuing from this provision or from its obligation to provide a healthy living environment are three:

- it shall regulate by law the substance and scope of the right to a healthy living environment,
- it shall regulate by law the conditions applicable to carrying out economic and other activities,
- it shall regulate by law the conditions for and scope of compensation to be provided by a person, who has damaged the living environment.

Slovenia included in this Article the basic international principle of environmental protection that is part of the so-called Rio Declaration[1]. Among the

1 Rio Declaration on Environment and Development, UN Conference on Environment and Development, Rio de Janeiro, 1992.

27 principles on sustainable development, the Declaration indicates in its Principle 16 that the polluter should, in principle, bear the cost of pollution. Slovenia has implemented this principle in the Slovenian legal order within EPA. The essence of requirement for sustainable development is in the provision of equal proportions between the developmental needs of an individual, the society as a whole and the need to preserve the environment. To this end, the legislator defined in EPA the basic objectives of environmental protection and the means and/or ways to achieve them.

The legislator provides in EPA the right to a healthy living environment in particular by defining the permissible limits of interventions into the environment. Under the Act (Article 5), intervention into the environment means any permanent and temporary human activity or omission of activity which, through its impacts, may pose a risk to or endanger health or environment and, as a consequence, bring about an artificial change of the environment, environmental nuisance or hinder its natural changes, and it applies in particular to:

- exploitation and use of natural good,
- interventions into the environment,
- products and other activities,
- trade in and use of goods,
- emissions into water, air or soil, waste disposal and accumulation, and other impacts on environment.

6. Article 73 defines the protection of natural and cultural heritage:

   Article 73: Everyone is obliged in accordance with the law to protect natural sites of special interest, rarities and cultural monuments.
   The state and local communities shall promote the preservation of the natural and cultural heritage.

The above Article 73 obligates anyone to protect natural sites of special interest, rarities and cultural monuments in accordance with the law. Protection of natural and cultural heritage is regulated under the Cultural Heritage Protection Act (CHPA, UL RS 7/1999) and the Nature Conservation Act (NCA, UL RS 56/1999). Nature Conservation Act defines natural heritage as natural worth, lays down a system of protection of natural worth, which includes procedures and methods of acknowledging the status of natural worth as well as implementing the protection thereof. Natural worth is defined as a rare, valuable or precious natural phenomenon and any other valuable phenomenon or constituent of the animate or inanimate nature.

7. Article 74 refers to free enterprise:

   Article 74: Free economic initiative shall be guaranteed. The conditions for establishing commercial organisations shall be established by law.

Commercial activities may not be pursued in a manner contrary to the public interest.

What is important in this Article for environmental protection is that any economic activity shall not be carried out contrary to public benefit. Nevertheless, limitations to free enterprise owing to public benefit shall be permissible, when they are in accordance with the principle of proportionality: an intervention shall be indispensable, when establishing the proportional balance between the curtailment of a constitutional right of an individual and the public benefit obtained thereby.

The Constitution of the Republic of Slovenia includes the basic elements of sustainable development, in particular in Article 67, laying down the appropriate spatial development planning.

3. Operationalisation of the basic principles of environmental protection in the environment protection act

Environment Protection Act (UL RS 32/1993) is the framework law in the sector of environmental protection and operationalises the basic principles of environmental protection in Slovenia[2]. It has unified the environmental protection principles and transferred the application of solely technical solutions for the restriction of pollution (end of the pipe technologies) within the context of provision of sustainable development.

Important for spatial planning are in particular the following principles, where the provisions of the Act are based upon:

Principle of integrity requires from the state and the narrower territorial units (municipalities and provinces) that at planning and programming, adopting regulations, rendering promotion and relief, issuing permits and adopting tax policy or public financing policy, control and other measures, an assessment of their impacts on environment should be made.

Principle of cooperation requires that at regulating and deciding on environmental protection issues a prior cooperation of all the involved and interested parties should be provided for.

Principle of preventive requires that the rules of handling, limit values and other standards should be designed, and each intervention in environment planned and implemented so as to cause the minimum possible change of environment and the minimum possible risk for environment, and so as to

2 Environmental protection in Slovenia is covered by numerous, mainly sectoral laws, which shall not be dealt with in this article owing to the extensiveness of topics.

minimise the use of space, substances and energy in the construction, production, trade and consumption and, thereby, to minimise environmental nuisance already at its origin.

Principle of responsibility of the person causing environmental nuisance is laying down that the person, who causes excessive environmental pollution should be held responsible under the penal law and should be liable to pay an indemnity. The same responsibility applies to the person, who has made it possible for the polluter or allowed him through illegal or irregular action to cause excessive pollution.

Principle of payment for environmental nuisance requires that the polluter should cover the total expenses of environmental nuisance caused.

Principle of promotion imposes on the state or local community to promote, within their respective competence, those interventions in the environment, which reduce the exploitation of the productive potentials of environment, reduce the consumption of substances and energy, prevent and restrict environmental nuisance below the level of permissible limits.

Principle of protection of rights is based on the obligation of everyone that, when deciding on an intervention in the environment and during the intervention itself, takes all the prescribed and other reasonable measures to provide for the realisation of the constitutional right of an individual to a healthy and clean living environment without any requirement of a compensation of any kind. Citizens as individuals or their societies, associations and organisations may by legal action require the competent court to order the holder of activity to suspend the intervention, when it may cause or has caused an immediate risk for the environment, a critical environmental nuisance of damage to the environment, or when it may cause or has caused an immediate public life or health risk, or to prohibit any beginning of interventions in the environment, where there is a great probability of causing such effects.

Environment Protection Act is based on the paradigm of sustainable development. It introduces three different forms of environmental assessment, which are the preventive forms of protection of environment and space, as follows:

- environmental impact assessment for the strategic decisions concerning spatial development,
- strategic impact assessment on environment in order to ensure the adequacy of spatial and sectoral plans, and
- assessments of impacts on environment at major projects.

Level of impact should be defined without the inclusion of the presupposed impacts of planned activities on the environment. Therefore, the environmental impact assessment represents only a very general basis for the spatial planning and is less fit for use in the decision-making in the concrete cases. Besides, it does not assess space by presenting the different levels of sensitivity of space, which would enable a comparison of the different variants of interventions in space (Plut, 2002, 109).

The Act focuses in particular on the solving and preventing problems ensuing from environmental pollution. Curative aspects do not get any special attention. It does mention the improvement of sanitary conditions, but mainly as measures to prevent problems that might be caused by future interventions in space.

In 1993, the Act laid the foundation of the modern environmental protection in Slovenia. In its preparation, a number of principles and institutes of the international environment protection treaties and declarations[3] were incorporated in it, as well as the fundamental principles and some requirements of Community legislation. The latter were not fully incorporated and were mainly intended as guidelines and not as legal commitments. When Slovenia decided to accede the European Union as a full Member State, and by adoption of several sectoral Acts in the Republic of Slovenia[4], and by analysis of this field in other countries[5], there arose the need for the adoption of a new Environment Protection Act, which is already in preparation[6].

4. Some other sectoral acts, strategies and national programmes in this field

In 1999, the Nature Conservation Act (UL RS 56/1999) was passed, which is based on the constitutional principles of sustainable approach or the sustainable conservation of nature[7]. By this Act, the field of nature conservation obtained a fully independent role not only in spatial assessment, but also in

3 UN Declaration on Environment and Development of Rio de Janeiro (1992), Basel Convention, Convention on Climatic Changes, Convention on Bio-Diversity.

4 Forest Act, Mining Act, Water Act, Marine Fisheries Act, Nature Conservation Act.

5 Environment Protection Act of 1993 simply nationalised the then existing social property on the individual natural sources (water, wild game and similar). In the present Draft Act, the definition of property on the nationalised natural sources is omitted. Environment Protection Act of 1993 introduced the term of ecological annuity, under which certain individuals should be entitled to a compensation owing to the fact alone that an intervention into environment has been executed, whether or not any damage has been caused thereby. The new Act will omit such a definition.

6 Draft Environment Protection Act, www.gov.si/mop.

7 Name of the Act is inadequate. It refers to the conservation of nature, and not to the protection of nature. Protection of nature is part of environmental protection, more accurately, it represents the protection of the natural environment. Conservation of nature is a term referring to the non-alteration of nature. Nature in itself is a dynamic system.

spatial planning. The Act provides directions for the harmonisation of different interests in space as well as the compulsory bases to be included in the spatial plan. As other sectors, also this sector tried to ensure by the law its competence for the decision making on the spatial development in certain areas, which is contrary to the presupposition that spatial planning represents an area, where different interests in space are harmonised.

Economic Development Strategy of 2001[8] is a strategic instrument of the state that by problem analysis defines the factors of economic development, long-term goals and targeted development scenario, and the basic directions of state operations in the individual fields. It is based upon the equal treatment of the economic, social and environmental extensions of welfare, thus providing a basis of integration of developmental activities. Sustainable development is defined on this basis. A result of sustainable development is therefore a balanced improvement of all these constituents of welfare. Welfare improvement is not restricted to the individual areas only in the country, and therefore, in addition to the social and environmental constituents of development, a particular focus is on the sustainable regional development. Strategy includes some principles of sustainable development, and space and environment are treated as a developmental and limiting factor at the same time. As proven already in the past[9], this strategy will, owing to its sustainably oriented substance, also represent a strong tool in the fight against the possible investment megalomania in Slovenia.

Economic Development Strategy from 1996 to 2001 was formally and legally less important than the sectoral national programmes and other strategies. When the over-ambitious and often environmentally disputable sectoral appetites needed to be limited, it was a strong tool of the non-governmental organisations (NGOs) in the fight against the continuation of energy policy as a sum of construction projects of power supply companies. It contributed to the non-adoption of the environmentally inadequate variant of the National Energy Programme , as the investment appetites were constantly confronted with the macroeconomic possibilities defined in the strategy.

By adoption of the National Programme of Environmental Protection (UL RS 83/1999) at the Government of the Republic of Slovenia in December 1998, the basic aims of environmental protection were confirmed as well as the methods of their realisation. Basic aim of the National Programme of Environmental Protection is a better living environment and the enforcement of environment as the limiting and promotional factor of development. Sustainable development is a priority of this programme, which is based on two levels:

---

8 www.gov.si/umar.

9 Environmental aspects and aspects of sustainable development in the economic development strategy of the Republic of Slovenia by 2006, www.gov.si/umar/

a) At the first level, aims and tasks are defined, which are mainly within the competence of the state, and tasks, the implementation whereof is closely connected with the tasks required by the approximation of laws to the European Union[10].
b) At the second level, directions are defined for the integration of environmental protection into the individual sectors, in accordance with the principle of responsibility.

Programme includes aims, directions, strategies of environmental protection and use of natural sources for a period of minimum 10 years and represents the fundamental strategic basis of spatial planning.

5. Spatial planning act

By 1 January 2003 there were still in force in Slovenia the laws from the past socio-economic system, i.e. the Spatial Planning Act (UL RS 18/1984) and Urban Planning and Other Forms of Land Use Act (UL RS 18/1984). A new Act was in preparation for more than a decade and in its substance superseded the above Acts as well as the Urban Land Act (UL RS 44/1997).

Conceptual bases for the preparation of new systemic solutions in the field of spatial planning and management represent the Spatial Development State Policy, and the Assessment of Development State in Space in the Republic of Slovenia, adopted by the Government in December 2001.

Spatial Development Policy in the Republic of Slovenia is a document on which basis Slovenia provides for the spatial development, defines the basic principles and objectives of spatial planning and management in order to create quality living and working environment and to harmonise spatial development with the wider European space.

Assessment of State in Space in the Republic of Sloveniais a document that contains a description and assessment of characteristics of the Slovenian space and factors influencing the development of the spatial planning system.

Objective of spatial planning and management under the valid Spatial Planning Act (UL RS 88/2002) is to provide for the sustainable spatial development by harmonising the economic, social and environmental aspects of development. Direction of developmental processes and linked spatial arrangements must be based on the balanced developmental needs, and space must be planned and managed so as to:

---

10 In addition to the National Programme of Environmental Protection, the Environmental Accession Strategy of the Republic of Slovenia for the Accession to the European Union was prepared, which is the ftmdamental document as regards the programme and deadlines, on which basis the efforts of Slovenia towards the accession to the EU were made in the field of environment.

1. provide for the sustainable spatial development and a rational use of space and the preservation of spatial capacity for the present and future generations;
2. provide for quality living conditions in the urban as well as rural settlements, a quality and humane development of urban and other settlements, and to provide for their supply;
3. achieve a spatially harmonised and interlinked distribution of the different activities in space;
4. provide for the spatial possibilities of a sustainable development of communities;
5. provide for the unhindered access of handicapped persons to all the facilities and to use them in accordance with the law;
6. provide for the environmental protection, conservation of nature and cultural heritage, to enable the sustainable use of natural resources and to preserve other qualities of the natural and living environment;
7. provide for the defense of the state and for the protection in case of the natural and other catastrophes.

A specific article, i. e. Article 4, is intended for the adaptation of developmental needs to the needs for protection. At spatial planning, it is necessary, through the treatment and confrontation of the different needs and interests of development in space, to ensure the harmonisation of the economic, social and environmental aspects, and to create the conditions required for the sustainable spatial development. Developmental needs of the individual activities that influence spatial planning and management shall in the process of preparation and adoption of spatial acts[11] be subjected to the assessment and harmonisation with the requirements of protection so as to enable the planning of spatial arrangements within the framework of the basic objectives of sustainable spatial planning.

Sustainable spatial development is ensured by spatial planning in such a way as to prevent any predomination of interests of the individual activities to the detriment of the balance of developmental needs and protection requirements and other basic objectives of spatial planning and management en environmental protection.

---

11 Under the Spatial Planning Act, spatial arrangements are determined and planned on the basis of spatial planning documents. Spatial acts include the national, municipal and common spatial planning documents. National spatial planning documents represent the strategy of spatial development of Slovenia, the spatial order of Slovenia and the national detailed plans. Municipal spatial planning documents represent the strategy of spatial development of the municipality on the basis of the conception of urban development and the conception of landscape development and protection, the municipal spatial order, and the local detailed plans. Spatial arrangements that are of common interest for the state and municipalities are planned jointly by the state and municipalities. A joint spatial planning document of the state and municipalities represents the regional conception of spatial development.

Basis for the harmonisation of developmental needs and protection requirements are the analyses of developmental possibilities of individual activities in space and theenvironmental impact assessment, and in case of harmonisation in the process of preparation of detailed plan, also guidelines for planning. Harmonisation is managed by the producer.

In case that developmental needs of individual activities cannot be harmonised in the process of preparation of spatial planning document, or when these needs cannot be harmonised with protection requirements, in case of preparation of a national or joint spatial planning document, the non-harmonised areas shall be decided by the Government, within its competence, on a proposal of the minister competent for environment, and in case of preparation of a municipal spatial development document, by the Municipality Council, within its competence, on a proposal of the mayor.

Under the Act, new development shall be directed into the settlement areas (Article 5). When directing settlement, rational use of land and facilities in settlements must be provided for by primarily exploiting the empty and insufficiently exploited surfaces in settlements through their activation and renewal and restoration of degraded areas in settlements.

Extension of a development shall be permissible only exceptionally. Any extension shall as a rule be directed to the land that is less important from the point of view of sustainable use of natural goods and the conservation of nature and preservation of cultural heritage. Construction of facilities shall be permissible on the best agricultural land only, when there is no land available that is less appropriate for the agricultural cultivation.

In certain matters there cooperate the different spatial planning stakeholders (Article 7). The state and local communities must cooperate in matters concerning spatial planning and management, in particular at planning the development of activities in space and their spatial arrangement, concerning the common use of natural goods, and common transport, power and public utility facilities, as well as other spatial arrangements concerning environmental protection and conservation of nature and preservation of cultural heritage.

It is pointed out that in implementing spatial planning and management activities, the spatial planning stakeholders must take into account public benefits and private interests, and study them in relation to and in accordance with the basic objectives of spatial planning and management, and where private interest shall not be pursued to the detriment of public benefit (Article 9).

6. Conclusion

The period after 1990 was for Slovenia a period of numerous changes. In addition to characteristics brought by the change of the socio-economic system it was necessary to take into account new principles of environmental protection, created on a global level, in order to prepare a new constitution and legislation. The Constitution of the Republic of Slovenia, and the Environment Protection Act of 1993 provided an appropriate basis for the enforcement of the principles of environmental protection also in the field of spatial planning. The Spatial Planning Act took a decade of preparation to enter into force on 1 January 2003. At present there are in preparation numerous implementing regulations, and the Spatial Development Strategy in Slovenia is at the final stages of preparation.

Environmental protection features in the legislation and other documents as an equal factor in the decision-making at interventions into space. It remains to be seen, how the promising Acts will be brought into practice and what consequences they bring for the society as a whole. In particular in the period, when the regional differences are increasing, the (non)quality of environment remains an element of social differentiation, and at the same, it is turning into a developmental factor.

Literature and Sources

Assessment of state and tendencies in space, 2001, Ministry of Environment, Spatial Planning and Energy, Ljubljana.

Constitution of the Republic of Slovenia, UL RS 33/1991.

Cultural Heritage Protection Act, UL RS 7/1999.

Draft Environment Protection Act, ZVO-1, www.sigov.si/mop.

Economic Development Strategy, 2001, www.gov.si/umar.

Environment Protection Act, UL RS 32/1993.

Envrionmental Aspects and the Aspects of Sustainable Development of the Economic Development Strategy of the Republic of Slovenia by 2006, www.gov.si/umar.

Forest Act, UL RS 30/1993.

Local Detailed Plan of the Highway Segment Klanec-Sermin, Volume VII, Assessment of Impacts on Environment, 1999, Urban Planning Institute of Slovenia, Ljubljana.

National Programme of Environmental Protection, 1999.

Nature Conservation Act, UL RS 56/1999.

Plut, D., 2002, Environmental Protection Aspects and aspects of spatial development in Slovenia, Scientific Institute of the Philosophical Faculty of Ljubljana, Ljubljana.

Professional bases for the preparation of Economic Development Strategy in Slovenia, www.gov/umar.si

Spatial Planning Act, UL RS 88/2002.

Spatial Planning and Management Policy in the Republic of Slovenia, 2001, Ministry of Environment, Spatial Planning and Energy, Office of the Republic of Slovenia for Spatial Planning, Ljubljana.

Šturm, L., 2002, Commentary to the Constitution of the Republic of Slovenia, Faculty of Postgraduate and National Studies, Ljubljana.

Isabelle Trinquelle

Le droit de l'environnement en Grèce ou les aléas d'un régime juridique de protection

Introduction

Couvrant un territoire de 131955 km^2, la Grèce bénéficie d'un environnement privilégié: climat méditerranéen clément, richesses naturelles et paysages de grande qualité, faible densité de population sur une grande partie du territoire d'où faible pression sur l'environnement, faibles pollutions industrielles... L'environnement en Grèce ne s'apparentait pas, jusqu'à récemment, à une «problématique». Et la population est encore peu sensibilisée au caractère fragile et limité des ressources naturelles.

Devenue membre de la Communauté européenne en 1981, la Grèce encore très agricole s'est engagé dans des réformes destinées à accélérer le développement économique et social, selon les critères communautaires. Les changements socio-économiques rapides entraînés par cette orientation ne sont pas sans conséquence sur l'environnement.

Engendrant de plus fortes pressions sur le milieu et les ressources naturelles, l'entrée du pays dans la Communauté a également eu des effets bénéfiques sur le développement d'une politique environnementale: exigence accrue du public quant à la qualité de l'environnement et meilleure prise en compte de l'environnement, particulièrement sensible dans le domaine législatif et institutionnel. Autre élément important lié à l'appartenance communautaire, la Grèce bénéficie des programmes de financements européens (Fonds de cohésion et autres aides communautaires) dont une partie doit servir à préserver l'équilibre entre développement économique et protection de l'environnement.

Néanmoins, les préoccupations environnementales vont croissantes. Pour en aborder les grandes lignes, il est possible de distinguer trois types de territoire, en fonction de leurs caractéristiques et des pressions qu'ils subissent:

- les métropoles urbaines d'Athènes et de Thessalonique
- les îles et les zones côtières
- le reste du territoire continental

Sur les près de 11 millions d'habitants de la Grèce, 50% habitent les deux principales villes du pays, Athènes et Thessalonique. Au-delà des avatars communs à l'ensemble des grandes métropoles d'Europe, on comprend qu'une telle concentration rende particulièrement aigus les problèmes d'environnement urbain: pollution de l'air due au trafic routier, problèmes de gestion des déchets, d'assainissement des eaux usées, urbanisation mal maîtrisée, etc. Deux exemples permettent d'illustrer la difficulté de compenser les effets de la croissance urbaine. En matière de pollution atmosphérique, Athènes continue d'être montrée du doigt en Europe pour les taux de pollution élevés de son air ambiant. Concernant l'eau potable, la capitale s'approvisionne depuis les lacs de Marathon et d'Iliki et la rivière Momos, mais la demande est telle que le ministère de l'Environnement et des Travaux publics a entrepris d'augmenter les sources d'alimentation avec l'eau de la rivière d'Evinos à plus de 250km de la capitale!

Autre milieu, les côtes maritimes grecques qui longent la Méditerranée sur 16600km. Elles concentrent 90% de la population de la Grèce et sont le lieu privilégié du développement d'une activité économique essentielle pour le pays: le tourisme. Ce secteur représente 6% du PIB du pays et repose sur l'attractivité de l'environnement hellénique. Simultanément, le milieu insulaire et côtier pâtit de la pression anthropiquesaisonnière engendrée par le tourisme. Elle génère des problèmes particuliers, notamment en matière de préservation des ressources hydriques et de gestion des déchets. La surexploitation de l'eau liée au changement du mode de vie et au tourisme entraîne une raréfaction de la ressource (épuisement des nappes phréatiques) et produit des phénomènes de désertification et, dans certains endroits, de salinisation des nappes aquifères littorales. La question des déchets pose des problèmes particuliers en milieu insulaire du fait du contexte géographique (espace disponible limité pour l'élimination, isolement du reste du territoire, habitat dispersé, etc.) et de la variation saisonnière importante de la production de déchets liée au tourisme. Les coûts de la gestion en sont accrus et il n'est pas rare que des solutions de fortunes, illégales, telles que les dépôts sauvages ou l'incinération à l'air libre soient usitées. Autre méfait de l'activité touristique: les constructions et extensions urbaines parfois anarchiques qui génèrent un «mitage» du paysageet renforcent, du fait de leur éparpillement, la difficulté de gérer les questions d'eaux usées, de déchets ou de voiries.

Le reste du continent hellénique, encore très agricole et peu densément peuplé, est essentiellement confronté aux problèmes liés à l'exode rural. L'abandon des villages et des petites exploitations de montagne engendre un manque de moyens locaux pour prendre en charge une bonne gestion de l'environnement. Il est synonyme d'un défaut d'entretien du patrimoine naturel et culturel (notamment architectural) et de modifications du rapport de l'homme à la nature. Il signifie aussi une concentration de la production dans des zones plus clémentes (plaines de Thessalie par exemple) et suivant des méthodes intensives, avec les conséquences connues sur l'environnement (surexploitation des ressources en eau pour l'irrigation, appauvrissement et érosion du sol, surpâturage, concentration de la pollution par les pesticides).

Pour répondre à l'ensemble de ces différents challenges et à ses obligations communautaires, la Grèce a mis en place un cadre législatif et institutionnel marquant des évolutions importantes (I). La réglementation environnementale s'est ainsi étoffée dans l'ensemble des secteurs de l'environnement (II). De nombreux obstacles persistent cependant, empêchant l'effectivité de ces outils et, ce faisant, la préservation efficace de l'environnement (III).

I. Le cadre juridique de la protection de l'environnement

La législation environnementale s'est nettement renforcée depuis les années 1980, sous l'influence évidente des directives européennes. Les institutions responsables de son développement et de sa mise en œuvre se sont mises en place essentiellement au niveau national.

1) L'article 24 de la Constitution

L'environnement bénéficie d'une protection de la loi suprême. La Constitution Grecque de 1975, telle qu'amendée en 1986, énonce dans son article 24 que la protection de l'environnement naturel et culturel est un devoir de l'Etat. Ce dernier doit donc prendre les mesures préventives et répressives nécessaires pour remplir cette mission.

Jusqu'à récemment, cet article de la Constitution ne faisait pas expressément mention d'un «droit à l'environnement». Cependant, le fait qu'il soit inscrit dans la partie du texte constitutionnel consacrée aux droits individuels et sociaux a rapidement permis à une partie de la doctrine et de la jurisprudence de lui conférer un rôle essentiel. L'article 24 a, de ce fait, aidé à la reconnaissance d'une obligation pour l'ensemble des pouvoirs législatifs, exécutifs et judiciaires de garantir un certain niveau de protection de l'environnement.

Le Conseil d'Etat et sa jurisprudence ont joué un rôle considérable en ce sens. L'intervention de la Section 5 du Conseil d'Etat grec – et sous l'influence de son Président d'alors, M. Decleris – est remarquable de par son originalité et sa détermination et a contribué à la définition de certains principes du développement durable et à leur intégration dans le droit. Parmi ces principes, on peut retenir:

- Le principe de l'ordre public environnemental qui suppose que la recherche d'un équilibre entre les systèmes anthropologiques et écologiques est de la responsabilité de l'Etat et de chaque individu. Sur ce fondement, le Conseil a interdit l'épandage de pesticide par voie aérienne sur les champs d'oliviers pour prévenir les atteintes à la santé et à la biodiversité (Conseil d'Etat 470/1993) ou la construction de projets qui menaçaient la beauté des côtes (IC 739/1994, 54/1994).
- Le principe de développement durable qui permet de condamner un développement économique qui va à l'encontre de la protection des écosystèmes. La cour suprême a énoncé ce principe pour la première fois

dans l'affaire Petrola. Le projet d'une raffinerie d'Elefsina de créer une unité de désulphurisation rencontrant l'opposition d'associations environnementales, le Conseil d'Etat a ordonné l'analyse du caractère durable du projet avant de se prononcer. En l'occurrence, la question était de savoir s'il y avait un risque d'accroître la pollution atmosphérique dans la zone concernée (Conseil d'Etat 53/1993, voir aussi 2755/1994, 5267/1995).

- Le principe de prévention qui suppose que l'action préventive de l'Etat est d'une importance majeure pour la protection de l'environnement (Conseil d'Etat 2759/1994).
- Le principe de planification qui reprend les termes de la Constitution et suppose que l'Etat est tenu d'élaborer une planification nationale, régionale et locale. En dehors de ces plans, une autorisation individuelle accordée pour un projet ou une activité quelconque peut être jugée illégale (Conseil d'Etat 304/1993, 2844/1993).
- Le principe de l'obligation de restaurer un écosystème endommagépar lequel, contrecarrant la théorie du fait accompli, le Conseil d'Etat refuse de faire perdurer des situations préjudiciables à l'environnement. Sur le fondement de ce principe, le Conseil a ainsi pu exiger la restauration de l'ensemble de la zone humide de Georgioupolis en Crète (Conseil d'Etat 306/1992, 296/1993).

La jurisprudence montre ainsi qu'il est possible de pallier les défaillances de la législation et de son application. Au stade du recours au juge, il ne peut cependant s'agir que de constater les dommages et de tenter de les réparer.

En avril 2001, la Constitution de l'Etat grec a été révisée. L'article 24 a connu quelques modifications qui font preuve d'un certain avant-gardisme et semble donner en partie raison au Conseil d'Etat: au devoir de l'Etat susmentionné, un réel «droit à l'environnement» a été adjoint et les politiques environnementales doivent désormais suivre le «principe du développement durable». La jurisprudence future montrera si de telles évolutions constitutionnelles offrent au citoyen un réel droit individuel à l'environnement susceptible d'être exercé contre les atteintes au milieu.

2) La Loi cadre de 1986 pour la protection de l'environnement

La loi Nationale 1650/86 pour la protection de l'environnement offre un cadre législatif pour la réglementation des différents secteurs de l'environnement. Ses objectifs sont très généraux puisqu'ils couvrent:

- La prévention de la pollution et de la dégradation de l'environnement;
- La préservation de la santé humaine;
- La promotion d'un développement équilibré respectant le renouvellement des ressources naturelles, les équilibres écologiques et les critères de protection de l'environnement.

Ils reprennent ainsi les objectifs fixés par la Communauté européenne en la matière dans le Traité (article 174 définissant la politique environnementale), en ajoutant toutefois la notion de respect des équilibres écologiques.

La loi de 1986 fixe des éléments essentiels du droit de l'environnement. Elle définit les notions d'environnement et de protection de l'environnement. Elle prévoit de limiter les effets négatifs de l'activité humaine via la procédure d'étude d'impact sur l'environnement pour certains projets et activités. Elle fixe les principes de base de la protection de la qualité de l'air, de l'eau, du sol et de la gestion des déchets, des substances dangereuses et du bruit. Elle prévoit les principaux outils de protection des espaces naturels (parcs nationaux, aires de protection spéciale). Enfin, la loi précise les sanctions pénales et administratives ainsi que la responsabilité civile pour dommage à l'environnement.

Les autorisations des rejets et émissions liées aux activités humaines comme l'utilisation des ressources doivent respecter l'environnement en se fondant sur :

- les connaissances scientifiques ;
- les meilleures technologies disponibles, économiquement viables ;
- le contexte régional, ses particularités humaines et environnementales et ses besoins en terme de développement ;
- la préexistance d'usage collectif d'une zone donnée ;
- les plans d'urbanisme et les plans de développement existants ;
- la sensibilité des citoyens sur les questions d'environnement, basée sur une information et une éducation appropriée.

On retrouve à nouveau, parmi ces critères, nombre de ceux inscrits dans le Traité auxquels sont adjointes des notions intéressantes telles que celles du respect des usages collectifs préexistants ou de la sensibilité des citoyens informés et éduqués.

Parallèlement à l'adoption de ce cadre législatif important, un cadre institutionnel s'est mis en place pour assumer les missions environnementales désormais identifiées.

3) Le cadre institutionnel en matière d'environnement

En 1980, un ministère de l'Environnement, du Logement et de la Planification était créé et prenait la responsabilité des questions d'environnement. Avec la loi 1032/1980, ce ministère fusionnait avec celui en charge des travaux publics pour devenir le ministère de l'Environnement, de la Planification et des Travaux Publics (YPEHODE), dont relève aujourd'hui l'essentiel du secteur de l'environnement. Cette unification des deux ministères était justifiée par la volonté d'intégrer les préoccupations environnementales dans le secteur « écologiquement sensible » des travaux publics et, plus généralement, de mieux concilier le développement économique du pays avec la protection de l'environnement.

L'YPEHODE est divisé en plusieurs directions générales en charge de l'environnement, de la planification urbaine, et de la planification et des projets. Elles traitent de la protection de la nature, des eaux intérieures, des déchets, de l'impact environnemental, du risque, de la qualité de l'air, du bruit et du trafic routier. Le ministère est en charge de la définition des politiques nationales en la matière, des plans et programmes nationaux et de la coordination et du contrôle de leur mise en œuvre.

Sur la base de la loi 1650/86 et des principes du développement durable énoncés tant au niveau communautaire qu'international (Sommet mondial de Rio en 1992), le premier Programme Opérationnel Environnemental de la Grèce a été adopté pour la période 1994-2000. Sorte de carnet de route du ministère, il fixe les objectifs en fonction des principaux problèmes identifiés et prévoit la création des infrastructures nécessaires à la gestion de l'environnement.

Des organismes spécialisés ont été créés pour assister l'YPEHODE dans certaines de ses tâches ou pour la mise en œuvre de programmes spécifiques découlant du Programme Opérationnel: le Fonds Spécial pour la Mise en oeuvre des Plan urbains, l'Organisation pour la planification et la protection de l'environnement d'Athènes, l'Organisation pour la cartographie et le cadastre grec, etc. En 1995, un Bureau pour le Réseau national d'information sur l'environnement et, plus récemment, un Centre National de l'Environnement et du Développement Durable doivent assurer la centralisation des données environnementales et le relais de l'information vers l'Union européenne (réseau EIONET et Agence Européenne de l'Environnement).

Bien entendu, d'autres ministères interviennent dans certains secteurs de l'environnement: ainsi en est il du ministère de l'Agriculture pour les forêts, du ministère de la mer Egée pour la protection de l'environnement dans les îles, du ministère de la Marine marchande pour les questions de pollution marine, ou encore du ministère de l'Industrie pour les pollutions industrielles. Cependant, l'administration grecque semble encore trop sectorisée et hiérarchisée pour que des coopérations puissent se mettre efficacement en place en faveur de l'environnement.

Concernant l'administration territoriale, la Grèce connaît trois niveaux de découpage: municipal, préfectoral et régional. Bien que la décentralisation soit en évolution, les compétences environnementales restent très centralisées.

Les 13 periferia (régions) sont dirigées par des représentants de l'Etat nommés par le Gouvernement et sont subordonnées au ministère de l'Administration Publique et de la Décentralisation et, pour les questions relatives à l'environnement, à l'YPEHODE. Elles ont essentiellement une compétence en matière de développement régional.

Le second échelon compte 57 nomós (préfectures). Les nomós sont dirigées par un préfet, directement élu, et un Conseil préfectoral. Leurs compétences environnementales concernent principalement:

- la définition et la mise en oeuvre de la politique de l'environnement sur leur territoire;
- la réglementation relative à l'environnement naturel (par voie de décision préfectorale ou de décision préfectorales jointes si le champ d'application de la réglementation dépasse les frontières d'une seule préfecture);
- la mise en œuvre des plans d'utilisation du sol élaborés par l'YPEHODE.

Depuis 1994, les préfectures sont en principe assistées dans leur travail par des Corps d'Inspecteurs de la Protection de l'Environnement. Ceux ci sont responsables du contrôle de la bonne application de la réglementation et du suivi des plaintes. En 2000, seule la préfecture d'Athènes bénéficiait d'un tel service...

Les Municipalités (villes de plus de 10000 habitants) et les Communautés (qui comprennent au moins 1000 habitants), échelon local de décentralisation, ont connu une réforme importante en 1997. Leur nombre est passé de 6350 à 1033 (900 municipalités pour 133 communautés). Simultanément, elles se sont vu reconnaître davantage de compétences en matière d'environnement (Loi 2539/97). Ces compétences recouvrent le traitement des déchets urbains et des eaux usées, l'approvisionnement en eau potable et la protection de l'environnement sur le territoire de la commune. Les maires ont une compétence propre en ce qui concerne la gestion du trafic automobile, la qualité de l'air et la planification de l'utilisation du sol. Diverses expériences d'intercommunalité permettent de faciliter la gestion de ces problématiques et de concentrer les efforts. Ainsi, l'Association des Communautés et Municipalités de la Région de l'Attique a pris en charge la gestion des déchets.

II. Le développement d'un arsenal législatif sectoriel

Passant en revue les principaux secteurs de l'environnement, le constate s'impose d'un développement certain des dispositions législatives et réglementaires, majoritairement issu de la transposition du droit communautaire. On se heurte cependant très rapidement au problème de sa mise en œuvre.

1) La maîtrise de l'impact des activités humaines sur l'environnement

La maîtrise de l'utilisation du sol

Bien que les politiques de planification régionale et urbaine datent des années 1960, la Grèce est particulièrement en retard en matière de gestion de l'urbanisation.

Selon la Constitution, l'Etat a la responsabilité d'élaborer un plan général pour le pays et de développer la planification des villes et des zones résidentielles (article 24 §2). L'objectif de cette planification est de promouvoir un aménagement du territoire et un développement urbain assurant les meilleures conditions de vie possibles. En application de cette disposition, la Loi 360/1976 sur la planification régionale et l'environnement enjoint le ministère en charge de l'environnement d'élaborer un plan national et des plans régionaux d'aménagement. Le plan national doit être voté par le Parlement, les plans régionaux approuvés par le Conseil National de la Planification Régionale et de l'Environnement. Une fois adoptés, ces plans sont opposables aux personnes privées comme aux personnes publiques. La Loi 1650/1976 prévoit un système de zonage pour les activités productives. Mais sa mise en œuvre repose sur les règlements qui doivent préciser les caractéristiques et l'usage des différentes zones. Pour la planification locale, la loi relative l'utilisation du sol (Loi 1337/1982) s'applique aux zones urbaines et aux petites villes. Le développement des zones rurales fait l'objet d'un autre texte (décret présidentiel sur la construction en zones situées en dehors des plans urbains officiels). Au-delà de 2000 habitants, l'utilisation du territoire doit faire l'objet d'un Plan général de gestion. En deçà, il est requis un plan d'utilisation du sol destiné à maîtriser les limites des villages.

Cependant, la pratique révèle des délais et lacunes importants dans le système prévu. Selon le rapport du CNEDD, en dehors des zones urbanisées, seuls 6 % de l'espace feraient l'objet d'un plan d'utilisation du territoire. Certes, pour les zones non couvertes par un Plan général de gestion, la loi permet de designer des «Zones de Développement Contrôlé». De plus, les règles de construction sont censées pallier à l'absence de planification. Cependant, ces dernières sont pour la plupart anciennes et trop laxistes pour permettre la maîtrise des constructions en dehors des zones déjà urbanisées.

Le tourisme étant devenu un facteur d'expansion incontrôlée de la construction, y compris dans des zones naturelles encore préservées, une réglementation spécifique a prévu le contrôle du développement des zones touristique. Selon ces dispositions, l'administration centrale peut identifier une zone comme étant saturée et y refuser de nouvelles installations touristiques.

Les études d'impact sur l'environnement

En application de la loi 1650/86 et de la Directive européenne 85/337 concernant l'évaluation des incidences de certains projets publics et privés sur l'environnement, deux décisions ministérielles datant de 1990 (69269/5387/1990 et 75308/5512/1990) mettent en place les procédures d'étude d'impact sur l'environnement et d'autorisation administrative pour les projets et activités ayant un impact significatif sur l'environnement. Autorisation et étude d'impact sont donc associées. Les projets et activités sont répertoriés selon diverses catégories en fonction du danger qu'ils représentent pour l'environnement. Les plus dangereux, tels que installations de l'industrie chimique, raffineries, centrale thermique, etc. sont soumis à autorisation (catégorie A). La procédure est allégée pour les autres catégories. Pour les projets et activités de catégorie A, la demande d'autorisation doit être précédée d'une «Approbation Préliminaire d'Attribution de Site» (sauf si le lieu d'implantation est situé sur une zone déjà affectée au type d'activité prévue) et d'une Etude d'Impact sur l'Environnement.

Complétant les dispositions législatives et réglementaires, le Conseil d'Etat a affiné à travers sa jurisprudence le cadre juridique spécifique aux procédures d'étude d'impact sur l'environnement. Il en ressort les principaux éléments suivants:

- L'approbation du choix d'un site est inconstitutionnel s'il entraîne une planification inorganisée, irrationnelle, violant l'article 24 §2 de la Constitution qui requiert un plan général (Conseil d'Etat 304/1993, 2844/1993).
- L'étude d'impact doit être soumise et approuvée non seulement avant le début des travaux ou avant l'autorisation du projet ou de l'activité, mais aussi avant tout autre acte préparatoire tel qu'une procédure d'expropriation (Conseil d'Etat 37/1993, 1040/1993).
- L'étude d'impact doit être soumise et approuvée également pour l'«Approbation Préliminaire d'Attribution de Site» sur lequel est envisagé le projet ou l'activité. L'étude doit consister en une étude complète et scientifique sous peine d'être annulée (Conseil d'Etat 1520/1993).
- Pour les travaux publics majeurs qui comprennent plusieurs constructions distinctes, des études d'impact spécifiques à chaque tranche du projet sont insuffisantes. Il convient d'analyser l'impact du projet dans son ensemble et non de façon fractionnée (Conseil d'Etat 2759 et 2760/1994).

Dans cette dernière affaire, la Cour suprême se prononçait suite à des plaintes déposées en 1992 et 1993 par plusieurs ONG environnementales. Celles-ci demandaient l'annulation de deux décisions d'autorisation administratives relatives à des infrastructures importantes visant à détourner la rivière Acheloos (canalisation de l'eau de la rivière Acheloos, barrage électrique). La raison invoquée était l'insuffisance de l'étude d'impact sur l'environnement. Dans ses décisions 2759 et 2760 du 20 juillet 1994, le Conseil d'Etat donnait raison aux associations en considérant que la diversion de la rivière Acheloos dans la plaine de Thessalie est une opération particulièrement complexe et majeure qui présente de sérieuses menaces sur l'écosystème. De ce fait l'étude de l'impact du projet sur l'environnement devait être globale et non morcelée, afin d'appréhender les effets sur l'ensemble du territoire concerné. Il est à noter que les travaux n'ont pas été interrompus suite à la décision du Conseil, ce qui révèle une considération limitée du gouvernement pour les préoccupations environnementales dans ce projet considéré comme extrêmement important pour le développement et supporté financièrement par la Communauté européenne.

La pression du système judiciaire communautaire est un autre élément d'évolution de la procédure d'étude d'impact. Suite à un recours de la Commission européenne devant la CJCE, la Grèce a ainsi adapté son dispositif national aux modifications apportées par la directive 97/11 amendant de façon sensible la directive 85/337.

2) La protection des ressources naturelles

Protection de la nature

5500 espèces de la flore et 900 espèces de la faune ont été identifiées en Grèce. Un programme d'action pour la gestion et la protection de la biodiversité prévoit de renforcer la connaissance de ces espèces afin de mieux en garantir la préservation. La protection des espèces passant par la préservation de leur habitat, divers dispositifs ont été mis en place en faveur des espaces naturels.

La forêt qui couvre près de la moitié du territoire grec fait l'objet d'une attention particulière(mention spéciale dans l'article 24 de la Constitution, Code Forestier, Loi n° 998 de 1979 sur les forêts et le territoire forestier). Les dispositions spécifiques à cet espace naturel sont mises en œuvre sous la responsabilité du ministère de l'Agriculture. En la matière, la gestion est facilitée par le fait que la plupart de la surface forestière appartient à l'Etat. Cependant, les orientations sont davantage axées sur la production que sur la préservation de la valeur écologique de ce patrimoine naturel. Ce qui entraîne les travers connus dans de nombreux Etats européens: reboisement avec des espèces à croissance rapide, introduction d'espèces exotiques, accroissement des voies d'accès, etc. Il y a évidemment une grande différence entre les forêts de montagne, au nord du pays, qui subissent une pression décroissante du fait de l'exode rural et les forêts de régions plus fréquentées telles que la Crète, soumises à une surexploitation génératrice d'érosion des sols. Cependant, globalement, et en dépit des feux subis chaque année, la surface globale d'espace boisée s'accroît. La question est alors de savoir comment et à partir de quelles espèces cette croissance s'effectue.

Si le premier Parc National, le Parc National du Mont Olympe, date de 1938, ce sont les directives européennes (directives «oiseaux» de 1979 et «habitat» de 1992) qui ont donné une véritable impulsion au réseau des zones naturelles protégées en Grèce.

La Loi 1650 de 1986 a apporté d'importants changements dans le domaine de la protection des ressources et espaces naturels. Elle permet notamment d'étendre la protection aux espaces marins et milieux aquatiques. De nouvelles catégories d'espaces protégés, tels les «paysages protégés» ou les «régions de développement écologique», ont vu le jour. Un premier Parc Naturel

Régional bénéficiant d'une structure de gestion a été créé en 1999 afin de préserver et développer le Parc Marin de Zakynthos, abritant des zones de ponte de la célèbre tortue caretta caretta. A l'image de cette tortue marine, d'autres espèces emblématiques, telles que le phoque moine (monachus monachus) au nord des Sporades, font l'objet d'une protection particulière par voie d'arrêté ministériel ou de décrets présidentiels. C'est alors, en principe, l'occasion de protéger tout le biotope constituant l'habitat de l'espèce en question.

La Grèce a également ratifié la plupart des conventions internationales de première importance en matière de protection de la nature comme la Convention sur le patrimoine mondial ratifiée en 1971, la Convention de Berne relative à la conservation de la vie sauvage et du milieu naturel de l'Europe ratifiée en 1983 ou encore la Convention de Ramsar sur les zones humides d'importance internationale qui couvre les zones côtières des deltas du Nestos, d'Evros et d'Axios, du golf d'Amvrakikos, des lacs de Mitrikou et de Vistonida, les lagons de Messolonghi et de Kotykhi, ainsi que les lacs de Prespa, de Yolvi-Lagada et de Kerkini.

Le total des aires protégées reste cependant faible et les plans de gestion des espaces protégés insuffisamment nombreux. Seul 2,5 % du territoire est couvert par les statuts de Parc national, Monument naturel ou Forêt de valeur esthétique, mis en place sur la base du Décret présidentiel N° 996 de 1971. Ils sont gérés par le Service Forestier Grec qui dépend du ministère de l'Agriculture. La situation devrait évoluer avec l'établissement progressif du réseau Natura 2000 : près de 20 % du territoire, soit 270 sites, ont été identifiés comme devant faire l'objet d'une protection spéciale.

Protection de l'air

L'accroissement de l'activité économique du pays et l'amélioration du confort vont de paire avec l'augmentation des rejets atmosphériques issus de la production d'énergie, des véhicules et d'autres activités humaines.

Depuis le milieu des années 1990, l'augmentation de la consommation d'énergie est parmi les plus importantes des pays de l'OCDE. Du fait de l'utilisation importante du charbon et du pétrole dans les installations de production d'énergie, cette croissance n'est évidemment pas sans effet sur la qualité de l'air. L'utilisation du gaz naturel (pour les centrales électriques, pour la ville d'Athènes cette année), de l'énergie solaire (pour les chauffes-eau) ou éolienne (notamment dans les îles) réduit un peu l'usage de combustibles polluants sans pouvoir cependant compenser les effets de leur consommation croissante. Divers programmes d'action et lois relatives à l'énergie promeuvent l'utilisation de technologies propres ou peu consommatrices d'énergie, ainsi que la production d'énergie renouvelable. Ainsi, la Loi de développement 1892/90 et la Loi relative aux incitations au développement économique

2601/98 prévoient des aides à l'investissement pour les entreprises de 40 à 50%. Le Programme Opérationnel en matière d'Energie fait quant à lui la promotion des énergies solaire, éolienne, géothermique, etc. Il prévoit également des subventions pour les investissements en ce sens. Un Plan National d'Action pour la Réduction du CO2 et autres gaz à effet de serre fixe des objectifs de réduction de la consommation d'énergie pour les secteurs de l'industrie, des transports, de l'habitat et des services. Dans le cadre de la politique communautaire relative au changement climatique, la Grèce s'est vu autoriser une augmentation de ses émissions de gaz à effet de serre de 25% jusqu'en 2008-2012 (par rapport aux niveaux de 1990), alors que l'UE dans sont ensemble s'est engagée à réduire de 8% ces émissions par rapport à 1990.

Les rejets dans l'air de SOx, NOx et CO2 sont encore très importants en Grèce. Ainsi, les émissions de dioxyde de souffre en provenance des grandes installations de combustion dépassent les seuils fixés par la réglementation européenne. Viennent désormais s'ajouter les émissions de composés organiques volatils et d'ozone en milieu urbain.

Ce problème de pollution atmosphérique est particulièrement préoccupant à Athènes, classée parmi les villes les plus polluées d'Europe et connue pour son nuage de pollution, le nefos. Certes depuis une dizaine d'années des mesures ont été adoptées pour améliorer la qualité de l'air et réduire les taux de certains composés tels que le dioxyde de soufre, le dioxyde d'azote, l'oxyde de carbone ou le plomb.

La circulation automobile étant l'un des principaux facteurs de pollution, une attention particulière a été portée à cette source de pollution. Aux mesures européennes visant l'amélioration de la qualité de l'essence et la mise en circulation de «voitures propres», sont venue s'ajouter des mesures locales comme la mise en place de circulation alternée en centre ville (selon les numéros pairs ou impairs des plaques d'immatriculation des véhicules), l'augmentation des zones piétonnes, l'amélioration des transport en commun (extension du réseau du métro, augmentation des bus en circulation, création d'un tram). La fermeture ou l'éloignement de certaines activités industrielles ou commerciales, comme l'aéroport international d'Athènes, ont également contribué à l'amélioration de la situation dans le centre de la capitale. En outre, la ville met en place un système de suivi de la pollution atmosphérique. Globalement cependant, les moyens de contrôle des niveaux de pollution et les mesures destinées à limiter les émissions sont encore insuffisants, en particulier face à la croissance exponentielle du parc automobile en circulation.

La Loi cadre de 1986 prévoyait la mise en place de mesures de surveillance et de prévention de la pollution de l'air. Diverses décisions ministérielles ont permis de déterminer des normes de qualité de l'air ambiant en fixant des valeurs limites pour différents composants tels que le SO2 (Décision 99/87), le plomb (Décision 98/87) ou le NO2 (Décision 98/88). Elles se fondent sur les normes prévues par les directives européennes. Quant au suivi de la qualité de l'air, une Décision ministérielle 84498/95 de 1995 prévoit que le contrôle de la qualité de l'atmosphère relève des directions de l'environnement de chaque préfectures (à Athènes, cette responsabilité revient à l'YPEHODE).

Protection de l'eau

La Grèce est un pays globalement bien pourvu en eau, bien que la ressource soit inégalement répartie, certaines îles de la mer Egée allant jusqu'à connaître des situations de pénurie et subissant des risques de désertification. Les réserves d'eau douce sont principalement constituées des eaux de surface (près de 90 %) et seulement 10 % d'eau souterraine (principalement dans les îles et pour l'eau d'irrigation). Dans quelques zones côtières, le surprélèvement aboutit à des phénomènes d'intrusion d'eau de mer dans les nappes phréatiques entraînant une salinisation de l'eau.

La pression sur les ressources s'est accrue principalement du fait de la consommation agricole et de celle des ménages. Le tourisme impose une pression supplémentaire en période estivale. Entre 1980 et 2000, le volume d'eau prélevé dans les nappes a doublé, sans qu'il ait été suffisamment tenu compte des risques de surexploitation de la ressource. L'irrigation agricole est le grand responsable de cette situation puisque 40 % de l'eau destinée à à cet usage est pompé dans les nappes phréatiques.

Un projet colossale de diversion de l'une des deux plus grandes rivières du pays, la rivière Acheloos, a été entrepris afin d'approvisionner certaines villes (Larissa, Trikala, Volos) et d'irriguer près de 400 000 hectares de cultures dans la plaine de Thessalie, tout en remédiant à la surexploitation de la nappe phréatique dans cette région. Suite aux contestations de plusieurs ONG environnementales et à une décision du Conseil d'Etat jugeant l'étude d'impact insuffisante (Conseil d'Etat 2760/1994), ce projet a été légèrement modifié.

En marge de ces équipements à fort impact sur l'environnement, la plupart du temps subventionnés par l'Union européenne, on trouve encore l'emploi de techniques traditionnelles de gestion de l'eau. Ainsi, les mini-barrages faits de pierres dans certaines îles des Cyclades sont destinés à retenir l'eau de pluie et lui permettre de s'infiltrer et de recharger les nappes avant qu'elle ne s'évapore. Malheureusement, ces techniques ancestrales n'ont pas toujours la capacité de pallier les effets de la surconsommation.

En ce qui concerne l'aspect qualitatif, la qualité des eaux de surface comme de la plupart des eaux côtières est estimée satisfaisante (les données manquent quant à la qualité des eaux souterraines). La pollution touche cependant des zones soumises à une forte pression, telles que certains lacs au Nord du pays, ou les eaux côtières à proximité des grandes villes comme Athènes, Patras ou Thessalonique. Les principales sources de pollution sont les rejets des eaux usées et ceux de l'activité agricole, du fait de l'usage accru de pesticides.

La Loi cadre de 1986 pose le principe de préservation de la qualité des eaux de surface et souterraines et prévoit le contrôle des rejets dans le milieu aquatique. La loi sur la gestion de l'eau de 1987 développe ces orientations et prend pour principe que les ressources en eau sont propriété de l'Etat. Des mesures telles que l'adoption de seuils limites de consommation, l'ajustement du prix en fonction de la consommation, l'établissement de zones protégées autour de sources d'eau potables ont été adoptées afin de mieux contrôler la pression anthropique.

En matière de contrôle de la consommation, un décret présidentiel de 1991 a établi un seuil limite de consommation par habitant en prenant en considération les données climatiques et hydrologiques de la région, la densité de la population et le type d'infrastructure d'approvisionnement en eau. La fourchette va de 100-250 litres par personne par jour, pour les habitants raccordés au réseau de distribution de l'eau, à 400-500 litres pour les hôtels de classe A.

Un système d'autorisation administrative a été mis en place pour la construction d'infrastructure, pour les prélèvements d'eau (pompage, raccordement, etc.), pour les rejets des industries en mer (loi de 1977 sur la protection de l'environnement marin) ou dans les rivières. Depuis la loi de 1986, l'étude d'impact sur l'environnement est obligatoire pour tous les projets majeurs d'infrastructure. Au cours de la procédure d'autorisation de rejets industriels, le ministère de l'Environnement ou la préfecture concernée élabore un «permis environnemental» en fonction de l'état et de l'usage du milieu aquatique concerné. Il faut constater que les industriels ne sont pas contraints par la loi de contrôler leurs rejets, quand bien même le non respect des normes de rejet peut entraîner amendes ou sanctions administratives.

La réglementation européenne en matière de protection de l'eau a eu un rôle moteur dans l'équipement en station d'épuration des communes et des industries. Mais l'ensemble du territoire n'est pas encore couvert et la mise en conformité avec la réglementation européenne demande des moyens additionnels même si les normes de qualité ont été adaptées à celles prescrites par les directives communautaires.

La loi de 1987 définit les autorités compétentes. Au niveau central, les responsabilités sont particulièrement dispersées: le ministère du Développement à la charge de l'administration et de l'allocation de la ressource, il supervise le développement des infrastructures (tunnels, barrages); le ministère de l'Environnement doit assurer le contrôle de la qualité de l'eau et la protection des écosystèmes; le ministère de l'Agriculture traite de l'utilisation de l'eau pour l'irrigation; le ministère de la Santé et des Affaires sociales met en œuvre la réglementation relative à la qualité de l'eau; le ministère en charge de l'Administration publique et de la Décentralisation assure l'approvisionnement en eau potable (hors Athènes et Thessalonique) et aide à l'équipement des communes en station d'épuration; même le ministère de la Culture intervient en la matière pour ce qui à trait aux usages de l'eau à des fins sportives. Ces intervenants doivent se rencontrer et trouver une approche commune au sein du Comité interministériel de l'eau qui définit la politique de gestion de l'eau. Pour gérer certains problèmes spécifiques, les ministères impliqués collaborent afin de mettre en place des programmes communs. Ainsi, les ministères de l'Environnement et de l'Agriculture doivent définir ensemble des programmes de préservation des eaux souterraines dans les zones sensibles soumises à une forte pollution par les nitrates, comme la plaine de Kapaida, en Grèce centrale.

Des évolutions significatives sont intervenues quant à l'administration territoriale de l'eau: désormais le pays est divisé en 14 bassins hydrographiques, dont la gestion a été confiée à autant de Comités régionaux de l'eau. Ceux-ci travaillent en liaison avec l'administration régionale et permettent de coordonner une gestion plus proche de la réalité du milieu. Ils élaborent des plans régionaux de gestion de la ressource en eau, soumis à consultation des principaux acteurs territoriaux.

Dans les villes de plus de 10 000 habitants, la gestion de l'eau et l'approvisionnement sont confiés à des entreprises communales (à une entreprise privée pour Athènes). Pour les plus petites communes, la collectivité locale est directement en charge de cette gestion. Le secteur privé est encore relativement absent dans ce domaine, comparativement à d'autres pays d'Europe.

3) La gestion des déchets

Dans ce domaine aussi, l'amélioration rapide du niveau de vie et les changements de mode de consommation sont synonymes de problèmes. La production annuelle de déchets municipaux est estimée à 4 millions de tonnes. Sur ce volume, 15 % est produit en zone isolée (montagne, îles) et ne fait pas l'objet d'une collecte régulière.

Concernant les déchets ménagers, un premier inventaire des sites d'élimination de ce type de déchets (autrement dit, des décharges) a été réalisée en 1988. Il dénombrait 3500 sites opérant sans mesures spécifiques pour la prise en compte de l'environnement et 1500 sites respectant certaines mesures de protection. La Décision ministérielle conjointe de 1990 (69269/5387/90) a mis en place une procédure d'autorisation et de contrôle des installations recevant des déchets. Ainsi, toutes les décharges doivent mettre en œuvre les dispositions environnementales prévues par l'administration, faire l'objet d'un contrôle régulier et d'une réhabilitation du site en fin d'exploitation, à la charge des autorités locales. Ce règlement n'a cependant pas permis de mettre fin au danger que représentent les nombreux dépôts préexistants et mal contrôlés. Selon l'OCDE, 91 % des déchets ménagers solides sont encore mis en décharge et le ministère a dénombré environ 1500 importants dépôts non contrôlés et de nombreuses petites décharges sauvages (on compterait 27 décharge illégales dans l'Attique, 157 dans le Péloponnèse et 64 dans les îles des Cyclades et du Dodécanèse). Les conséquences de cette situation sont importantes : pollution de l'eau par écoulements, pollution de l'air par la dioxine ou le méthane (les dépôts de déchets représentent la seconde source d'émission de méthane en Grèce), risque d'incendie, etc.

Une Décision ministérielle 69728/824/96 a mis en place le cadre de la gestion des déchets. Elle est mise en œuvre par deux autres textes : la Décision 113944/1016/97 qui précise dans un Plan National pour la Gestion des déchets les priorités de réduction des déchets, de recyclage et réutilisation et d'amélioration des technologies ; la Décision 114218/97 qui pose les lignes directrices de la gestion des déchets solides.

Des progrès ont été réalisés en matière d'équipement : ouvertures de nouvelles décharges contrôlées, stations de transfert, stations de tri et unités de compostage (à Kalamata et dans l'Attique). L'YPEHODE met en place des « Sites d'enfouissement sanitaires légaux » répondant aux normes européennes (36 sites ouverts depuis 1994). Ces améliorations sont en partie co-financées par le Fonds de Cohésion. Deux programmes, l'un au nord du pays, l'autre au sud, sont développés pour sélectionner de nouveaux sites et entamer les travaux.

Mais ces efforts sont insuffisants face à l'accroissement continu de la production de déchets et les objectifs prioritaires de recyclage et de réutilisation sont loin d'être atteints. La Commission européenne a récemment tiré le signal d'alerte et sorti la menace d'un recours devant la Court de Justice Européenne du fait de la violation de la directive de 1999 relative à la mise en décharge. Ces dispositions ne sont pas respectées par environ 1450 dépôts de déchets en Grèce. De plus, la Commission a spécifiquement insisté sur les cas des sites illégaux de Paeania (Est de l'Attique), de Maroulas (Crète) et de Psittalia. Près d'Athènes, la décharge d'Ano Liosia qui reçoit les déchets de la capitale et de sa région aura atteint sa capacité limite en 2005. Le ministre de l'Environnement a désigné plusieurs sites qui pourrait recevoir de nouvelles installations. Cependant la décharge d'Ana Liosia a fait l'objet d'aménagement et aujourd'hui le nouveau site d'une capacité de 11.000.000 m^3 reçoit toujours la quasi totalité des déchets de l'Attique. L'installation n'est pas capable d'absorber un tel volume de déchets et la municipalité d'Ano Liosia a plusieurs fois décidé de fermer le site pour cause de saturation, entraînant l'apparition de montagnes de déchets sans destination possible dans la capitale…

Partout les projets de nouvelles installations nécessaires se heurtent au même problème : le refus de la population et des élus locaux concernés de voir de nouveaux sites se mettre en place près de chez eux. Bien entendu, les questions de coût des services et installations requis, de la formation du personnel et de la sensibilisation du public en général sont d'autres obstacles majeurs à l'amélioration rapide de la gestion.

Quant aux déchets dangereux, leur volume est limité par la part modeste du secteur industriel dans ce pays comparativement à d'autres Etats de l'Union européenne, et notamment de l'industrie lourde. Cette situation ne signifie pas pour autant que la gestion de ce type de déchets soit sans faille. Des cas de déchets industriels mélangés aux déchets ménagers, l'insuffisance des données relatives aux sites contaminés, le manque de planification sont des lacunes importantes.

Certes, des progrès ont été réalisés. Au total, plus de 76% des déchets dangereux sont en principe stockés dans des sites contrôlés, sur leur lieu de production, et près de 23 % sont recyclés. Le reste est en grande partie exporté pour être acheminé vers des installations adéquates (c'est la cas par exemple des déchets contenant des PCB, des pesticides, etc.). Ces transferts s'effectuent selon les règles prévues par la Convention de Bâle pour le contrôle des transports transfrontaliers de déchets dangereux, ratifiée par la Grèce par la Loi 2203/94.

Diverses décisions ministérielles ont transposé les directives européennes spécifiques à certains types de déchets nécessitant la mise en place d'un circuit prévention-récupération-traitement spécial. Ainsi, les décisions ministérielles conjointes 72751/3054/85 et 19396/1546/97 relatives aux déchets toxiques et dangereux et à l'élimination des PCDF, PCTF transposant les Directives 78/319, 76/403 et 91/689; la décision 73537/1438/95 relative à la gestion des batteries usagées contenant des substances dangereuses; les décisions 71560/3053/85 et 98012/2001/96 pour le traitement des huiles usées (Directives 75/439 et 87/101). Concernant ces dernières, un rapport de l'OCDE faisait valoir que près de la moitié des huiles usées produites en Grèce était collectée, sur laquelle, 55% étaient régénérés, 20% étaient réutilisés sans régénération, 20% utilisés comme combustibles et 5% éliminés sans contrôle.

Globalement, les taux de recyclage sont considérés comme étant les plus faibles de l'Union européenne. Le recyclage ne concerne que 9% des déchets et couvre surtout le verre, le papier et l'aluminium. Les taux de recyclage ont évolué entre 1989 et 1997 mais de façon insuffisante: ils sont passés de 24 à 28% pour le papier, de 19 à 28% pour le verre, de 22 à 35% pour les cannettes en aluminium. En 1998, une loi pour la Gestion alternative des déchets solides a été adoptée pour transposer la Directive européenne relative aux déchets d'emballage dans le droit national. Elle pose les objectifs de réduction de la production et de recyclage des déchets solides jusqu'à 2004. La mise en œuvre de cette loi a pris du retard en dépit du plan développé pour inclure des campagnes d'information, la création d'un organisme national chargé de mettre en place le système de recyclage des emballages, les aides financières, etc. En 2001, une nouvelle loi relative à la gestion des déchets a été adoptée (Loi n° 2939) qui insiste sur l'objectif primordial de réduction des déchets. Des décrets présidentiels sont attendus pour sa mise en oeuvre jugée trop lente notamment par les associations de défense de l'environnement. Une société a cependant été créée pour la mise en œuvre du programme de recyclage des emballages. Essentielle pour assurer la participation des consommateurs au système qui doit se développer, l'éducation de la population vis-à-vis des circuits de recyclages fait défaut.

La perspective des Jeux Olympiques, et de l'afflux supplémentaire et soudain de déchets qu'ils occasionneront, rendent plus que jamais nécessaire une rapide amélioration des circuits de récupération.

III. Les obstacles à l'effectivité du droit de l'environnement

Le problème de l'effectivité de la réglementation relative à l'environnement se pose dans tous les pays de l'Union européenne et au dehors. Ce n'est pas (plus !) que le droit de l'environnement manque de reconnaissance, mais plutôt que les intérêts qu'il protège sont toujours en concurrence avec d'autres intérêts mieux défendus. En Grèce, les défaillances dans la mise en œuvre de ce droit s'expliquent certainement par la jeunesse des politiques environnementales qui ont encore du mal à s'imposer face aux exigences du développement économique. C'est aussi l'accumulation de facteurs bloquants qui contribue à alourdir ce handicap : résistances locales, effets de la bureaucratie, manque de suivi des politiques et des textes de lois adoptés, manque de sensibilisation du public, manque de moyens humains, techniques et financiers, etc.

1) A l'épreuve des faits

Quelques exemples sont particulièrement évocateurs des difficultés rencontrées lors de la mise en œuvre des principes et obligations sur le terrain.

Constructions et urbanisme : une vision à courts termes

Sur l'ensemble du territoire, la question de la maîtrise de l'urbanisation pose un problème évident. L'ensemble du dispositif législatif s'est souvent montré insuffisant à maîtriser l'utilisation du territoire. Les instruments existants sont mal utilisés et les circonstances locales compliquent leur application : propriété foncière morcelée, système fiscal non adapté, développement d'activités concurrentes (tourisme et agriculture par exemple), manque de moyen de l'administration (insuffisance du registre cadastral, du système de classification de l'utilisation du sol, manque de coordination entre les outils de planification, etc.).

Un projet important d'élaboration du cadastre sur l'ensemble du territoire (seules quelques régions en étaient dotées jusque là) a été engagé dans les années 1990 avec le soutien de la Communauté européenne. Cet instrument important du suivi et de maîtrise du foncier n'a cependant pas pu aboutir à ce jour. Un scandale lié à l'insuffisance des résultats de la société chargée de la mise en œuvre du projet a porté préjudice à l'ensemble de la démarche.

L'exemple le plus flagrant de l'échec de l'outil législatif semble être celui de la Loi 360/76 qui pose des restrictions à l'urbanisation et fixe les sanctions pénales pour les constructions illégales. En effet, les dispositions destinées à contrôler l'utilisation du sol rencontrent une opposition importante de la part des petits propriétaires privés, habitués à user de leur droit de propriété comme ils le souhaitent. L'administration elle-même ne semble pas prête à faire valoir une vision à longs termes intégrant la préservation de l'environnement et la notion de développement durable.

Le Rapport sur l'Etat de l'Environnement publié en octobre 2001 par le Centre National pour l'Environnement et le Développement durable (CNEDD) dénonce le fait qu'entre 1950 et 1980 la quasi-totalité des espaces urbanisés incorporés dans les plans urbains étaient constitués d'espaces urbanisés en dehors de toute légalité, à la périphérie des villes. Ainsi, le développement anarchique des constructions de l'après guerre a été légalisé et incorporé dans les plans urbains. Même après cette période et jusqu'à aujourd'hui, les constructions illégales ont cours.

Cette année, le Gouvernement a annoncé un projet de loi sur les «constructions sauvages». Ce projet prévoit la légalisation de nombre de constructions illégales (600.000 personnes sont concernées), selon un classement fondé sur l'usage du bâtiment. De plus, étant donné que les populations concernées sont parmi les plus déshéritées, des réductions concernant le montant des amendes pour construction sans permis sont prévues. Le ministère de l'Environnement et des Travaux publics précise cependant que les constructions se trouvant en forêt, sur des sites archéologiques et en zones protégées, ne seront pas légalisées. Certaines démolitions ont même été ordonnées à l'encontre de constructions illégales sur le mont Pendeli, près d'Athènes. Mais la tendance générale reste trop favorable à la tolérance vis-à-vis des constructions illicites.

Le territoire côtier et insulaire subit de façon particulièrement flagrante cet état de fait: le tourisme étant devenu dans de nombreux cas l'une des principales sources de revenu, les autorités nationales comme locales semblent incapables de justifier et d'imposer des restrictions dans le développement de constructions destinées à l'accueil des vacanciers, comme on le voit sur les îles ou à Athènes à l'occasion des Jeux Olympiques. En effet, bien que la région de l'Attique ait été déclarée saturée en matière de capacité hôtelière, les autorités ont suspendu la restriction s'appliquant à la construction de nouveaux hôtels du fait de la perspective des Jeux de 2004. On peut se demander ce que deviendront ces installations après la courte période de l'événement sportif...

Les moyens défaillants de la protection de la nature

En 1999, le Parc Maritime de Zakynthos a été mis en place, premier parc national marin en Grèce, avec pour objectif de protéger la tortue marine Caretta caretta, espèce figurant parmi celles de l'annexe IV de la Directive Habitat 92/43 pour lesquelles l'article 12 du texte communautaire prévoit un régime de protection strict.

Quelques années plus tôt, le Conseil d'Etat avait annulé le permis de construire délivré par l'autorité administrative autorisant une compagnie privée à construire deux résidences sur l'îlot de Marathonisi, à l'endroit même où les tortues ont l'habitude de venir pondre (Décision 2301/1995). La commune de Lithakia sur l'île de Zakynthos et les associations à l'origine du recours (WWF, Greenpeace Hellas, Sea Turttle Protection Society) ont obtenu gain de cause, le Conseil d'Etat se fondant sur la Loi 1335/1983 portant ratification par la Grèce de la Convention de Bern relative à la conservation de la vie sauvage et du milieu naturel de l'Europe et sur le Décret présidentiel du 5 juillet 1990 établissant une zone de protection spéciale sur le territoire concerné afin de protéger la Caretta caretta .

Pourtant, en dépit de la mise en place du Parc maritime, un jugement de 2002 de la Cour de Justice Européenne demandait l'établissement et la mise en œuvre d'un système de protection stricte de cette espèce durant la période de ponte, sur plusieurs plages de l'île (CJCE 30 janvier 2002-aff.C103/00). Il s'agissait du premier arrêt de la Cour impliquant l'article 12 et le régime de protection strict.

En septembre de la même année, la Commission européenne menaçait à nouveau la Grèce d'un nouveau recours devant la Cour en cas de non respect de ses obligations. Elle constate que si la création du parc constitue une avancée certaine, de nombreuses insuffisances empêchent une protection réelle du site : absence de fonds disponibles pour des réalisations dans le parc et pour le bon fonctionnement de son agence de gestion; pas de compensation financières mise en place pour dédommager les propriétaires affectés par les mesures de protection; construction de nouvelles routes dans les limites du parc… Entre les ambitions affichées et les moyens de les assumer, trop de distances demeurent.

L'affaire du Kouroupitos : l'ingérable opposition des acteurs locaux

Ce cas concerne la Crète, dans la préfecture de La Canée où, depuis les années 1960, des déchets sont rejetés dans un ravin qui abrite le lit de la rivière Kouroupitos, à 200 mètres de la mer. On y trouve des déchets ménagers mais aussi des déchets issus de l'industrie ou des déchets hospitaliers. En 1987 (la Grèce étant membre de l'UE depuis 1981), la Commission européenne commence à recevoir des plaintes émanant d'une association locale et entame une discussion avec les autorités grecques à ce sujet. En effet, la situation met en danger la santé humaine et l'environnement et la Commission constate l'absence de plan de gestion des déchets dangereux pour cette région. Le Gouvernement grec fait valoir que la population locale s'oppose à la construction d'un site adéquate et qu'un désaccord persiste entre les autorités locales et les autorités étatiques.

Jugeant ces arguments insuffisants, la Commission entame une procédure en manquement d'Etat devant la Cour de Justice Européenne (CJE) en 1990. Le premier jugement rendu par la Cour le 7 avril 1992 constate que la Grèce ne respecte pas les obligations découlant des directives «déchets» et enjoint cet Etat de remédier à ces violations évidentes des règles de bonne gestion (articles 4 et 6 de la directive 75/442/CEE relative aux déchets et article 12 de la directive 78/319/CEE relative aux déchets toxiques et dangereux). Cependant, en 1997, cette décision restant sans effet, la Commission récidive son action devant la CJE.

Dans son arrêt du 4 juillet 2000, la Cour déclare qu'en n'ayant pas pris les mesures nécessaires pour assurer que les déchets seront éliminés sans mettre en danger la santé de l'homme et sans porter préjudice à l'environnement, la Grèce n'a pas pris les mesures indispensables que comporte l'exécution de l'arrêt du 7 avril 1992 et a manqué aux obligations qui lui incombent.

La Grèce se voit ainsi infliger une astreinte de 20000 euros par jour de retard dans la mise en oeuvre des mesures nécessaires pour se conformer à l'arrêt précité. Se faisant, elle crée un précédent peu glorieux: pour la première fois un Etat membre est condamné au paiement d'une amende, en application de l'article 228 du traité, pour non respect des dispositions communautaires environnementales.

La Commission dut encore envoyer périodiquement aux autorités des lettres exigeant le paiement de l'astreinte journalière. Enfin, en mars 2001, le site a été fermé et les déchets envoyés vers une installation adaptée. Le dossier était ainsi clôt, après que les contribuables grecs aient du payer une somme représentant un total de 540000 euros!

Cette affaire n'a bien entendu pas résolu l'ensemble des défaillances dans l'application des directives relatives aux déchets. Ainsi, par exemple, alors que le dossier était encore en cours, la CJE était à nouveau été saisie en 2000 pour une décharge non contrôlée à Epitalio dans le Péloponnèse. Pendant l'année 2002, la Commission poursuivait la Grèce, avec d'autres Etats membres, pour non conformité ou mauvaise application de la directive 75/439/CEE concernant le traitement des huiles usagées. Quant à l'opposition des autorités locales face à de nouvelles installations, le cas de la gestion des déchets de la capitale précédemment évoqué, montre que c'est un phénomène récurant.

2) Au delà de l'outil législatif

La nécessité d'une vision politique

Les exemples qui précèdent sont évocateurs des facteurs les plus fréquents de l'ineffectivité du droit. Ils ne devraient pas faire oublier cependant l'étendu de l'arsenal législatif et réglementaire grec en matière de protection de l'environnement. La Grèce fait d'ailleurs figure de bon élève parmi les Etats membres si l'on s'en réfère aux derniers rapports de la Commission européenne sur l'application du droit communautaire de l'environnement[1].

Certes la Commission entame régulièrement des recours devant la CJE contre l'Etat grec pour faire constater le manquement à certaines dispositions communautaires : ainsi a-t'elle décidé de saisir la Cour en raison de l'absence de conformité, notamment dans la région industrielle d'Elefsina, à la directive 91/271/CEE relative au traitement des eaux urbaines résiduaires qui impose aux États membres de veiller à ce que les agglomérations disposent de systèmes de collecte et de traitement des eaux usées. En 2002, la Grèce a été condamnée par la CJE pour n'avoir pas correctement transposé en droit interne la directive 96/61/CE relative à la prévention et à la réduction intégrées de la pollution.

Le problème majeur ne semble cependant pas se situer au niveau du dispositif législatif mais plutôt dans l'absence d'approche globale des questions d'environnement et de leur intégration dans le développement des diverses politiques.

Pourtant les chiffres parlent d'eux-mêmes en faveur d'une amélioration de la prise en compte de l'environnement. Le tourisme en est un exemple. Ce secteur a oublié le rôle joué par le patrimoine naturel dans sa propre survie, il a négligé le caractère épuisable des ressources naturelles, y compris de l'espace et des paysages. L'approche commerciale du tourisme qui a trop longtemps joué la carte quantitative, l'objectif étant d'attirer un maximum de touristes, montre aujourd'hui ses limites : la demande touristique évolue vers des critères de qualité, les recettes par touriste ont diminué et le rythme d'accroissement du secteur touristique est stagnant (6% du PIB depuis 1994). La politique touristique tente désormais de s'orienter vers davantage de critères de qualité et de diversité de l'offre (développement d'une forme d'éco-tourisme ou de tourisme rural en Epire ou en Crète par exemple).

De même en matière d'environnement urbain, l'intérêt économique et social d'une amélioration qualitative n'est plus contestée. Depuis le milieu des années 1980, des projets de restauration des centres urbains ont été développés pour améliorer l'environnement urbain et la qualité de vie en général. A Athènes, les transports en commun se sont développés, de nombreuses zones pédestres ont été aménagées, les sites archéologiques mis en valeur, la qualité de l'eau et de la gestion des déchets améliorée, le suivi de la qualité de l'air renforcée, etc. L'échéance des Jeux Olympiques de 2004 et les investissements énormes qui y sont rattachés sont une nouvelle occasion d'amélioration de l'environnement dans la capitale. Ainsi, les services de nettoyage et de gestion des ordures devraient être intensifiés dans diverses villes; des in-

---

1 Voir : Troisième étude annuelle sur la mise en œuvre et le contrôle de l'application du droit communautaire de l'environnement publié le 1.10.2002, SEC(2002) 1041 et Quatrième étude annuelle publié le 7.7.2003, SEC(2003) 804.

vestissements importants ont été fait pour améliorer la qualité de l'eau de baignade dans l'Attique et des travaux d'infrastructures de transports bénéficient aussi aux transports en commun (nouveau tram du centre ville aux quartiers longeant le littoral, train rapide jusqu'à l'aéroport ou extension du métro). Cependant, par d'autres aspects, l'organisation des Jeux Olympiques d'Athènes sont décriés du fait des atteintes à l'environnement qu'ils engendrent. Les installations des bassins destinés à l'aviron dans la zone humide, réserve naturelle, de Schinia près de Marathon en sont un exemple.

La nécessité d'intégrer les préoccupations environnementales dans les projets de développement est connu des décideurs politiques et peu à peu mise en pratique. Il est dommage que les politiques menées n'aillent pas jusqu'au bout de cette démarche et que des décisions contradictoires soient régulièrement adoptées – les unes allant dans le sens d'une prise en compte de l'écologie, les autres en faisant totalement fi. Le manque de cohérence entre les diverses politiques et à l'intérieur même d'un programme amenuisent les efforts entrepris dans le bon sens.

Les outils non réglementaires

Parallèlement à l'outils réglementaire, la Communauté européenne promeut d'autres instruments, tels que les mesures économiques ou les démarches volontaires et contractuelles. Ils viennent compléter la réglementation et participent à l'intégration de l'environnement dans les autres politiques.

Il existe quelques exemples allant dans ce sens en Grèce telles que les aides aux mesures environnementales (pour les technologies propres, pour les économies d'énergie) et les pénalités infligées en cas de pollutions marines, collectées dans un fonds spécial (Fonds Bleu) destiné à réduire la pollution. Cependant l'outils économique est encore peu usité. Ainsi, les tarifs de l'eau et de l'énergie tels qu'ils sont pratiqués ne permettent pas d'internaliser (autrement dit de prendre en compte) les coûts environnementaux et sociaux.

Les démarches volontaires du secteurs privés sont relativement limitées et sont le fait d'un nombre restreint d'entreprises parmi les plus importantes. Ainsi, les entreprises de l'industrie de l'aluminium ont mis en place en 1992 un organisme chargé de la récupération et du recyclage des emballages en aluminium et la principale entreprise hôtelière grecque, Grecotel, a développé une politique interne de gestion environnementale.

De même, en matière de sensibilisation du public, des progrès ont été accomplis, mais l'administration comme le grand public ne sont pas assez familiarisés avec le concept de développement durable, ni avec les principes de protection de l'environnement. Le réseau associatif dans ce domaine, même s'il s'épanouit, est encore trop restreint pour pouvoir efficacement participer à la sensibilisation des citoyens. Seules les associations ayant pignons sur rue, tel WWF ou Greenpeace, ont une action critique et d'opposition visible.

L'une des conséquences de cette situation est que la démarche participative et les processus décisionnels ouverts ne sont pas encore entrés dans les usages de l'administration.

Une approche territoriale de l'environnement

La Grèce présente une variété et de forts contrastes qui doivent nécessairement être pris en compte dans la politique environnementale: régions très rurales (Thessalie, Péloponnèse) ou très urbaines (Athènes, Thessalonique), zones côtières touristiques et intérieur des terres agricole, près de 3000 îles constituant un environnement spécifique, etc. Chacune des ces zones présentent des caractéristiques économiques, sociales, culturelles et écologiques qui appellent une approche particulière.

Développer une politique environnementale efficace suppose une adaptation au cas par cas de l'intégration des préoccupations écologiques dans les autres politiques, en suivant les principes directeurs que le droit grec de l'environnement a clairement établis. Ainsi, la gestion des déchets dans les îles des Cyclades et à Athènes doit poursuivre les mêmes objectifs de réduction, recyclage et traitement dans des conditions sûres et saines. L'une et l'autre sont cependant dépendantes de facteurs différents (par exemple, variation saisonnière du volume des déchets dans les îles touristiques et concentration quotidienne d'importants volumes à Athènes) qui requièrent des moyens différents. De même, le développement de l'activité touristique dans une zone agricole de montagne et dans l'Attique engagent des acteurs différents, supposent des infrastructures, des contraintes environnementales, des dimensionnement de projets distincts.

Une telle approche, si elle paraît évidente, demande davantage d'implication des acteurs locaux (pouvoirs locaux, entreprises, public) et une décentralisation des moyens humains et financiers. Il revient à l'Etat de maintenir la cohérence des actions menées avec le reste du territoire et avec les politiques nationales. Il est donc regrettable de constater que les principales défaillances concernent précisément l'information et l'éducation des acteurs locaux, les outils d'aménagement du territoire et de mise en cohérence de l'action de l'Etat.

Conclusion

Les dernières années ont vu la Grèce s'engager dans la croissance économique, supportée par les subventions européennes. Les activités qui se sont développées n'ont pas toujours, presque jamais, été conçues dans le cadre d'un programme de développement concerté et durable. Les projets même les plus importants n'ont que trop rarement fait l'objet d'évaluation coût-avantage et d'évaluation des impacts sur l'environnement. Les efforts n'ont pas suffisamment portés sur le long terme et la recherche d'effets démultiplicateurs.

Certes la Grèce peut encore être considérée comme bénéficiant de ressources naturelles et paysagère de qualité. Mais les signaux d'alerte sont nombreux dans tous les secteurs de l'environnement.

Propulsé par le droit communautaire, le droit grec de l'environnement s'est nettement développé depuis le milieu des années 1980. Les institutions et procédures administratives de contrôle et de suivi ont accompagnées cette évolution. Le cadre ainsi constitué a porté ses fruits notamment en terme de réduction des émissions dans l'air, d'amélioration de l'environnement urbain ou de gestion des déchets.

Cependant, de nombreux blocages et lacunes contredisent ces améliorations et certaines situations, en matière de gestion de l'eau ou des déchets par exemple, sont particulièrement inquiétantes. Parmi les obstacles les plus visibles, on note l'opposition des acteurs locaux, peu enclins à faire valoir les dispositions environnementales; le manque de cohérence de l'action de l'Etat, susceptible de méconnaître lui-même la législation en la matière ou de négliger les moyens nécessaires à sa mise en œuvre ; et encore, l'insuffisante sensibilisation et participation du public.

Aujourd'hui en Grèce, comme ailleurs dans le «monde occidental», l'approche classique du développement est remise en question et l'orientation vers un développement durable s'impose. Les jeunes générations s'impliquent davantage que les précédentes dans des démarches citoyennes en faveur de l'environnement. Et l'on redécouvre d'anciennes pratiques de certains milieux ruraux, témoignage de relation homme-nature fondée sur une gestion raisonnable et respectueuse des ressources.

Le droit de l'environnement peut évidemment contribuer aux évolutions positives vers le développement durable. L'arsenal législatif et réglementaire demande encore à être amélioré en ce sens. Mais c'est certainement sur l'application effective de la législation existante qu'il convient de renforcer les efforts. Celle-ci passe par la mise en œuvre de politiques plus globales pour chacun des secteurs d'activité (tourisme, industrie, développement urbain, agriculture, etc.): globales quant aux objectifs fixés, en intégrant l'aspect environnemental, globales dans le temps, en adoptant une vision à long terme, et globale dans l'espace, en optant pour une approche territoriale cohérente.

Principales références

Decleris M., The Law of Sustainable Development, European Commission, 2000

Kouvelis S., Natural resources and Development in Greece , http://www-penelope.drec.unilim.fr

National Center for the Environment & Sustainable Development, Greece, The State of the Environment, Athens, 2001
OECD, Environmental performance reviews – Greece , Paris, 2000.
Public Environmental law in Greece , http://www-penelope.drec.unilim.fr

NICOLAI VOLOVITCH

Ökologie und Grundsteuerreform in Rußland

Unter den Bedingungen der hohen Wirtschaftsmonopolisierung haben die grössten russischen Eigentümer (in anderer Terminologie – finanziell-industrielle Gruppen oder Oligarchen) legale Möglichkeiten, ihre Steuerzahlungen über das System der inneren Preise zwischen den Holdingunternehmen und Off-Shore-Firmen im Ausland zu minimieren. Die Oligarchen benutzen ihren politischen und wirtschaftlichen Einfluß, um zum Nachteil der allgemein anerkannten gesellschaftlichen Interessen einen maximalen Gewinn zu erhalten. In den letzten Jahren wenden sie sich oft gegen die mit dem Anfang der russischen Reformen in den Jahren 1991-1992 gesetzlich festgelegten Grundsteuern und Umweltzahlungen, die von den in den Bilanzen der Unternehmen gezeigten Ergebnissen der finanziell-wirtschaftlichen Tätigkeit nicht abhängen und in der Tat in die lokalen und regionalen Haushalte gehen. Da die Bodengesetzgebung noch ziemlich jung ist (das neue Bodengesetzbuch der RF wurde am 25.10.2001 angenommen), gibt es keine Antwort auf die Frage, wie effektiv die Regelung der Verhältnisse im Bereich der Bodennutzung unter den Marktbedingungen die gesellschaftlichen Interessen realisieren kann.

Es sind nur zwei Beispiele anzuführen: Nach dem Antrag der Aktiengesellschaft „Norilskij Nickel", (sie ist der weltweit größte Produzent von Palladium, Nickel, Kupfer und anderen Buntmetallen), die Regierungsverordnung Nr. 632 der Russische Föderation (RF) vom 28.08.1992 (über die Feststellung des Verfahrens für die Gebühren und deren Grenzwerte für die Umweltverschmutzung, die Unterbringung von schädlichen Abfällen und andere Arten der schädlichen Einwirkung) für ungesetzlich zu erklären, hat das Oberste Gericht der RF am 28.03.2002 diese Verordnung für ungesetzlich erklärt, da die Regierung nach der Meinung des Obersten Gerichtes nicht berechtigt ist, die tatsächlichen Steuerzahlungen festzulegen. Und erst nach der Entscheidung des Verfassungsgerichtes der RF wurden diese Zahlungen (deren Gesamthöhe betrug 2002 ca. 260 Mio. $) am 12.06.2003 wieder ermöglicht. Die Regierung der RF hat mit der Verordnung Nr. 344 ein neues Verfahren für die Festsetzung der Gebühren und deren Grenzwerte für die Umweltverschmutzung erlassen.

Ein weiteres Beispiel: Die breite öffentliche Diskussion über den Wunsch der Unternehmenseigentümer, Grundstücke, die sie heute nach dem Recht auf ständige (unbefristete) Nutzung oder nach dem Mietrecht besitzen, ihnen kostenlos zu übergeben. Die Eigentümer führen in diesem Fall verschiedene Beweisgründe an und benutzen ihren politischen Einfluss, um die Regelung, dass die Interessenten für die von ihnen besetzten Grundstücke einen Preis, der dem 3- bis 17-fachen Bodensteuersatz (für Moskau und St. Petersburg ist der Preis das 30-fache) entspricht, kaufen müssen, abzuschaffen. Als Hauptargument wird angeführt: bei einem solchen Abzug von Unternehmensmitteln würde weiteres bedeutendes industrielles Wachstum nicht möglich sein.

Unter diesen Bedingungen werden die regionalen und die lokalen Behörden und in ihrer Person auch die Bevölkerung zu Geiseln der Interessen einer kleinen Gruppe von Oligarchen. Die Situation kann sich ab dem 01.01.2004 weiter verschärfen. Ab diesem Tag werden die lokalen Behörden nach dem neuen Steuergesetzbuch der RF, das eine Reihe von Umsatzsteuern aufhebt, die Möglichkeit haben, für ihre Haushalte nur die Bodensteuer und -miete sowie andere Vermögenssteuern zu erheben. Gegenwärtig ist die Bodensteuer (auf deren Grundlage auch die Bodenmietsätze für Grundstücke sowohl des staatlichen als auch des städtischen Eigentums festgestellt werden) für jeden Bevölkerungstyp und jede Kategorie der Bodennutzung festgelegt, wobei die städtischen Behörden besondere Hebesätze anwenden dürfen. Dieses System ist aber veraltet.

Die künftigen Prinzipien der Erhebung von Abgaben für den Boden sind noch unklar. Laut der angenommenen Strategie wird geplant, in der Zukunft zu einer einheitlichen Immobiliensteuer und in den nächsten Jahren zur Erhebung einer Bodensteuer im prozentualen Verhältnis zum Verkehrswert des Grundstücks (Art. 65-66 des Bodengesetzbuches) überzugehen.

Im großen und ganzen kann man bestätigen, dass die Möglichkeiten der neuen Bodengesetzgebung noch nicht in vollem Maße zutage getreten sind, daher kann man von verschiedenen Varianten ihrer Anwendung sprechen. Die Behörden suchen ihrerseits nach Varianten einer vollständigen Erhebung der Bodenrente sowie einer rechtmässigen Besteuerung des Bodeneigentums. Dafür wird die Aufgabe gestellt, den ökologischen Faktor bei der Planung beim Städtebau zu berücksichtigen. Moskau ist in der Tat die einzige Region Russlands, wo bei der Regelung der Bodennutzung die ökologische Komponente von Bedeutung ist, was durch den hohen Bodenwert bedingt ist. Bei dem Bau und der Modernisierung von Objekten zahlen die Investoren bedeutende Summen für die Begrünung und die Ausstattung der Grundstücke mit allen nötigen Versorgungseinrichtungen. Gleichzeitig wurden erhöhte Mietsätze für den Boden auf dem Territorium des „Naturkomplexes" der Stadt festgelegt. Es wurde beschlossen und es wurden Mechanismen erarbeitet, um die umweltschädlichen Produktionen aus dem Zentrum und aus den Wohnvierteln in Industriezonen zu verlegen. Dabei existiert die Gefahr, dass die ökologischen Forderungen für die aktive Neuaufteilung der Einflussbereiche zwischen unterschiedlichen finanziell-industriellen Gruppen benutzt werden, und die Eigentümer und Investoren selbst sind vor Willkür der Stadt (Korruption) nicht geschützt.

Die Möglichkeiten, das System der Besteuerung konfliktlos für den Haushalt zu verändern, werden vorläufig durch die Größe des staatlichen Bodeneigentums selbst bestimmt. Innerhalb der letzten zwei Jahre hat der größte

Teil von Bodennutzern in den Regionen die genutzten Grundstücke als Eigentum gekauft.

Die neuen Eigentümer wollen die Bodenbesteuerung reduzieren. Deswegen ist es erforderlich, die Varianten der Bodenbesteuerung und der Mietzinsen zu präzisieren.

Gegenwärtig sind diese Anforderungen auf der föderalen Ebene nur in der Form verankert, dass es vorgeschrieben ist, um die umweltschädlichen Produktionsstandorte zusätzliche sanitäre und Schutzzonen (SSZ) zu bilden (für besonders gefährliche Verschmutzer – bis 2 km von der Betriebsgrenze). Bei Zahlungen für ökologische Schäden geht man von dem Qualitätsstandard der Umwelt (von der maximal zulässigen Konzentration von Schadstoffen – MZK) aus. Diese Hauptstandards der Umweltqualität (keine Gefahr für die Gesundheit des Menschen und für die Umwelt) wurden noch in der UdSSR erarbeitet, einige Standards über die maximal zulässigen Konzentrationen der schädlichen Stoffe in der Luft, im Wasser und im Boden basieren auf den Empfehlungen der Weltorganisation für Gesundheitswesen (WHO). Wenn die Standards an der Grenze der SSZ nicht erreicht werden, ist theoretisch der Bau von solchen Betrieben unmöglich.

Die Standards der MZK sind hoch, der natürliche Schadstoff-Hintergrund ist in einigen Regionen höher. Auf der Grundlage der MZK und für deren Einhaltung werden die maximal zulässigen Normen der Schadstoffemission für die Betriebe (ZNS) festgestellt. Die Betriebe bekommen entsprechende Genehmigungen für die Schadstoffemissionen, die für eine bestimmte Zeit gültig sind. Für einige Verschmutzer wurden provisorische Normen eingeführt (BCB) – annähernd ungefährliche Einwirkungsniveaus und provisorisch abgestimmte Emissionsniveaus. Die BCB ist höher als die ZNS – sie werden in Entsprechung mit der besten alternativen Technologie festgelegt und basieren auf technologischen Verfahren.

Insgesamt wurden nach der Regierungsverordnung Nr. 344 vom 12.06.2003 die Basissätze für 210 Luft- und für 142 Wasserverschmutzer sowie für die Entsorgung von 5 Grade der Schädlichkeit von Abfällen festgelegt. Viele sprechen von der Notwendigkeit, diese Listen zu ergänzen.

Aber solche hohen Standards wurden früher, und werden auch heute oft nicht realisiert. Der tatsächliche ökologische Zustand Russlands ist schlecht. Die Verschmutzung der Umwelt ist extrem. Die Anzahl der Städte, wo die MZK um einen Faktor von 10 überschritten wird, stieg im Jahre 2001 auf 40. Luft- und Wasserverschmutzer sind hauptsächlich Erdölprodukte. Die mit Abfällen der Betriebe verschmutzten Bodenflächen betragen 62 Mio. Hektar. Gelagert wurden 90 Mrd. t Abfällen (darunter sind 2.6 Mrd. t giftig). Das Durchschnittsniveau der Wiederverwertung von Abfällen ist 36 %.

Die Gebühr für die Verschmutzung heisst im neuen Steuergesetzbuch „ökologische Steuer“.

Wird der ZNS eingehalten, so gilt der Basis-Steuersatz. Wird mit einer BCB gearbeitet, ist die Steuer 5 mal höher. Wird auch die BCB nicht eingehalten, beträgt die Steuer des 25-fache. Diese Summen reichen aber nicht aus, um alte Verschmutzungen zu beheben und die Gefahr für die Umwelt zu minimieren. Weil Umsatz u. Export mit geringen Innere-Firmenpreisen laufen, zahlen die Holdings (u. damit die ganze Wirtschaft) niedrige Gewinn- und Umsatzsteuern. Damit leiden besonders die regionalen und örtlichen Haushalte. Mit geplanten Reformen in der Energie- und Eisenbahnbranche (in Richtung Privatisierung) besteht große Gefahr, dass sich diese negativen Prozesse der weiteren Monopolisierung vergrössern werden. Andere Steuern, mit denen etwas für die Umwelt getan werden kann, gibt es nicht.

Daraus resultiert u. a.:

1) Russland möchte in der Reihe der ökonomisch starken Länder sein. Der Präsident hat vor kurzem das Ziel genannt, in 8 Jahren Russlands Inlandprodukt zu verdoppeln. Deshalb können ökologisch schädliche Betriebe, die Positionen auf dem Weltmarkt sichern, nicht geschlossen werden.
2) Die Regierung will die Umsatzsteuerarten senken, und teilweise durch Vermögensteuern ersetzen. Besonders die lokalen Haushalte verlieren 2004 fast alle eigenen Steuern, außer der Vermögen- u. Grundsteuer.

Wir haben ein Dilemma – wir sind von umweltschädlicher Industrie abhängig, aber möchten die Situation durch Umweltschutz verbessern.

Die Lösung muss für verschiedene Regionen u. Städte unterschiedlich sein.

Es gibt große Unterschiede auf den Bodenmärkten in Moskau und in den Provinzen. Wenn in den Provinzen die umweltschädlichen Betriebe in der Regel der einzige Beschäftigungsort für die Bevölkerung sind, so haben solche Gesellschaften in Moskau ihre Hauptbüros und Wohnungen für die Eigentümer und Topmanager. Wegen der Nachfragekonzentration ist der Marktwert von Grundstücken in Moskau sehr hoch und wird immer höher. Deshalb hat Moskaus Stadtverwaltung verschiedene Lösungen praktiziert.

Vor ein paar Jahre kauften Spekulanten schädliche Betriebe und stoppten die Produktion. Statt für die Industrieproduktion benutzen die neuen Inhaber die Grundstücke als Baustelle für Wohnungen oder Büros und Handelszentren. So hat Moskau einige wichtige Betriebe, die ihren Sitz in der Stadtmitte hatten, verloren.

Moskau strebt an, die Gewerbebetriebe zu erhalten, weil die Betriebe 25% des Einkommens der Stadt bringen. Zuerst wurde als Strategie betrachtet, mit Hilfe von privaten Investoren die Betriebe in spezielle Gewerbegebiete umzusiedeln. Dazu übte die Regierung administrativen Druck aus, damit die Betriebe keine schädliche Produktion in Wohngebieten (z. B. in der Stadtmitte) weiter führen dürfen. Parallel dazu gilt es, Wettbewerb zwischen den Investoren, die Rechte kaufen, um Betriebe in bestimmte Stadtteile überzusiedeln, und um selbst die ehemaligen Industriegrundstücke neu zu bebauen.

Aber seit diesem Sommer hat die Stadtregierung eine neue Strategie. Zuerst werden Grundstücke am Stadtrand für die Industrie neu erschlossen. Dann werden den Betrieben für die Übersiedlung dorthin Kosten erstattet. So hofft die Stadtverwaltung, mehr Geld durch die Vergabe der Stadt gehörender Grundstücke (zeitlich beschränktes Baurecht) nach der Übersiedlung für das Budget der Stadt zu gewinnen.

Solche Modelle kann man z. Z. nur in Moskau und in geringer Zahl in anderen Zentren realisieren, wo die Grundstückspreise hoch sind. Auch hilft in Moskau die städtische Bodenpolitik, die die Privatisierung ausschließt, so dass Grundstücke nur vermietet werden. Das ist im Grunde ein Verstoß gegen die russische Gesetzgebung. In den Regionen haben die Grundstücke der Industriebetriebe, wo die Technologien seit den 60-er Jahren unverändert blieben, einen niedrigen Marktwert (um so mehr, als es verboten wurde, diese stark mit Schadstoffen belasteten Grundstücke zu privatisieren). Die Komplikation besteht darin, dass das Verfahren für die Bestimmung des Preises von Grundstücken, die für die Bebauung zur Verfügung gestellt werden, keine verbindliche Kompensation des zugefügten ökologischen Schadens vorsieht. Deswegen bedarf die Berücksichtigung dieses Faktors bei der Bewertung der Steuerbasis einer ernsthaften wissenschaftlichen Begründung.

Dabei verfolgen die Regionen oft nur fiskalische Ziele. Es ist nötig, eine Reihe von Ergänzungen und Änderungen der Gesetzgebung zu verabschieden. Der Minimalverkaufspreis sollte dabei den Kompensationswert des möglichen ökologischen Schadens in sich einschliessen.

Wie versucht man, in der Provinz Probleme zwischen Ökologie und Wirtschaft zu lösen? Nehmen wir als Beispiel ein Kombinat in Sibirien – Atschinsktonerdekombinat (AGK) – ein „ Global Player" der Branche.

Es produziert pro Jahr mehr als 1 Mio t- Tonerde und 0.6 Mio t kalziniertes Soda (Kaustik Soda) und andere Produkte – wichtige Rohstoffe für die Aluminiumindustrie. Vor zwei Jahren produzierte das Kombinat auch Zement, jetzt ist diese Produktionanlage vollig für Tonerde umgestaltet aus Gründen der Rentabilität.

AGK gehört seit 2000 zur Holding „Russisches Aluminium", mit 10% der Weltaluminiumproduktion. Die Inhaber der AGK sind meist in Zypern registriert. Der wichtigste Inhaber ist aber bekannt als Gouverneur der Provinz Tschukotka.

Im AGK sind zehntausend Menschen beschäftigt. Für die Stadt mit 120 Tausend Einwohnern spielt der Betrieb eine besondere Rolle, so wie auch für ganze Region. Aber er bringt für das Budget wenig. Schon seit drei Jahren stehen in den Bilanzen, die die Manager nach westlichem Standard führen, 10-13 Mio $ Verluste, weil das ganze Geschäft durch eine Firma aus Tschukotka mit inneren Preisen läuft.

Die regionale Regierung steht vor der Frage, wie kann das Einkommen für Stadt und Region vergrößert werden? Das fehlende Steuervolumen versuchen sie teilweise durch die Grundsteuer zu ersetzen. Ein Weg ist es, die Grundsteuer zu erhöhen. Aber seit 1991 gibt es ein Gesetz, in dem für jede Stadt nach Größe und Lage durchschnittliche Grundsteuersätze als konkrete Ziffern vorgegeben sind. Oft beschliessen für solche Betriebe örtliche Macht-organe höhere Mietzinsen oder spezielle Hebesätze für die Grundsteuer. Z. B. gilt laut Gesetz in Atschinsk der durchschnittliche Grundsteuersatz von 8 RR pro qm. Aber in Kreis Atschinsk gilt ein Steuersatz von 0.26 RR pro qm. Deshalb zahlt AGK Miete, deren Satz 20 mal höher als der Grundsteuersatz ist. Viele Betriebe haben in den letzten 2 Jahren Grundstücke gekauft (wie auch AGK – siehe Tabelle 1). Dadurch zahlen sie keine Miete, nur Grundsteuer. Und zweitens – das neue russische Steuerrecht schreibt vor, dass die Grundsteuer maximal 2 % des Katasterwerts (der am Verkehrswert orientiert sein soll) sein darf.

Wie im Fall des AGK das Problem gelöst werden soll, ist aber völlig unklar.

1) Ein Residualverfahren geht nicht, weil der Betrieb schon 30 Jahre funktioniert.
2) Für das Vergleichswertverfahren gibt es keine ähnlichen Betriebe in der weiteren Umgebung.
3) Mit Erträgen kann auch nicht gerechnet werden, weil der Betrieb offiziell nur Verluste macht.

Dann bleibt nur die Möglichkeit, mit Kosten zu rechnen und einen Preis deduktiv zu bestimmen. Wir haben versucht, einen „Sachwert" zu kalkulieren.

Tabelle 1. Grundstücke des Atschinsktonerde Kombinates

Rechte	Fläche, ha	Darunter				
		In der Stadt				Im Kreis
			Einschliesslich			
		insgesamt	Betrieb	Technologiezone (Reinigungssee u. Abfall)	Schutzzone	(Tagebau u. Abfall)
Privates Eigentum seit Ende 2002	390	390	390°	-	-	-
Mietrecht (25–49 Jahre)	1550	940	-	590	350	610
Insgesamt	1940	1330	390	590	350	610

° darunter 57,5 ha, die 2002 zusammen mit einem Eisenbahnunternehmen gekauft wurden.

Wir konnten bei unserer Untersuchung nur 4 Arten von Kosten prüfen:

1. Bezahlung des Grundstücks an die Landwirtschaft oder Forstwirtschaft

Es gibt einen Erlass der Russischen Regierung, dass für jedes Grundstück, das von der Landwirtschaft oder Forstwirtschaft für eine Bebauungng abgegeben wird, gewisse Summen bezahlt werden müssen. Für die Grundstücke der AGK bedeutet das (nach der Qualität des Bodens), dass pro qm 11.6 bis 31.4 RR von AGK an die öffentliche Hand bezahlt werden mussten.

Das Grundstück der AGK, das von Bauern oder Anderen enteignet wurde, hat die Größe von 3231,4 ha, darunter 51 % Ackerland (z. Z. 31.4 RR pro qm). 1590 ha (ohne die fehlende Schutzzone) nutzt der Betrieb heute noch. Dafür muss voll gezahlt werden. 1641 ha wurden nur zeitweilig genutzt und zurückgegeben. Dafür sind nur 50 % zu bezahlen. Insgesamt wären für 1590 ha heute 32.6 RR zu zahlen.

Die Bebauung von AGK dauerte 10 Jahre, der Tagebau konnte nach 4 Jahren genutzt werden. Werden 14 % Zinsen gerechnet (10 % Liegenschaftszinsen, 18 % Kreditzinsen – durchschnittlich 14 %), so ergibt sich für 10 Jahre ein Faktor von 3,707 – also 120,8 RR pro qm für AGK, und ein Faktor von 1,68896 für den Tagebau – also 55 RR pro qm.

2. Erschließung mit Elektrizität und durch die Eisenbahn

AGK hat zwar ein eigenes Kohlekraftwerk und eigene Wasserversorgung und Entsorgung, braucht aber auch fremden Strom.

Der Eisenbahnanschluß hat einen heutigen Wert von 258 Mio RR, für 980 ha Betriebsfläche zusammen mit der technischen Zone sind das 26.3 RR pro qm. Für die Elektroversorgung von außen wären noch 129 RR pro qm zu zahlen.

3. Ökologische Forderung einer Schutzzone

Die Schutzzone existiert nicht. Für AGK sollte sie 493 ha (Bauprojekt) groß sein. Das wären 47 RR pro qm (Grundstückserwerb und Schutzwaldaufforstung).

Das Grundstück des Betriebes kann damit für die Grundsteuer mit dem Katasterwert von 323 RR pro qm bewertet werden – etwa 10 Euro pro qm. Das bedeutet unter anderem, dass Industriegrundstücke überall einen Wert haben, der sich allerdings großenteils aus der Erschließung ergibt.

Aber wenn das Gesetz als Grundsteuer 2 % vom Katasterwert ermöglicht, ist die mögliche Grundsteuer 6,44 RR – weniger als heute (8 RR). So kann die Katasterbewertung zu Verlusten für Regionen u. Städte führen. Z. Z. erhält die Stadt von AGK 95 Mio RR, der Kreis 35 Mio RR. Zusammen sind das 4.2 Mio $. Diese Summen teilen lokale Budgets mit der Region. Deshalb ist jede Senkung gefährlich.

Weitere Möglichkeit, die Grundsteuer mindestens zu erhalten, wäre die Bewertung der ökologischen Folgen. Für die ökologischen Folgen bezahlt AGK etwa 3 Mio $ pro Jahr. Weil der Betrieb besonders ökologieschaedlich ist, gelten spezielle Normen für die Schadstoffe (BCB sind zeitlich begrenzt bis 2007), deshalb zahlt das Kombinat das 5-fach des Basissatzes als Oekosteuer.

Die Grundsteuer könnte noch erhöht werden, wenn der Eingriff in die natürlichen Ressourcen auch bewertet würde.

Das könnte z. B. für den Wasserverbrauch gelten. Der Betrieb entnimmt für eigene Nutzung ca 50 Mio Kubikmeter Wasser vom Fluß. Außerdem entnimmt er noch ca 50 Mio Kubikmeter, die stark verschmutzt zurückgegeben werden (hauptsächlich Schwermetalle und Ölrückstände). Auch die Luft wird erheblich verschmutzt. Jährlich fallen als Abfall mehr als 6 Mio Tonnen an – direkte Folge der Fabrik.

Wie das zu bewerten ist, das müssen neue Forschungen ergeben. Die möglichen Ziffern liegen in einer riesigen Spanne.

Die Möglichkeit, aus dem Fluss 120 bzw. 50 Mio. Kubikmeter Wasser entnehmen zu können, ist eine Eigenschaft des Grundstücks, das seine Grundrente erhöht – so wie Adam Smith bei Farmen auf den Shetlandinseln die Möglichkeit des Fischfangs zur Grundrente zählte. Ohne die Möglichkeit der Wasserentnahme wäre die Produktion unmöglich. Ein Bewertungsmaßstab lässt sich über Kosten finden, wobei noch zu prüfen sein wird, ob mit innerrussischen Kosten oder Weltmarktkosten gerechnet werden müsste.

Die ermöglichte Luftverschmutzung wird schon durch einen höheren Steuersatz (Ökosteuer) bezahlt. Diese soll aber als Strafsteuer den Schadstoffausstoß mindern. Trotzdem ist zu prüfen, ob die Möglichkeit, ein Werk zu betreiben, das solche Belastungen auch nach dem erlaubten Standard mit sich bringt, als Teil des Grundstückswerts angesehen werden kann, oder ob durch Schutzzonen der Schaden aufgefangen wird bzw. werden sollte. Deren Kosten werden aber eine Erhöhung des Grundstückswerts wohl ausschließen.

Hinsichtlich des Abfalls kann von einem Einfluss auf den Bodenwert kaum ausgegangen werden, wenn der Abfall nach den anerkannten Regeln der Technik mit entsprechenden Kosten entsorgt wird.

Die Fragen der Bewertung von Grundstücken, auf denen natürliche Ressourcen in Anspruch genommen werden können – nicht nur die Bewertung von Umweltschäden –, bedürfen noch weiterer wissenschaftlicher Forschung. Das gilt sicher nicht nur für Russland.

HAGEN HENRŸ

From Land Legislation to Participatory Forms of Preserving Nature in Sub-Saharan Africa – New Approaches or Old Ones?

I. Introduction

1. The European Faculty of Land Use and Development is mainly concerned with developments in Europe. Why then speak about Sub-Saharan Africa?

At its 32nd symposium in Strasbourg last year I argued that developments in Europe have an incidence on Sub-Saharan Africa. Land legislation in Sub-Saharan Africa, for example, is copied from Western models. What is decided in the Western hemisphere will find its way into the statute books of Africa. There it lives on as phantom laws. These phantom laws unfold weak legal, but strong extra-legal negative effects.

2. The title of my contribution "From Land Legislation to Participatory Forms of Preserving Nature in Sub-Saharan Africa" reflects a much acclaimed shift in this respect when it comes to the preservation of nature: Supposedly a shift from mimicry to participation. The subtitle "New Approaches or Old Ones?" questions, however, the pertinence of this claim. Indeed, I shall try to demonstrate that 'participation' is just another in an inflationary series of catchwords. It was invented by a well intentioned, but western-centric international community of donors, researchers and politicians. They are little aware of the ever growing loss of influence of the majorities of the peoples of this globe in the decision making processes concerning the preservation of nature. For the rest, the concept of participation – if such it may be named? – is a thing of the past. As we meet, ecological intervention is the word. And, furthermore, the debate on the preservation of nature operates with a reference to 'resource management', ignoring the unchanged legal framework, which consists mainly of private ownership rights, including intellectual ownership rights related to natural resources.[1]

3. Concerned mainly with aspects of public international legal procedures and the preservation of nature, I possibly commit a sacrilege in this assembly and I reduce the wider concern for the well being of oikos in an unscientific

1 Dall'Aglio, Andrea, Ingérence écologique: un débat, in: Nouveaux Cahiers de l'IUED, no. 3 (Ecologie contre nature. Développement et politiques d'ingérence), Paris: PUF et Genève: IUED 1995, 169-191 (172).

way. May I justify, however, this sacrilege and this reduction with the need to ask whether such noble goals as the preservation of nature in its biodiversity may be pursued without having regard to the ways leading thereto? The answer is clearly: No! I shall sketch – for more I am not able to do in the space allocated – the line of diminishing participation by the peoples of Sub-Saharan Africa in the development of their countries. This line starts with land legislations and ends with the adoption of recipes for natural resource management, including land, through local communities. These local communities are being invented[2] by development funding agencies, far away from the so much cherished "local people" they claim to serve.

4. The contribution is a report on an ongoing research into the ways public, and indeed public international affairs, like the preservation of nature, have become the affair of ever fewer decision makers. It is – therefore – partly hypothetical in nature … and it is provocative in style: As long as worldwide news headlines on a short electrical power cut in North America[3] are to capture the minds of Africans, Asians, Meso- and South Americans, for whom electrical power supply remains an unfulfilled dream, and as long as westernized people around the globe use more than 250 l of drinking water per capita and day[4], participation remains an empty shell. In a world of finite resources, economic growth[5] equals fewer opportunities to participate for most people.

---

2 Cf. Gallon, Thomas-Peter, Mythos oder Methode bei der Planung von Partizipation – Die verklärte "Fokonolona"-Tradition und "Soziale Integration" in einem madegassischen Dorf, in: afrika spectrum 91/2, 181-197; Henrÿ, Hagen, Boden(rechts)-ordnungen in Afrika südlich der Sahara. Am Ende der Skandale ein Entwicklungsskandal?, in: Verfassung und Recht in Übersee. Law and Politics in Africa, Asia, and Latin America 1998, 196-214; Merten, Peter, Anspruch und Wirklichkeit zielgruppenorientierter Entwicklungsplanung: Das Beispiel integrierter ländlicher Regionalentwicklung im Handeni-Distrikt im Nordosten Tanzanias, in: afrika spectrum 91/2, 199-220 (199): "Direkte Zielgruppenbeteiligung konnte in Handeni (Tanzania) nicht erreicht werden; stattdessen übernahm … ein sozialökonomisch und -anthropologisch orientierter Forscher die Rolle des Stellvertreters der Zielgruppe im Planungsprozess".

3 In August 2003 the electrical power supply broke down in most of the northeastern parts if the USA and of Canada for a couple of hours. Worldwide news coverage of this event amounted to about the length of the power cut.

4 Cf. Helsingin Sanomat, 11.8.2003, B 1.

5 Economic growth remains the means by which development is thought to be "reached". It is based on an equation which does not take into consideration natural resources. Neither the so-called Brandt Report, nor the so-called Brundtlant Report (critical of that latter report, which emphasizes economic growth as a prerequisite of development at the wake of the UN Rio Conference on Environment and Development, Grinevald, Jacques L'ingérence des écologistes dans les affaires internationales, in: Nouveaux Cahiers de l'IUED, no. 3 (Ecologie contre nature. Développement et politiques d'ingérence), Paris: PUF et Genève: IUED 1995, 121-142 (132))

II. From Land Legislations in Sub-Saharan Africa to Natural Resource Management: A Chain of Diminishing Participation

5. By the time European powers took hold of Sub-Saharan Africa, European land law differed radically from the social relationships referring to land in Sub-Saharan Africa. Sub-Saharan African societies exclude foreigners from access to land or, where they grant that access, it can be withdrawn under certain circumstances. The colonial powers thought it inadmissible that property/ownership rights to land, which had contributed to the development of Europe, were to be withheld from African soils. They therefore introduced the notion of private individual, registered ownership of land.

Participation in the decision on this policy was not an issue. By and large, this policy failed.[6]

6. After independence, the Sub-Saharan African States continued this policy, unchanged. The State became the owner of all unregistered land – still today some 97 to 99 % of the total land areas – and was empowered to grant temporary use rights.

Participation was not an issue for those States. But it became an issue for the international community. What used to be normal colonial practice and expected by the international community during the first development decade in terms of an interventionist state, came to be strongly criticized in the 1980ies in terms like 'lack of democracy', 'lack of endogenous development and participation', 'Human Rights abuses' etc..

7. The latest intervention into the ways in which the majorities of the African peoples conceive the social relationships referring to land is the establishment of so-called protected areas, totally under the control of the State.[7] Ac-

---

dealt with the challenge raised in the early 1970ies with the Report by the Club of Rome. Cf. Steppacher, Rolf, L'ingérence écologique et la globalisation de l'économie de marché, in: Cahiers de l'IUED, no. 3 (Ecologie contre nature. Développement et politiques d'ingérence), Paris: PUF et Genève: IUED 1995, 99-114 (106, 106/7) as for the impossibility of rapid economic growth – and hence participation! – based on finite natural, not renewable resources.

6 It is estimated that not more than 1-3 % of all land is registered. Cf. LeRoy, Étienne, Actualité des droits dits "coutumiers" dans les pratiques et les politiques foncières en Afrique et dans l'Océan Indien à l'orée du XXIe siècle, in: Cahiers d'Anthropologie du droit 2002, 237-264 (257).

7 Barume, Albert Kwokwo, Heading Towards Extinction? Indigenous Rights in Africa: The Case of the Twa of the Kahuzi National Park, Democratic Republic of Congo, Kopenhagen: International Work Group for Indigenous Affairs 2000; Juma (Ibrahim H., Extinction of Customary Land Rights in the Wildlife Conservation Areas of Tanzania: The Case of Mkomazi Game Reserve, in: Recht in Afrika. Law in Africa. Droit en Afrique 2000, 133-172) speaks of "the demise of customary land

cording to United Nations (Environmental Program (UNEP)) estimates[8], some 100 protected areas around the world cover a surface equaling that of China and India. Around 10% of the territories of the Southern African States are under strict protection measures, including protection laws. In Tanzania, the size of these areas exceeds that of the territories of the Netherlands, Slovakia and Switzerland combined. If one adds to these areas the so-called buffer zones[9], the protected areas in these countries cover a considerable portion of the state territory, subtracted from popular participation.

The buffer zone concept, which takes the protected areas as an undisputed given, stems from the 1968 UNESCO "Man and Biosphere Program"[10]. While Sub-Saharan African States might exclude the majorities of their populations from the design of their land law policies, not the least because the so-called international community so required, the "Man and Biosphere Program" excludes both the peoples and the governments from the decision making process.[11] Furthermore, its positive effects for the environment are not proven in all cases.[12]

8. This latest concept of land law policy in Sub-Saharan Africa converges with the international community's concern for the preservation of nature. The 1992 United Nations Conference on Environment and Development in Rio de Janeiro symbolizes the starting point of 'resource management' aiming to preserve the bio-diversity, whilst having regard to social justice. For the sake of this concern, the international community sanctions positively the means by which Sub-Saharan Africa States have tried since independence to consolidate their control over the totality of land comprised within their ahistorical boundaries. The administrative structures which these states inherited from colonial times, and on which they have to rely on for the sake of their own continuity, lack mechanisms through which democratic or participatory decision making procedures could be furthered. Under these circumstances

---

rights in the wildlife conservation areas of Tanzania", despite all lip service paid to the recognition of customary law. Dall'Aglio. op. cit., 177, referring to Cameroon.

8 Reported in Pretoria News, 17.9.2003, 10.

9 I.e. the zones adjacent to the protected areas and where human activity is restricted to those which maintain the ecological security of the protected areas while providing benefits to local communities. Cf. Neumann, Roderick P., Primitive Ideas: Protected Area Buffer Zones and the Politics of Land in Africa, in: Development and Change 1997, 559-582.

10 Cf. Culmsee, Heike, Das UNESCO-Biosphärenreservatkonzept als Instrument der Entwicklungszusammenarbeit, in: WeltWissen. Entwicklungszusammenarbeit in der Weltgesellschaft, Herausgeber Markus Kaiser, Bielefeld: transcript 2003, 163-181.

11 Cf. Dzingirai, Vupenyu, The New Scramble for the African Countryside, in: Development and Change 2003, 243-263.

12 Schmidt-Soltau, Kai, Conservation-related Resettlement in Central Africa: Environmental and Social Risks, in: Development and Change 2003, 525-551.

the Sub-Saharan African States carry on with homogenizing the vast variety of political orders[13] – kingdoms, republics, chieftaincies, acephalous subsistence societies – to be found on most of their territories, in order to fit them into the European-type state pattern.[14] Attempts[15] aiming to allow for more participation through decentralization and federalization have failed. It remains to be seen whether new approaches, known under the name of 'new constitutionalism' will allow for more participation.[16]

This homogenization of political orders equals less participation. It accounts for much of the loss of valuable knowledge which could help to preserve nature.[17] The introduction of culturally foreign land law destroys ecological knowledge.[18]

---

13 Cf. for example Magnant, Jean-Pierre, Peuple, ethnies et nations: Le cas du Tchad, in: Droit et Cultures 8/1984, 29-50.

14 Englebert, Pierre, The contemporary African state: neither African nor state, in: Third World Quarterly 1997, 767-775; Okafor, Obiora Chinedu, After Martyrdom: International Law, Sub-State Groups, and the Construction of Legitimate Statehood in Africa, in: Harvard International Law Journal Vol. 41, 2000, 503-528.

15 Cf. for example Dezentralisierung, Demokratisierung und die lokale Repräsentation des Staates. Décentralisation, démocratisation et les représentations locales de la force publique, herausgegeben von Jakob Rösel / Trutz von Trotha, Köln: Rüdiger Köppe 1999; Stiel, Philipp C., Dezentralisierung in Marokko und Burkina Faso, in: Recht in Afrika. Law in Africa. Droit en Afrique 1999, 133-152.

16 Cf. for example Gonidec, Pierre-François, Constitutionnalismes africains, in: Revue Africaine de Droit International et Comparé. African Journal of International and Comparative Law 1996, 12-37; Nahum, Fasil, Ethiopia: Constitution for a Nation of Nations, in: The Review (International Commission of Jurists) no. 60 (1998), 91-101; Tshiyembe, Mwayila, L'Afrique face au défi de l'État multinational, in: Le monde diplomatique, septembre 2000, 14 f.. Cf. also van Eijk, Ryan, The United Nations and the Reconstruction of Collapsed States in Africa, in: African Journal of International and Comparative Law. Revue Africaine de Droit International et Comparé 1997, 573-599; Henrÿ, Hagen, Political Order in Subsaharan Africa – the European View of African States at Stake, in: Kansainoikeus. Ius Gentium 1988, 131-161; Scholler, Heinrich, Vom kolonialen zum modernen afrikanischen Staat, in: Verfassung und Recht in Übersee. Law and Politics in Africa, Asia and Latin America 1983, 103-117; Sur, Serge, L'État entre l'éclatement et la mondialisation, in: Revue belge de droit international 1997/1, 5-20; Wunsch, James S., Refounding the African state and local self-governance: the neglected foundation, in: The Journal of Modern African Studies 2000, 487-509.

17 Cf. Morin-Labatut, Gisèle and Akhtar Shahid, Traditional Environmental Knowledge: A Resource to Manage and Share, in: Development – Journal of the Society for International Development 1992:4, 24-30; Linden, Eugene, Lost tribes, lost knowledge, in: Time, 23.9.1991, 50-58, cited by Schulte-Tenckhoff, Isabelle / Horner, Stephen, Le Bon Sauvage, nouvelle donne, in: Nouveaux Cahiers de l'IUED, no. 3 (Ecologie contre nature. Développement et politiques d'ingérence), Paris: PUF et Genève: IUED 1995, 21-39 (22, fn. 4).

18 As for this general phenomenon cf. Raffestin, Claude, Les ingérences paradoxales

9. Some Sub-Saharan African countries have started to privatize – at least the management of – their protected areas.[19] Such redefinition of public goods reduces the space of the res publica. The ill-conceived "Tragedy of the Commons"[20] laid the ground for this definition of all[21] natural resources as commodities and their inclusion in the "common heritage of mankind"[22] or "natural heritage of mankind"[23]. The purpose of this definition is the application of the same remedy which was to allow for an escape from the "Tragedy of the Commons", i. e. the allocation of private ownership rights relating to natural resources[24], by definition exclusive of participation.

Thus, one may observe in Sub-Saharan Africa a double movement: From plural and local political orders to the homogenizing and centralizing state and from the extension of the private to the dismantling of the public, and hence the dismantling of the locus for participation.

10. Most Sub-Saharan Africans live in acephalous subsistence societies with few ties to the centralized state structures. To them, the principle of the rule of law and of subsidiarity, decentralization laws and democratization programs[25],

---

de la pensée écologique, in: Nouveaux Cahiers de l'IUED, no. 3 (Ecologie contre nature. Développement et politiques d'ingérence), Paris: PUF et Genève: IUED 1995, 41-51 (42 ff.).

19 For example privatization of 5 % of the Krüger National Park in the Republic of South Africa; Hufty, Marc / Razakamanantsoa, Andry / Chollet, Martial, Néo-colonialisme vert à Madagascar?, in: Nouveaux Cahiers de l'IUED, no. 3 (Ecologie contre nature. Développement et politiques d'ingérence), Paris: PUF et Genève: IUED 1995, 143-148.

20 Hardin, G., The Tragedy of the Commons, in: Science no. 162, 1968, 1242-1248. Negatively critical of Hardin, Muttenzer, Frank, La mise en œuvre de l'aménagement forestier négocié, ou l'introuvable gouvernance de la biodiversité à Madagascar, in: Cahiers d'Anthropologie du droit 2002, 135-186 (173).

21 Such definition has been applied to land ever since the so-called physiocrats (Quesnay, Turgot et al.) lost their influence in economics. As for the merchandisation of nature cf. also Steppacher, op. cit., 104.

22 The recent invitation to the Vth World Park Congress by the Republic of South Africa and the World Conservation Union (IUCN), 8-17th September 2003 in Durban, advertised the congress with slogans like "Conservation is everyone's business. Protected areas are our collective heritage", cf. Pretoria News, 17.9. 2003, 10. The slip from "common" to "collective" is no coincidence.

23 Dall'Aglio, op. cit., 174.

24 Hufty / Razakamanantsoa / Chollet, op. cit., 148 speak of a "guerre commerciale pour l'accès aux ressources génétiques comme matières premières".

25 Obviously, democracy is a culture specific concept, cf. Pannikar, Raimundo, in: InterCulture no. 136 (1999). Obviously also, the western type representative democracy is not the only one, cf. Poirier, Jean, Décentralisation, démocratie et éthocratie, in: Dezentralisierung, Demokratisierung und die lokale Repräsentation des Staates. Décentralisation, démocratisation et les représentations locales de la force publique,

introduced in a great number of Sub-Saharan African countries, are meaningless.[26] Their order of custom needs no rule of law.[27] Tribes without rulers[28] do not know of centralized government which would call for decentralization, subsidiarity or the like. Sustainable resource management[29] has been an integral part of their 'land law'. Even more: The traditional land law has included, by design, the wider concept of sustainable ways of living.[30]

Most Sub-Saharan African States use the principles of the rule of law and subsidiarity, decentralization laws and democratization programs to divest themselves of their responsibilities. They transfer these responsibilities to underfinanced and understaffed sub or non-state levels, while at the same time they continue clinging to their centralized power structures.[31]

11. The channeling of funds through NGOs to finance nature preservation measures has contributed to maintaining the illusion of popular participation

---

herausgegeben von Jakob Rösel / Trutz von Trotha, Köln: Rüdiger Köppe 1999, 79-88.

26 Cf. for example LeRoy, Etienne, Problèmes de légitimité dans les théories contemporaines de l'Etat de Droit en France et en Afrique, in: Legitimation von Herrschaft und Recht. La légitimation du pouvoir et du droit, herausgegeben von Wilhelm J. G. Möhlig and Trutz von Trotha, Köln: Rüdiger Köppe 1994, 43-54.

27 Reference is made here to the title of an article written by Diamond (Stanley, The Rule of Law Versus the Order of Custom, in: Social Research 38, 1971, 42-72.

28 Reference is made here to a classic edited by John Middleton and David Tait (Tribes Without Rulers. Studies in African Segmentary Systems, London: Routledge & Kegan Paul 1958). As concerns traditional ruling by consensus cf. Wiredu, Kwasi, Democracy and Consensus in African Traditional Politics: A Plea for a Non-Party Polity, in: Philosophy from Africa, ed. by P. H. Coetzee and A. P. J. Roux, Oxford: Oxford University Press, 374-382.

29 Concept first developed by Carl von Carlowitz, Sylvicultura Oeconomica. Die naturmäßige Anweisung zur Wilden Baum-Zucht, Leipzig 1713. Cf. also Fievet, Gilles, Réflexions sur le concept de développement durable: prétention économique, principes stratégiques et protection des droits fondamentaux, in: Revue belge de droit international 2001/1, 128-184.

30 Similar patterns are reported from India (cf. Gadgil, Madhav, Social Restraints on Exploiting Nature: The Indian Experience, in: Development – Journal of the Society for International Development 1987:1, 26-30), which leads to suggest that the split is not between South and North but within countries. Cf. also Schulte-Tenckhoff/ Horner, op. cit., 28. Mistakenly, the 'chef de terre' has been considered as the guardian of the access to land (as we conceive it) only. In fact, he has a much wider role.

31 Critical Muttenzer, op. cit., 141 f., 180 f.; Rochegude, Alain, Foncier et décentralisation. Réconcilier la légalité et la légitimité des pouvoirs domaniaux et fonciers, in: Cahiers d'Anthropologie du droit 2002, 15-43 (19) and Diop, Moustapha, Légimitation, paradoxe et contradictions du caractère "public" du foncier en Afrique: du monopole foncier de l'Etat à la décentralisation des ressources foncières, in: Cahiers d'Anthropologie du droit 2002, 85-133 (111 f.) referring mainly to Rochegude.

in the design and implementation of programs and projects aiming to preserve nature in Sub-Saharan Africa. Not used to having a say in these affairs, knowing the limits imposed by public debt burdens[32] and trained in satisfying the needs of so-called donors for an ever faster changing development and ecology rhetoric [33], local partners, in close collaboration with unskilled international NGO representatives, uphold this image.[34]

However defective national law making might be in terms of participation of the people, it provides at least a possibility. These mechanisms do not.

The Rio Agenda 21 has been followed by a great number of declarations[35], which all emphasize the need for having (autochthonous) people participate in the decision making processes concerning the preservation of nature. However, States, international governmental and non governmental organizations as well as transnational enterprises may generally be assured that the link of any programs or projects to their funding guarantees maximum "agreement" of the concerned people with the imposed rules of the game.[36]

The moral legitimacy of the aims of these NGOs and funding agencies is out of question. But law is to prevent private moral from being imposed on other people and peoples.

The language used by these publicly funded NGOs demonstrates the confusion created in public international law. To cite a news paper report on the recent V[th] World Parks Congress in Durban/South Africa: "the conference

---

32 As for the link between the funding of so-called development projects and programs and commercial interests, disguised as ecological interests, cf. Puckett, Jim, Déchets sans frontières: les enjeux de la Convention de Bâle, in: Nouveaux Cahiers de l'IUED, no. 3 (Ecologie contre nature. Développement et politiques d'ingérence), Paris: PUF et Genève: IUED 1995, 75-85 (75); Steppacher, op. cit., 109: "chaînes causales qui démontre comment les acteurs globaux produisent de l'ingérence écologique dans le local"; cf. also 110.

33 Secher Marcussen (Henrik, Burkina Faso: Systèmes de (mé)connaissance et politiques environnementales, in: Nouveaux Cahiers de l'IUED no. 10 (Sciences sociales et coopération en Afrique: les rendez-vous manqués, Paris: PUF et Genève: IUED 2000, 147-165) writes: "… enchaînant un mot clé après l'autre sans aucun fondement scientifique."; similar p. 160. Cf. also Steppacher, op. cit. 100.

34 More than 800 NGOs are active for example in Cambodia, a country with some 12 million inhabitants.
The clash between NGOs, whose democratic legitimacy is rarely established, and sovereign states is obvious. Cf. Tanaka, Maki, Bridging the Gap between Northern NGOs and Southern Sovereigns in the Trade-Environment Debate: The Pursuit of Democratic Dispute Settlements in the WTO under the Rio Principles, in: Ecology Law Quarterly 2003, 113-188. As for numerous unresolved constitutional questions having an impact on the (financial) viability of so-called civil society entities cf. Muttenzer, op. cit., 163 f.

35 Cf. Schulte-Tenckhoff/Horner, op. cit., 30.

36 Dall'Aglio, op. cit., 176.

will conclude with … the Durban Accord and Action Plan, which the 2500 (!) delegates (… the world's foremost environmental experts …) will ratify …".[37] It is not the first time in history that self-proclaimed expertise replaces democratic solution finding. Are we about to live in ecological expertocracies?[38] "Que le souci écologique soit aujourd'hui la source, et aussi l'alibi, des prétentions totalitaires d'un colonialisme environnemental de la part des plus puissants, c'est une évidence."[39]

12. Interestingly, the shift of the locus of law making from national and, through them, international institutions, to mechanisms that have no democratic legitimacy was criticized in its early phase when transnational companies misused their transnationality to escape from the reach of lawmakers[40] and started to impose through that ability their will on national and international lawmakers. Since then, NGOs have adopted the same behavior.[41] Their experts on and technicians of global ecology determine, together with the equally democratically not legitimized funding agencies the course of preservation measures.[42] It appears that ever fewer democratically legitimized actors determine the goals of and the actions concerning the preservation of nature. Their stemming from different parts of the world camouflages their belonging to the same geo-cultural space.[43]

They have forgotten about concepts like sovereignty over national resources.[44] That concept was one of the pillars of the equally forgotten New

37 Marshal, Leon, Conservation Africa, in: Pretoria News, 17.9.2003, 17.

38 Secher Marcussen, op. cit., 147 f. and Cleaver (Frances, Reinventing Institutions: Bricolage and the Social Embeddedness of Natural Resource Management, in: The European Journal of Development Research, Vol. 14, no. 2, 2002, 11-30 concerning the relationship to institutional formation) are rather skeptical about the "knowledge" underlying environmental policies.

39 Comeliau, Christian, Nations sans frontières, in: Nouveaux Cahiers de l'IUED, no. 3 (Ecologie contre nature. Développement et politiques d'ingérence), Paris: PUF et Genève: IUED 1995, 118-120 (118).

40 Cf. Finger, Mathias, Promoteurs globaux – résistance locale: l'ingérence au quotidien, in: Nouveaux Cahiers de l'IUED, no. 3 (Ecologie contre nature. Développement et politiques d'ingérence), Paris: PUF et Genève: IUED 1995, 87-98 (87, 90).

41 Hufty / Razakamanantsoa / Chollet, op. cit., 143: "Voir les industriels faire bon ménage avec les environnementalistes et les représentants des indigènes …".

42 Finger, op. cit., 91 f., is even of the opinion that the whole after Rio process escapes political (and democratic) control and has been usurped by the Global Environmental Facility (GEF) controlled by the Worldbank. Similar Dall'Aglio, op. cit., 170.

43 Sabelli, Fabrizio, L'éthique globalitaire et l'esprit du nouveau colonialisme, in: Nouveaux Cahiers de l'IUED, no. 3 (Ecologie contre nature. Développement et politiques d'ingérence), Paris: PUF et Genève: IUED 1995, 13-20 (17 f.); Dall'Aglio, op. cit., 170.

44 Dall'Aglio, op. cit., 169.

International Economic Order. Its legacy is the Human Right to Development, as expressed, for example, in the right to external and internal[45] self-determination. Land is a crucial factor for any form of self-determination.[46]

13. This lack of participation is increasingly being justified in public international law by a supposed "right of ecological intervention"[47] should the "common heritage of resources" be endangered. This is not a new paradigm, but rather the logical evolution of the relationship between unequal partners. It started with a supposed right of economic intervention. Then, it evolved into a supposed right of humanitarian intervention. We are now at the stage of a supposed right of ecological intervention and public international lawyers have already invented a right of political intervention. The principle of state sovereignty over natural resources which was to do away with the vestiges of colonial type international relations has never been weaker than now. Its misuse by many a state after independence is much to blame for its being forgotten as a means to provide at least for some degree of participation.

Nothing may demonstrate better how unbalanced international relations are than the supposed right of demographic intervention[48] ... in the countries of the South, of course.

It is to make us believe that the resource problem, created by the North, will be solved by simply dividing the finite quantity of resources by a lesser number of people in the South.

14. The farthest-reaching twist in legal terms operated by the supposed right of ecological intervention is that it operationalizes what the Human Rights instruments stipulate as principles. Where the latter leave the implementation to national and international organizations, thus offering participation at least a chance, the supposed right of ecological intervention excludes that possibility.[49]

---

45 So far only the right to external self-determination has been recognized. Cf. Gonidec, Pierre-François, Conflits internes et question nationale en Afrique: Le droit à l'autodétermination interne, in: Revue Africaine de Droit International et Comparé. African Journal of International and Comparative Law 1997, 543-572; Gusy, Christoph, Selbstbestimmung im Wandel. Von der Selbstbestimmung durch den Staat zur Selbstbestimmung im Staat, in: Archiv des Völkerrechts 1992, 385-410.

46 Cf. Schulte-Tenckhoff/Horner, op. cit., 25.

47 Cf. UN Res. 43/131 (8.12.1988) and 45/100 (14.12.1990). The notion of "ecological intervention" was first used by the French foreign minister, Michel Rocard, at the so-called Rio Earth Summit in 1992, cf. Tamiotti, Ludivine, Ingérence écologique: un concept, in: Nouveaux Cahiers de l'IUED, no. 3 (Ecologie contre nature. Développement et politiques d'ingérence), Paris: PUF et Genève: IUED 1995, 159-168 (159 f.). Sabelli, op. cit., 15 f.

48 Agreed upon at the Cairo UN Population Conference in 1994.

49 Similar Tamiotti, op. cit., 161 f.

15. Such ecological intervention could only be justified if there were a universal understanding of nature. That is, however, not the case.[50] In the international debate, nature continues to be defined in scientific and western terms.[51] In a world of many cultures this translates into exclusion, the very contrary of participation.

The 1992 United Nations Conference on Environment and Development in Rio de Janeiro, by recognizing the "... traditional and holistic scientific knowledge of autochthonous people of their lands..."[52], paradoxically proves how desperate we are in trying to put those bits and pieces together, which in Sub-Saharan Africa never separated.

III. Conclusion

16. If indeed three generations of Human Rights[53] haven't produced but "rights to intervention", then they have produced little.

17. Domination, i. e. the contrary of participation, has many masks. The latest one is knowledge and information[54] on the right way to preserve nature.[55]

---

50 As for the different concepts, conceptions and conceptualizations of nature in time and space cf. Höffe, Otfried, Sieben Gesichter der Natur. Ein Versuch über Anthropologie und Ökologie, in: Neue Zürcher Zeitung 5./6.4. 2003, 78; InterCulture no. 132 (1997) and no. 137 (1999); Leidig, Guido, Contribution to the 32nd Symposium of the European Faculty of Land Use and Development, in: Social Strategies, Vol. 38 (in print); Neumann, op. cit., 560; O'Neill, Catherine A., Risk Avoidance, Cultural Discrimination, and Environmental Justice for Indigenous Peoples, in: Ecology Law Quarterly 2003, 1-57; Radkau, Joachim, Natur und Macht. Eine Weltgeschichte der Umwelt, München: Beck 2000; Raine, Peter, Au-delà de l'universalisme. Le chaman et l'écologiste. Un horizon toujours ouvert, INTERculture no. 140 (2001); les sept points cardinaux. orientations éco-logiques, Cahier de l'IUED no. 7, Paris: PUF et Genève: IUED 1978.

51 Similar Grinevald, op. cit., 122, 124, 136.

52 Cf. Chap. 26 of the Report on the United Nations Conference on the Environment and Development, doc. A/CONF, 151/4, Part III, Action principle 26.1 (zitiert bei Schulte-Tenckhoff/Horner, op. cit., fn. 13.

53 1st generation: civil and political rights; 2nd generation: economic and social rights; 3rd generation: solidarity rights, right to development.

54 Cf. Clerc, Alain, Les oscillations paradoxales de la pensée écologique, in: Nouveaux Cahiers de l'IUED, no. 3 (Ecologie contre nature. Développement et politiques d'ingérence), Paris: PUF et Genève: IUED 1995, 115-117 (117); Hufty/Razakamanantsoa/Chollet, op. cit., 147.

55 N'Guessan, Cacou Eloi Koudio IV, Les éléphants, les lamantins et le développement, in: Nouveaux Cahiers de l'IUED, no. 3 (Ecologie contre nature. Développement et politiques d'ingérence), Paris: PUF et Genève: IUED 1995, 149-153 (152); Tamiotti, op. cit., 167.

But: "Notre ingérence (par la monopolisation de l'information) … est en train de détruire la socio-diversité intellectuelle dont nous pourrions tirer parti".[56]

This not only concerns North/South relations but more and more urban/non-urban relations. The world over, societies are urbanizing and industrializing. Urbanized societies are out of contact with nature, but they have taken over the knowledge and information systems concerning nature.[57] Again, Sub-Saharan Africa partakes little in this development because it is the only place in the world where peasants have not yet been conquered by the modes of life that go with urbanization and industrialization.[58]

18. But how ensure participation of the majorities in the making of decisions that affect the globe – if that is what we want? The original principle of sustainability, i. e. ecological balance, must be extended to include social justice[59], economic security and political stability through participation. Democracy has to be redefined as a way of making decisions along these lines.[60] The institutions within which and/or the mechanisms through which this could happen are unknown. I do not believe that another hierarchy[61] will do. Too little attention has been paid to the process of decision making related to development. This concerns also public international law-making.[62]

19. An unresolved matter is whether we should maintain in law, because of the very nature of our law[63], our anthropocentric outlook on nature or whether a biocentric[64] vision, and hence a different vision of law all together, is needed.

---

56 Raffestin, op. cit., 50 f. Cf. also Finger, op. cit., 93: "socio-cultural erosion". Same words used by Steppacher, op. cit., 111, who links the issue to lessening participation.

57 Écologie contre nature. Développement et politiques d'ingérence. Nouveaux Cahiers de l'IUED, no. 3, Paris: PUF et Genève: IUED 1995, propos, p. 12; Raffestin, op. cit., 44.

58 Hydén, Göran, Das Kleine ist mächtig: Die strukturelle Anomalie des ländlichen Afrika, in: Das Afrika der Afrikaner. Gesellschaft und Kultur Afrikas, herausgegeben von Rüdiger Jestel, Frankfurt a. M.: Suhrkamp 1982, 205-246.

59 Comeliau, op. cit., 120.

60 Similar Grinevald, op. cit., 128.

61 As suggested for example by Dall'Aglio, op. cit., 185.

62 For the importance of the procedural aspect cf. Henrÿ, Hagen, Vom Entwicklungsrecht zum Menschenrecht auf Entwicklung – vielfältige Einheit anstelle von Einheitlichkeit, in: Zeitschrift für Rechtsvergleichung 1994, 3-29; Joas, Hans, Wertevermittlung in einer fragmentarisierten Welt, in: Universitas 7/2003, 667-686 (678 ff.); Retour du foncier, Cahiers d'Anthropologie du droit 2002, Éditorial, 7-13 (9).

63 Dall'Aglio, op. cit., 172 f., is right in rejecting attempts to treat nature as a legal entity. Tamiotti, op. cit., 168, consequently calls upon lawyers to first recognize a Human Right to environment before an ecological intervention may be justified.

64 Cf. Grinevald, op. cit., 135. A considerable part of public international law relates to sustainable development. For an overview cf. Luff, David, An Overview of Interna-

20. Which nature are we protecting since gene-modified plants and animals – and why not humans –? will soon populate our globe? The ecosystem/s we are protecting have long ceased to be natural. We project our modern culture into something we call nature.[65]

21. What kind of participation are we talking about if patenting the results of research built on biodiversity[66] destroys that very biodiversity and the cultural diversity, and hence participation, that goes with it?[67] What participation are we talking about in a world where the economic decision making power rests in the hands of ever fewer people[68] and where the production of expert knowledge follows these footsteps?[69]

Privatization of nature, be it in the form of exclusive ownership rights[70], in the form of patents on the result of research benefiting from biodiversity[71] or in the form of the privatization of public services comes at the prize of less participation.

22. The impasse is our incapacity to share. The well off – who, by the way, do not only live in the North and not all Northerners are well off! – waste 250 l of drinking water per day and capita. Populations in urban centers of the South explode. Each side is pointing its finger to the other.[72] The Rio Convention on

---

tional Law of Sustainable Development and a Confrontation between WTO Rules and Sustainable Development, in: Revue belge de droit international 1996/1, 90-144.

65 Are we at the end of nature? Cf. Luke, Timothy W., International and Interenvironmental Relations: Reassessing Nations and Niches in Global Ecosystems, in: Alternatives 2003, 393-422; Raffestin, op. cit., 47.

66 Sabelli, op. cit., 19: "Traquer le chaman pour le faire parler de ses plantes est ainsi devenu un véritable sport …"; Schulte-Tenckhoff/Horner, op. cit., 36 ; Hufty/Razakamanantsoa / Chollet, op. cit., 145 f., 147.

67 Patent rights are exclusive. Cf. Berlan, Jean-Pierre et Richard C. Lewontin, La menace du complexe génético-industriel, in: Le Monde diplomatique, décembre 1998, 1 and 22 f.

68 Cf. Steppacher, op. cit., 111.

69 Similar Sabelli, op. cit., 19.

70 The UN Declaration on the Right to Development was co-adopted with a Declaration on the right to private ownership.

71 Dall'Aglio, op. cit., 179.

72 Similar, Auroi, Claude, Politiques démographiques et écologie, in: Nouveaux Cahiers de l'IUED, no. 3 (Ecologie contre nature. Développement et politiques d'ingérence), Paris: PUF et Genève: IUED 1995, 61-74 (74); Steppacher, op. cit., 107, 113; Comeliau, op. cit., 119; Yonoudjoum, Clément, L'éthique globalisante et l'esprit du renouveau, in: Nouveaux Cahiers de l'IUED, no. 3 (Ecologie contre nature. Développement et politiques d'ingérence), Paris: PUF et Genève: IUED 1995, 154-157 (155).

Biodiversity holds States responsible for the "conservation and sustainable utilization of their biological resources"[73]. For many states this means burden sharing without profit sharing, since some have little left of their bio-diversity but they hold on to their intellectual property rights on knowledge produced thanks to the free access to the bio-diversity in other countries.

Sharing is less about the technical use of natural resources. It is rather about sharing ideas, notions, concepts and conceptualizations concerning these resources.[74] We ignore the participatory value of sharing because we do not know how to deal with the radically different.

73 Zitiert bei Dall'Aglio, op. cit., 175.
74 Similar Dall'Aglio, op. cit., 181.

Verzeichnis der BeiträgerInnen

AKSAMITAUSKAS, ČESLOVAS, Assoc. Prof. Dr., Department of Geodesy and Cadastre, Vilnius Gediminas Technical University, Vilnius, LT

ALEKNAVIČIUS, MARIUS, MA and Dipl. in International Business Management, Lithunian Institute of Agrarrian Economics, Vilnius, LT

ALEKNAVIČIUS, PRANAS, Assoc. Prof. Dr., Advisor to the Minister of Agriculture, Vilnius, LT

ANTUCHEVIČIENĖ, JURGITA, Department of Building Technology and Management; Vilnius Gediminas Technical University, Vilnius, LT

BAJEROWSKI, TOMASZ, Department of Land Planning and Management, Faculty of Geodesy and Land Management, University of Warmia and Mazury in Olsztyn, Olsztyn, PL

BITINAS, ALBERTAS, Geological Survey of Lithuania, Vilnius, LT

CYMERMAN, RYSZARD, Prof. Dr., Department of Land Planning and Management, Faculty of Geodesy and Land Management, University of Warmia and Mazury in Olsztyn, Olsztyn, PL

DÉJEANT-PONS, MAGUELONNE, Docteur en Droit, Head of the Spatial Planning and Landscape Division, Chef de la Division de l'aménagement du territoire et du paysage DG IV, Conseil de l'Europe/Council of Europe

DÉNAS, ZYDRUNAS, Vilnius University, Vilnius, LT

DEVEIKIS, STEPONAS, Président de l'Association Lituanienne des Experts en Evaluation des Biens, Viceprésident de la Corporation «Matininkai», Sarl, Lecteur à Vilniaus Gedimino Technikos Universitetas, Vilnius, LT

HENRŸ, HAGEN, Dr., Governments' Counsellor, Helsinki, SF

KAKLAUSKAS, ARTURAS, Prof. Dr. Ing. habil., Vilnius Gediminas Technical University, Vilnius, LT

KOWALCZYK, CEZARY, Department of Land Planning and Management, Faculty of Geodesy and and Land Management, University of Warmia and Mazury in Olsztyn, Olsztyn PL

KRAUS, WILLY, Prof. Dr. rer. pol. Dr. h.c. mult., Institut für Entwicklungsforschung und Entwicklungspolitik, Ruhr-Universität Bochum, Bochum, D

KUMETAITIS, ALGIRDAS, Vilnius Gediminas Technical University, Vilnius, LT

KUMPIENE, JURATE, PhD-student, Division of Waste Science & Technology, Lulea University of Technology, Lulea, SE

LEIDIG, GUIDO, Dr. rer. pol., Sekretär FESF Strasbourg, F

LENDI, MARTIN, Prof. Dr. jur., Eidgenössische Technische Hochschule, Zürich, CH

MALIENÉ, VIDA, Assoc. Prof. Dr. Ing., Department of Construction Economics and Property Management, Vilnius Gediminas Technical University, Vilnius, LT

MATTSON, HANS, Prof. Dr., Real Estate Planning and Land Law, Department of Infrastructure, Royal Institute of Technology, Stockholm, SE

NEUHOFER, HANS, Prof. Hofrat Dr. jur., Prof. an den Universitäten Linz und Wien, Direktor des Oberösterreichischen Gemeindebundes i.R., Wels, A

NOWAK, ANDRZEJ, Prof. Dr., Department of Regional Planning and Land Development, Faculty of Geodesy and Land Landmangement, University of Warmia and Mazury in Olsztyn, Olsztyn, PL

NOWAK-RZASA, MAGDALENA, Department of Land Planning and Management, Faculty of Geodesy and Land Management, University of Warmia and Mazury in Olsztyn, Olsztyn, PL

PALABINSKA, AINA, Dozent Mag. oec., Lehrstuhl für Bodenordnung und Geodaesie, Agraruniversität Lettland, Jelgava, LV

PALCIAUSKASKAITE, RITA, Head of the Territorial Planning and Landscape Architecture Division of the State Land Survey Institute, Republic of Lithuania National Service under the Ministry of Agriculture, Vilnius, LT

PETROŠKEVIČIUS, PETRAS, Department of Geodesy and Cadastre, Vilnius Gediminas Technical University, Vilnius, LT

RASLANAS, SAULIUS, Assoc. Prof. Dr. Ing., Vilnius Gediminas Technical University, Vilnius, LT

SLIAUPA, SAULIUS, Institute of Geology and Geography of Lithuania, Vilnius, LT

STALIUNAS, SILVESTRAS, The Director, Republic of Lithuania National Service under the Ministry of Agriculture, Vilnius, LT

STANIONIS, ARMINAS, Department of Geodesy and Cadastre, Vilnius Gediminas Technical University, Vilnius, LT

ŠUBIC KOVAČ, MARUŠKA, Assist. Prof. Dr., Municipal Economics Institute, Faculty of Civil and Geodetic Engineering, University of Ljubljana, Ljubljana, SI

SURVILA, ROMUALDAS, Deputy Director of the Land Survey Institute, Republic of Lithuania National Service under the Ministry of Agriculture, Vilnius, LT

TRAPPE, PAUL, Prof. Dr. phil., FESF, President of the Board, Strasbourg, F

TRINQUELLE, ISABELLE, Governments' Counsellor, Athens, GR

VOLOVITCH, NIKOLAI WLADIMIROWITCH, Dr., Institut Ozenki Sobstwennosti, Tomsk, RU

WEISS, ERICH, Prof. Dr.-Ing., Dr. h.c. mult., Lehrstuhl für Bodenordnung und Bodenwirtschaft, Institut für Städtebau, Bodenordnung und Kulturtechnik der Rheinischen Friedrich-Wilhelms-Universität Bonn, D

ZAKAREVIČIUS, ALGIMANTAS, Prof. Dr. Ing. habil., Department of Geodesy and Cadastre, Vilnius Gediminas Technical University, Vilnius, LT

ZAVADSKAS, EDMUNDAS KAZIMIERAS, Prof. PhD, D. Sc., Dr. h.c., Vice Rector Vilnius Gediminas Technical University (VGTU), Department of Building Technology and Mangement, Vilnius, LT

ŽEBROWSKI, WOJCIECH, Prof. Dr., Department of Land Planning and Management, Faculty of Geodesy and Management, University of Warnia and Mazury in Olsztyn, Olsztyn, PL

Social Strategies

Monographien zur Soziologie und Gesellschaftspolitik

Monographs on Sociology and Social Policy

Editor: Prof. Dr. Paul Trappe

Social Strategies ist eine Reihe für Monographien und Proceedings. Sie behandelt soziologische Grundsatzfragen und gesellschaftspolitische Problemstellungen, die zur sozialen Wirklichkeit einen engen Bezug haben. Probleme, die sich heute stellen, sollen analysiert und Wege zur Überwindung aus sozialwissenschaftlicher Sicht aufgezeigt werden. Damit soll diese Reihe einen Bereich des sozialen Wandels erfassen, der zwischen Aktualität und «utopischem Übermorgen» liegt und der im neueren soziologischen Schrifttum oftmals nicht berücksichtigt wird. Zukunftsgerichtete Konzeptionen können dabei als Orientierungshilfe dienen.
In der Reihe können Beiträge in deutscher, englischer und französischer Sprache veröffentlicht werden.

Social Strategies is a series of monographs and proceedings. It deals with fundamental sociological questions and socio-political problems closely related to social reality. It analyses current day problems and shows, from a sociological viewpoint, how they can be overcome. The series covers the sphere of social change between the present day and the 'Utopian future' – something lacking in more recent sociological writings. Ideas directed towards the future may serve as guidelines. Contributions to the series can be published in German, English and French.

La collection **Social Strategies** regroupe des monographies et travaux scientifiques traitant des questions sociologiques de base ainsi que des aspects socio-politiques de la réalité sociale. Elle analyse les problèmes contemporains et montre comment les sciences humaines peuvent les surmonter. *Social strategies* aborde le domaine des changements sociaux ente aujourd'hui et les «surlendemains utopiques». En se servant de conceptions tournées vers l'avenir comme point de repère, elle comble les lacunes sur ce sujet de la littérature sociologique récente. Des textes en allemand, anglais et français peuvent être publiés dans cette collection.

Verlag / Editions Peter Lang AG
Bern, Berlin, Bruxelles, Frankfurt/M., New York, Oxford, Wien
Hochfeldstrasse 32, Postfach 746, CH-3000 Bern 9
info@peterlang.com, www.peterlang.com, www.peterlang.net
Postgirokonto CH: Bern 30-2699-7
Postgirokonto BRD: D-60288 Frankfurt/M. Kto. 300593-604

Social Strategies – Monographien zur Soziologie und Gesellschaftspolitik

Vol. 1: **Christian Giordano und Robert Hettlage** Mobilisierung oder Scheinmobilisierung? Vergriffen

Vol. 2: **Heli Aurich** Partizipation an der Stadtplanung. Basel 1976. VIII + 156 S. + 18 Grafiken. SFr. 20.-

Vol. 3: **Adolf Gasser** Staatlicher Grossraum und autonome Kleinräume. Vergriffen.

Vol. 4: **Verena Burkolter** The Patronage System. Theoretical Remarks. With an Introduction by Prof. Samuel N. Eisenstadt/Jerusalem. Basel 1976. XII + 54 Seiten. SFr. 8.-

Vol. 5: **MaxWeidmann** DasDrogenverhalten von Basler Schülern. Eine Studie zur Epidemiologie und Soziogenese des jugendlichen Drogenkonsums. Mit einer Einleitung von Prof. Paul Kielholz/Basel. Basel 1977. VIII + 124 Seiten + Anhang von 78 Seiten. SFr. 20.-

Vol. 6: **Paul Trappe** Development from Below as an Alternative. A Case Study in Karamoja/ Uganda. With a Preface by Donald V. McGranahan and Prince Sadruddin Aga Khan/ Geneva. Basel 1978. VI + 102 p. SFr. 12.-

Vol. 7: **Peter Abplanalp** Marktkommunikation und Konsumentenverhalten. Basel 1978. XIV + 252 Seiten. SFr. 24.-

Vol. 8: **Klaus Schrape** Theorien normativer Strukturen und ihres Wandels, Teil I: Theoretische Einführung, Rekonstruktion und Kritik der Theorie von Jürgen Habermas. Basel 1977. XV + 260 Seiten. SFr. 28.-

Vol. 9: **Klaus Schrape** Theorien normativer Strukturen und ihres Wandels, Teil II: Zur Rekonstruktion und Kritik der Theorien von Talcott Parsons, Ralf Dahrendorf und Niklas Luhmann. Basel 1978. XVI + 304 Seiten. SFr. 28.-

Vol. 10: **Klaus Schrape** Theorien normativer Strukturen und ihres Wandels, Teil III: Ein eigener Ansatz. Basel 1979. XVIII + 382 Seiten. SFr. 28.-

Vol. 11: **Kurt Eichenberger, Walter Buser, Alexandre Métraux, Paul Trappe** (Herausgeber) Grundfragen der Rechtssetzung. Basel 1978. XVI + 520 Seiten. SFr. 32.-

Vol. 12: **Paul Trappe** Entwicklungssoziologie Herausgegeben und eingeleitet von Hans Werner Debrunner. Basel 1984. XVIII + 711 Seiten. SFr. 60.-

Vol. 13: **Roland Harri Wettstein** Über die Ausbaufähigkeit von Rawls' Theorie der Gerechtigkeit, Vorüberlegungen zu einer möglichen Rekonstruktion. Basel 1979. 60 Seiten. SFr. 8.-

Vol. 14: **Harro von Senger** Partei und Staat: Parteinormen und staatliches Gesetzesrecht in der Volksrepublik China. Basel 1981. 77 Seiten. SFr. 12.-

Vol. 15: **Lotti Gujer, Edith Hunziker, Ruth Hungerbühler** Basler Frauenuntersuchung. Einleitung von Paul Trappe. Basel 1982. VIII + 609 Seiten. SFr. 48.-

Vol. 16: **Klaus M. Leisinger** Health Policy for Least Developed Contries. With a Foreword by Carl E. Taylor/Baltimore. Basel 1984. XII + 444 Seiten. SFr. 30.-

Vol. 17: **Ueli Mäder** Gewaltfreie Revolution in Entwicklungsländern. Basel 1984. VIII + 308 Seiten. SFr. 30.-

Vol. 18: **Michael Haller** Was sollen wir mit den neuen Medien machen? (Forschungsprojekt im Auftrag des Schweizerischen Wissenschaftsrats). Basel 1986. 60 Seiten. SFr. 12.-

Vol. 19: **Paul Trappe** Soziale Breitenwirkung einer Entwicklungsintervention. «Lac Alaotra – Grenier de Madagascar.» Basel 1987. VIII + 272 Seiten. SFr. 45.-

Vol. 20: **Raphael Laube** Maasai-Identität und sozialer Wandel bei den Maasai. Mit einem Vorwort von Prof. Meinhard Schuster. Basel 1986. VIII + 213 Seiten. SFr. 30.-

Vol. 21: **Martin Kolb** Gewalt gegen Frauen. Eine Untersuchung im Auftrag des Justizdepartementes Basel-Stadt. Basel 1986. X + 77 Seiten. SFr. 18.-

Social Strategies – Monographien zur Soziologie und Gesellschaftspolitik

Vol. 22: **Paul Trappe**, ed., Politische und gesellschaftliche intermediäre Gewalten im sozialen Rechtsstaat. Basel 1990. X + 340 Seiten. SFr. 42.–

Vol. 23: **Ueli Mäder et al.**, Armut im Kanton Basel-Stadt. Basel 1991. XII + 672 Seiten. SFr. 60.–

Vol. 24: **Marco Nese** Soziologie und Positivismus im präfaschistischen Italien (1870–1922). Basel 1993, IX + 184 Seiten. SFr. 33.–

Vol. 25: **Urs Fazis und Jachen C. Nett**, eds., Gesellschaftstheorie und Normentheorie. Symposium zum Gedenken an Theodor Geiger, 9.11.1891–16.6.1952. Basel 1993. VIII + 319 Seiten. SFr. 48.–

Einführung:
Karl Pestalozzi, Basel
Paul Trappe, Basel

Teil I: Position und Trends zu Gesellschaftstheorie und Normentheorie:
Niklas Luhmann, Bielefeld
Annemarie Pieper, Basel
Ota Weinberger, Graz
Gerhard Schmid, Basel
Henning Ottmann, Basel
Horst Baier, Konstanz

Teil II: Beiträge zur aktuellen Geiger-Forschung:
Eugen Buss, Hohenheim
Sven Eliaeson, Karlstad
Urs Fazis, Basel
Athanasios Gromitsaris, Athen
Friedrich Lachmayer, Wien
Petra Leuenberger, Basel
Jachen C. Nett, Basel
Hans Oswald, Berlin
Horst Poettker, Frankfurt/Main
Heinrich Popitz, Freiburg/i. Br.
Klaus Schrape, Basel
Hubert Treiber, Hannover
Erich Zalten, Basel
Wolfram Zitscher, Kiel

Vol. 26: **Urs Fazis** «Theorie» und «Ideologie» der Postmoderne. Studien zur Radikalisierung der Aufklärung aus ideologiekritischer Perspektive, Basel 1994. IV + 322 Seiten. ISBN 3-908243-01-7 SFr. 48.–

Vol. 27: **Karlheinz Hottes, Werner Gocht, Paul Trappe**, eds., Krisenkontinent Afrika – Ansätze zum Krisenmanagement, IAfEF-Klausurtagung 1994, Basel 1995, VII + 320 Seiten. ISBN 3-908243-02-5 SFr. 48.–

Vol. 28: **Christoph Bosshardt**, ed., Beiträge zu Transformationsprozessen und Strukturanpassungsprogrammen, i. A. des Interdisziplinären Arbeitskreises für Entwicklungsländerforschung (IAfEF), Basel 1997; VIII + 362 Seiten. ISBN 3-908243-04-1 SFr. 52.–

Vol. 29: **Christoph Bosshardt**, ed., Problembereiche interdisziplinärer Forschung; 30 Jahre IAfEF; Klausurtagungen in Freiburg i. Br. 1997 und Wien 1998, Bern 1999; VI + 380 Seiten. ISBN 3-906763-29-3 SFr. 65.–

Vol. 30: **Ueli Mäder**, Subsidiarität und Solidarität, Bern 2000; 304 Seiten. ISBN 3-906765-36-9 SFr. 65.–

Vol. 31: **Hans Neuhofer, Victoria Jäggi-Torra**, eds., Aktuelle Probleme der Stadt- und Landplanung; Bodenordnung in städtischen Verdichtungsgebieten und alpinen Regionen; im Auftrag der Europäischen Fakultät für Bodenordnung, Bern 2000; 260 Seiten. ISBN 3-906765-45-8 SFr. 61.–

Vol. 32: **Willy Kraus, Paul Trappe**, eds., Nachhaltige räumliche Entwicklung auf dem europäischen Kontinent – Interdisziplinäre Ansätze; im Auftrag der Europäischen Fakultät für Bodenordnung (FESF) und des Interdisziplinären Arbeitskreises für Entwicklungsländerforschung (IAfEF), Bern 2000; X + 458 Seiten. ISBN 3-906766-25-X SFr. 89.–

Social Strategies – Monographien zur Soziologie und Gesellschaftspolitik

Vol. 33: **Christoph Bosshardt**, Homo Confidens; Eine Untersuchung des Vertrauensphänomens aus soziologischer und ökonomischer Perspektive, Bern 2001; 285 Seiten.
ISBN 3-906767-12-4 SFr. 69.–

Vol. 34: **Piotr Sałustowicz**, ed., Civil Society and Social Development; Proceedings of the 6th Biennial European IUCISD Conference in Krakow 1999, Bern 2001; 422 pages.
ISBN 3-906767-21-3 / US-ISBN 0-8204-5343-9 SFr. 89.–

Vol. 35: **Victoria Jäggi, Ueli Mäder und Katja Windisch**, eds., Entwicklung, Recht, Sozialer Wandel; Festschrift für Paul Trappe zum 70. Geburtstag, Bern 2002; 701 Seiten.
ISBN 3-906767-96-5 SFr. 98.–

Vol. 36: **Erich Weiss und Tanja Zangger**, éds, 31ᵉ Symposium International FESF Strasbourg; Les régions en face de l'aménagement du territoire, du droit foncier et de la protection de l'environnement / Functions of the regions in the realms of spatial planning, landed property and environmental protection / Aufgaben der Regionen im Hinblick auf Raumplanung, Bodenrecht und Umweltschutz, Bern 2002; 473 Seiten.
ISBN 3-906769-38-0 / US-ISBN 0-8204-5936-4 SFr. 99.–

Vol. 37: **Michael S. P. Freudweiler**, Soziale Normen in der multikulturellen Gesellschaft, Bern 2003; 286 Seiten.
ISBN 3-906770-60-5 SFr. 71.–

Vol. 38: **Henrÿ Hagen und Ruşen Keleş**, éds, 32ᵉ Symposium International FESF Strasbourg; Des anciens et des nouveaux droits fonciers dans leur contexte culturel / Old and new land tenure rights in their cultural context / Zur Systematik alter und neuer Bodenrechte im kulturellen Kontext, Bern 2004; 257 Seiten.
ISBN 3-03910-224-9 / US-ISBN 0-8204-7005-8 SFr. 69.–

Vol. 39: **Katja Windisch**, Gestalten sozialen Wandels; Die Entwicklungssoziologie Richard F. Behrendts, Bern 2005; 259 Seiten. ISBN 3-03910-366-0 SFr. 65.–

Vol. 40: **Edmundas Kazimieras Zavadskas**, ed., 33ᵉ Symposium International FESF Strasbourg; Des développements récents dans la protection de l'environnement / Recent Developments in Environmental Protection / Neuere Entwicklungen im Umweltschutz, Bern 2005; 395 Seiten.
ISBN 3-03910-612-0 / US-ISBN 0-8204-7161-5 SFr. 86.–